Synthesis Lectures on Mathematics & Statistics

Series Editor

Steven G. Krantz, Department of Mathematics, Washington University, Saint Louis, USA

This series includes titles in applied mathematics and statistics for cross-disciplinary STEM professionals, educators, researchers, and students. The series focuses on new and traditional techniques to develop mathematical knowledge and skills, an understanding of core mathematical reasoning, and the ability to utilize data in specific applications.

Deguang Han • Qianfeng Hu • Bei Liu • Rui Liu

Banach Space Aspects of Frame Theory with Applications

 Springer

Deguang Han
Department of Mathematics
University of Central Florida
Orlando, FL, USA

Qianfeng Hu
School of Science
Hebei University of Technology
Tianjin, China

Bei Liu
Department of Mathematics
Tianjin University of Technology
Tianjin, China

Rui Liu
School of Mathematical Sciences
Nankai University
Tianjin, China

ISSN 1938-1743   ISSN 1938-1751   (electronic)
Synthesis Lectures on Mathematics & Statistics
ISBN 978-3-032-10173-0   ISBN 978-3-032-10174-7   (eBook)
https://doi.org/10.1007/978-3-032-10174-7

This Springer imprint is published by the registered company Springer Nature Switzerland AG
The registered company address is: Gewerbestrasse 11, 6330 Cham, Switzerland

If disposing of this product, please recycle the paper.

Preface

The main purpose of this book is to give a novel introduction to the theory of functional analytic aspects of frames for Banach and Hilbert spaces with focuses on the Banach space aspects of the theory and applications. The theory of Hilbert space frames, a useful tool in mathematics and engineering, has a natural generalization in the Banach space settings, namely Banach space frames (or framings). This generalization is deeply connected with some basic theory of Banach spaces including the theory of various types of approximation properties. In fact, Banach space techniques play an essential role and are necessary even in the investigation of Hilbert space frame theory. This Banach space aspect of the frame theory is particularly demonstrated by the recent investigation on the geometric properties of framings, operator-valued measures, and bounded linear maps. This recently developed general dilation theory not only unifies the dilation theory for framings, arbitrary operator-valued measures, and bounded linear maps on both commutative and noncommutative domain algebras (such as Banach algebras, C^*-algebras and von Neumann algebras), but also presents an insight on the intrinsic connections between the dilation theory and many other important areas of research, including quantum information theory and Kadison's simlarity problem for bounded linear maps between C^* algebras.

We begin Chap. 1 with a short introduction to Hilbert space frame theory by focusing on discrete frames and their basic properties. Then followed by brief discussions on several related topics, including continuous frames, Gabor frames and more general group representation frames. In Chap. 2 we present various generalizations of Hilbert frames to Banach spaces including the concepts of atomic decompositions, Schauder frames, and framings and discuss their connections with the reflexivity property of Banach spaces. We also introduce and present some recent work related to the concepts of the completely bounded frames for operator spaces and Lipschitz frames for Banach spaces. Approximation properties and their relations with frames are the topics of Chap. 3. In this chapter, we utilize the dilation techniques to show that a Banach space admits a frame if and only if it is isomophic to a complement subspace of a Banach space with a basis. This is related to the characterization of Johnson, Rosenthal, Zippin, and Pełczyński on the Banach space bounded approximation property. We characterize the operator spaces with the completely bounded approximation property and the Lipschitz bounded approximation property, and then recover the famous Godefroy-Kalton theorem, which states that the Lipschitz bounded approximation property and the bounded approximation property are equivalent for every Banach space. Chapter 4 is devoted to the Banach space dilation theory for operator-valued measures and bounded linear maps. We first present our general dilation theory, including the minimal and maximal dilations, for operator-valued measures and bounded linear maps on Banach algebras. In the C^*-algebra or von Neumann algebra case, it leads to a natural generalization of the classical dilation theorem of Naimark and Steinspring from positive operator-valued measures and completely bounded linear maps to arbitrary operator-valued measures and bounded linear maps that are not necessarily completely bounded. One important case of this theory, the dilation of quantum measures (i.e., operator-valued measure on the projection lattices of von Neumann algebras) and weak operator topology continuous linear maps, will also be a focus of this chapter. In Chap. 5, we investigate nonself-adjoint dilation for linear systems and present a generalized Choi-Kraus dilation and several structural theorems for this generalized dilation theory. Chapter 6 presents some applications of frame theory in quantum information theory with focuses on the quantum detection problem and phase retrieval problem by frame-induced positive operator-valued measure (POVM). In addition to presenting some basic results on injective frames, we also present some of our recent work on exact phase-retrievable frames and phase retrievable group representation frames. In Chap. 7, we present another application of Banach space frame theory by investigating the problem of sampling and signal recovering in mixed Lebesgue spaces. We discuss the conditions, reconstruction algorithms and stability on exact reconstruction for signals in shift-invariant subspaces by nonuniform samples with appropriately sufficient density.

This book presents some of the main results developed in the last decades on the theory of frames, dilations, operator-valued measures, and some of their applications. We do not intend to provide a comprehensive review of this topic, but instead to offer a new perspective of the frame theory to demonstrate that frame theory is heavily entangled with the theory of Banach spaces and it provides a central and effective tool in dealing with many problems in other areas of mathematics and engineering. We hope that the materials presented in this book can help the reader to get a quick understanding and rapidly access to the related research topics.

Acknowledgments

We wish to thank a number of our friends and colleagues for useful comments and suggestions during the preparation of this book. Firstly, we are extremely thankful to our friend, mentor, and collaborator David R. Larson who has guided us in the last decade on a series of research topics on frames and related dilation theory. The following is the list of friends and colleagues who have supported our research projects related to the topics of this book, and whose motivational and critical comments inspired us and helped us far more than they probably realize, including Pete Casazza, Lixin Cheng, Ole Christensen, Longyun Ding, Hans Feichtinger, Daniel Freeman, Karlheinz Gröchenig,

Don Hadwin, Hua He, Rui Li, Franz Luef, Fusheng Lv, Chi-Keung Ng, Zhong-Jin Ruan, Thomas Schlumprecht, Jie Shen, Wenchang Sun, Jingsong Wu, Quanhua Xu, Qingyue Zhang, Bentuo Zheng, etc.

Deguang Han acknowledges partial support from NSF grant DMS-2105038. Qianfeng Hu is partial supported by National Natural Science Foundation of China (Grant Nos. 12471131, 12301162 and 12071230) and Natural Science Foundation of Hebei Province of China (Grant No. A2024202039). Bei Liu acknowledges support from Tianjin Natural Science Foundation Project (24JCYBJC00480). Bei Liu and Rui Liu acknowledge support from National Natural Science Foundation of China (Grant Nos. 12471131, 12071230, 11971348). All authors contribute equally and the names of the authors are listed in alphabetical order according to their family names.

Orlando, FL, USA   Deguang Han
Tianjin, China   Qianfeng Hu
Tianjin, China   Bei Liu
Tianjin, China   Rui Liu
July, 2025

Contents

1 Frames in Hilbert Spaces 1
  1.1 Preliminary 1
  1.2 Frames and Basic Properties.. 4
    1.2.1 Frames.. 4
    1.2.2 Frame Operators and Dual Frames................................. 7
    1.2.3 Geometric Characterizations 12
    1.2.4 Continuous Frames 13
  1.3 Structured Frame Theory 15
    1.3.1 Gabor Frames 15
    1.3.2 Frame Representation 18

2 Frames in Banach Spaces 33
  2.1 Schauder Frames in Banach Spaces... 33
    2.1.1 Atomic Decompositions 33
    2.1.2 Duality, Reflexivity and Schauder Frames....................... 39
  2.2 Completely Bounded Frames for Operator Spaces 45
    2.2.1 Operator Spaces and Completely Bounded Maps.................. 45
    2.2.2 Completely Bounded Frames (Cb-Frames) for $C_r^*(\mathbb{F}_2)$ 46
  2.3 Lipschitz Frames in Banach Spaces... 53

3 Frames and Approximation Properties... 59
  3.1 Approximation Properties 60
  3.2 Cb-frames and Completely Bounded Approximation Property............. 69
    3.2.1 Cb-Frames for Operator Spaces...................................... 70
    3.2.2 Subspaces of Operator Spaces with Cb-basis 76
  3.3 Lipschitz Frames and Lipschitz Bounded Approximation Property........ 77
    3.3.1 Lipschitz Bounded Approximation Property for Operators 77
    3.3.2 Linearization for Lipschitz Bounded Approximation Property...... 81

4   Banach Space Dilations 87
   4.1 Classical Dilation Theorems.. 88
      4.1.1 Naimark's Dilation Theorem.. 88
      4.1.2 Stinespring's Dilation Theorem 89
   4.2 Dilations for Operator-Valued Measures................................... 91
      4.2.1 Minimal and Maximal Dilations 92
      4.2.2 Imprimitive Systems... 103
      4.2.3 OVMs with Bounded p-Variations 106
      4.2.4 Quantum Measures 111
   4.3 Homomorphism Dilation for Bounded Linear Operators 116
      4.3.1 Banach Algebra Case.. 116
      4.3.2 Commutative von Neumann Algebra Case 119
      4.3.3 Noncommutative von Neumann Algebra Case...................... 128

5   Algebraic and Generalized Choi-Kraus Dilations............................... 131
   5.1 Algebraic Dilation for Linear Systems..................................... 131
      5.1.1 Principle and Universal Dilations 132
      5.1.2 Structural Theorems 135
   5.2 Generalized Choi-Kraus Dilations 141
      5.2.1 Generalized Choi-Kraus Dilations 142
      5.2.2 Classifications for Linear Maps on Matrix Algebras M_n 150

6   Quantum Detection and Phase Retrieval Frames 161
   6.1 Quantum States and Observables 161
   6.2 Quantum Detection from Frame POVMs 163
      6.2.1 Injective Discrete Frames 164
      6.2.2 Injective Continuous Frames 168
      6.2.3 Injective Finite Multiwindow Gabor Frames 179
   6.3 Phase Retrieval for Finite Frames 182
      6.3.1 Phase-retrievable Frames on Finite Abelian Groups 185
      6.3.2 Exact Phase-retrievable Frames and Subspaces 190

7   Samplings in Mixed Lebesgue Spaces 197
   7.1 Mixed Lebesgue Spaces 197
   7.2 Shift-Invariant Subspaces.. 205
   7.3 The $L^{p,q}$-Stability of the Shift-invariant Subspaces....................... 218
      7.3.1 Principal Generated Case... 219
      7.3.2 Finitely Generated Case 233
   7.4 Nonuniform Samplings 236
      7.4.1 The $L^{p,q}$-Stability of Nonuniform Samplings 236
      7.4.2 The Reconstruction Algorithm...................................... 237

7.5   Samplings for Non-decaying Signals 241
       7.5.1   Samplings in Non-decaying Shift-Invariant Subspaces 243
       7.5.2   Ideal Samplings 255

References 261

Notations

The following is a list of standard notations that will also be used in this book.

$\mathbb{N}$	the set of all positive integers
$\mathbb{Z}$	the set of all the integers
$\mathbb{Q}$	the set of all the rational numbers
$\mathbb{R}$	the real number field
$\mathbb{C}$	the complex number field
$\mathbb{R}^n$ (resp. $\mathbb{C}^n$)	the n-dimensional real (resp. compex) vector space
$M_{n,m}(\mathbb{C})$	the space of all the $n \times m$ complex matrices, abbreviated as $M_{n,m}$ ($M_n = M_{n,n}(\mathbb{C})$)
X, Y, Z	Banach spaces over the scalar field $\mathbb{K} = \mathbb{R}$ or $\mathbb{C}$
X^*	the dual space of a Banach space X. The elements of a dual space X^* will typically be denoted by f, φ, or x^* etc.
$\mathcal{H}$	a separable Hilbert space over $\mathbb{K}$
$\langle \cdot, \cdot \rangle$	the inner product of a Hilbert space $\mathcal{H}$. For Banach spaces, we also use the notation $\langle \cdot, \cdot \rangle$ to denote the bilinear functional given by $\langle x, f \rangle = f(x)$ for all $f \in X^*$, $x \in X$
$B(X, Y)$	the space of all bounded linear operators from a Banach space X to a Banach space Y, and write $B(X) := B(X, X)$
I	the identity operator on a Banach space X The standard notations c_0, $\ell^p (1 \leq p \leq \infty)$, and c_{00} are adopted for the spaces that are commonly used in the literature
$C(K)$	the continuous function space on a compact Hausdorff space K
$L^p(\mu)(1 \leq p \leq \infty)$	the space of all measurable functions that are p-integrable
wot	the weak operator topology of $B(\mathcal{H})$
sot	the strong operator topology of $B(\mathcal{H})$

Frames were introduced by Duffin and Schaeffer in 1952 in the context of nonharmonic Fourier series [69]. After a rapid development in the past decade, frame theory has been widely used to develop innovative techniques in tackling many challenging problems in engineering applications including signal analysis and image processing. It also plays a central role in some theoretical developments in other areas of mathematics including wavelet analysis [60], time-frequency analysis [102], sampling theory [4], operator theory and representation theory [122], nonlinear sparse approximation [62], pseudo-differential operators [90], and quantum computation [74]. The recently settled Kadison-Singer Problem [47, 58] shows that frame theory also has deep connections with many other theoretical problems in mathematics. We refer to [51, 117] for related materials of frame theory and its applications. In this chapter, we briefly introduce some basic properties of frames and their connections with time-frequency or Gabor analysis, wavelet analysis and representation theory.

1.1   Preliminary

Orthonormal bases and Riesz bases for Hilbert spaces are central concepts that are essential in many areas of analysis. Recall that if $\{e_n\}_{n=1}^{\infty}$ is an orthonormal basis of a Hilbert space $\mathcal{H}$, then for every $x \in \mathcal{H}$, we have

$$x = \sum_{n=1}^{\infty} \langle x, e_n \rangle e_n.$$

In other words, orthonormal bases allow us to represent every element explicitly in a simple manner. Recall that every Riesz basis has the form of $\{Te_n\}$, where $\{e_n\}$ is an

© The Author(s), under exclusive license to Springer Nature Switzerland AG 2026
D. Han et al., *Banach Space Aspects of Frame Theory with Applications*, Synthesis
Lectures on Mathematics & Statistics, https://doi.org/10.1007/978-3-032-10174-7_1

orthonormal basis and $T \in B(\mathcal{H})$ is bounded invertible. Riesz bases also allow us to represent every element in a Hilbert space uniquely and stably in terms of series expansions. These concepts have natural extensions to Banach spaces.

Definition 1.1 A sequence of elements $\{x_n\}_{n=1}^{\infty}$ in an infinite-dimensional Banach space X is said to be a *Schauder basis* (or simply, *basis*) of X if for each $x \in X$ there is a unique sequence of scalars $\{a_n\}_{n=1}^{\infty}$ such that

$$x = \sum_{n=1}^{\infty} a_n x_n,$$

where the series converges in norm. A sequence $\{x_n\}_{n=1}^{\infty}$ is called an *unconditional basis* of X if $\{x_{\pi(n)}\}_{n=1}^{\infty}$ is a Schauder basis for every permutation $\pi : \mathbb{N} \to \mathbb{N}$.

Example 1.1.1 Let $e_n = (0, \ldots, 0, 1, 0, \ldots)$ be the sequence where the single nonzero entry is in the n-th coordinate. It is easy to check that $\{e_n\}_{n=1}^{\infty}$ is a basis of ℓ^p for $1 \leq p < \infty$, and of c_0, which is called the *standard basis*. $\square$

Example 1.1.2 Let $X = C[0, 1]$ and consider the Schauder system:

$$\{\chi, \ell\} \cup \{s_{n,k}\}_{n \geq 0, k=0,\ldots,2^{n}-1},$$

where $\chi = \chi_{[0,1]}$, $\ell(t) = t$, and $s_{n,k}$ is the continuous function given by

$$s_{n,k}(t) = \begin{cases} 1, & t = \frac{k+1/2}{2^n}, \\ \text{linear}, & \text{on } [\frac{k}{2^n}, \frac{k+1/2}{2^n}] \text{ and on } [\frac{k+1/2}{2^2}, \frac{k+1}{2^2}], \\ 0, & \text{otherwise} . \end{cases}$$

Some of the elements in this system are demonstrated as follows (Fig. 1.1):

Fix any $f \in C[0, 1]$. We can choose scalars a, b such that the function $g = f - a\chi - b\ell$ satisfies $g(0) - g(1) - 0$. From [51], there exist scalars $c_{n,k}$ such that $g = \sum_{n=0}^{\infty} \sum_{k=0}^{2^n-1} c_{n,k} s_{n,k}$ with uniform convergence of the series. So

$$f = ax + b\ell + g = ax + b\ell + \sum_{n=0}^{\infty} \sum_{k=0}^{2^n-1} c_{n,k} s_{n,k}$$

where the series converges in the uniform norm. Furthermore, this representation is unique. So the Schauder system is a basis for $C[0, 1]$. $\square$

To better understand the bases or frames in a Banach space, we also list the following useful facts [5]. Given a basis $\{x_n\}$ for a Banach space X and $x = \sum_{n=1}^{\infty} a_n x_n$. If we define

Fig. 1.1 Some elements of the Schauder system: $\chi, \ell, s_{0,0}, s_{1,0}, s_{1,1}$ and $s_{2,0}$

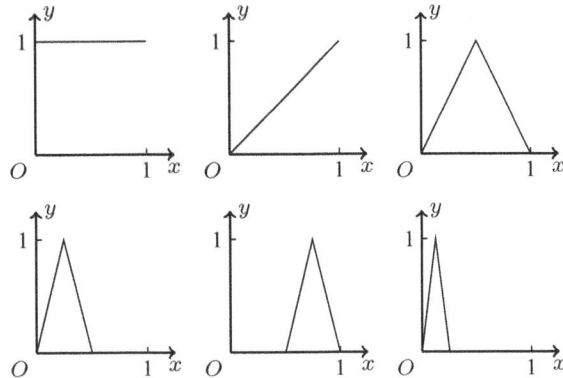

$f_n(x) = a_n$, then f_n is linear functional on X, which is called the *coefficient functionals* for the basis $\{x_n\}$. That is,

$$x = \sum_{n=1}^{\infty} f_n(x)x_n.$$

Theorem 1.1 *Suppose that $\{x_n\}_{n=1}^{\infty}$ is the basis for Banach spaces X and $\{f_n\}_{n=1}^{\infty}$ are the corresponding coefficient functionals. Then $f_n \in X^*$. Moreover, there is a positive constant M such that*

$$1 \le \|x_n\|\|f_n\| \le M, \forall n.$$

Immediately, $f_n(x_m) = \delta_{n,m}$, where $\delta_{n,m}$ is the Kronecker delta defined by $\delta_{n,m} = 1$ if $n = m$, and $\delta_{n,m} = 0$ if $n \ne m$, and the functionals $\{f_n\}$ are called the *biorthogonal functionals* associated to the basis $\{x_n\}$. We also use the pair of sequences $\{x_n, f_n\}$ to denote the basis of X. The following characterization of bases is well known [5].

Theorem 1.2 *A sequence $\{x_n\}_{n=1}^{\infty}$ of nonzero vectors is a basis for a Banach space X if and only if the following conditions are satisfied:*

1. $\{x_n\}$ has dense linear span in X,
2. There is a constant K such that

$$\left\|\sum_{i=1}^{n} a_i x_i\right\| \le K\left\|\sum_{i=1}^{m} a_i x_i\right\|$$

for all scalars (a_i) and all $n < m$.

The number K is called the *basis constant* of the basis $\{x_n\}$. A basis with the basis constant 1 is sometimes called a *monotone basis*. And, the sequence $\{x_n\}$ is called a *basic sequence* if (2) holds.

Definition 1.2 Let $\{x_n\}_{n=1}^\infty$ be a basis for a Banach space X and $\{y_n\}_{n=1}^\infty$ be the basis for a Banach space Y. We say that $\{x_n\}$ is equivalent to $\{y_n\}$ if the convergence of $\sum_{n=1}^\infty c_n x_n$ implies the convergence of $\sum_{n=1}^\infty c_n y_n$ and vice versa for any scalar sequence $\{c_n\}$. If $X = Y$, we write $\{x_n\} \sim \{y_n\}$ to denote this equivalence.

It follows from the Closed Graph Theorem that two bases $\{x_n\}$ and $\{y_n\}$ are equivalent if and only if there exists a constant $0 < C < \infty$ such that

$$C^{-1}\left\|\sum_{i=1}^\infty a_i y_i\right\|_Y \leq \left\|\sum_{i=1}^\infty a_i x_i\right\|_X \leq C\left\|\sum_{i=1}^\infty a_i y_i\right\|_Y$$

for any $(a_i) \in c_{00}$. That is, $\{x_n\}$ and $\{y_n\}$ are equivalent if and only if the basis-to-basis map $x_i \mapsto y_i$ extends to an isomorphism between X and Y. So we get:

Theorem 1.3 *Two bases $\{x_n\}_{n=1}^\infty$ and $\{y_n\}_{n=1}^\infty$ for a Banach space X are equivalent if and only if there is an isomorphism $T : X \to X$ such that $T x_n = y_n$, for all $n \in \mathbb{N}$.*

1.2   Frames and Basic Properties

1.2.1   Frames

Recall that each orthonormal basis $\{e_n\}_{n=1}^\infty$ for a separable infinite-dimensional Hilbert space $\mathcal{H}$ satisfies the Plancherel identity:

$$\sum_{n=1}^\infty |\langle x, e_n\rangle|^2 = \|x\|, \ \forall x \in \mathcal{H}.$$

However, there is a sequence without being orthonormal or a basis that can still satisfy Plancherel equality. Here is a simple example.

Example 1.2.1 Let $\mathcal{H} = \mathbb{R}^2$ and set

$$x_1 = (0, \frac{\sqrt{6}}{3}), \quad x_2 = (-\frac{\sqrt{2}}{2}, \frac{\sqrt{6}}{6}), \quad x_3 = (\frac{\sqrt{2}}{2}, \frac{\sqrt{6}}{6}).$$

Then it is easy to check that $\sum_{n=1}^3 |\langle x, x_n\rangle|^2 = \|x\|^2$ for every $x \in \mathcal{H}$. $\square$

Fig. 1.2 The three vectors of
the Mercedes frame

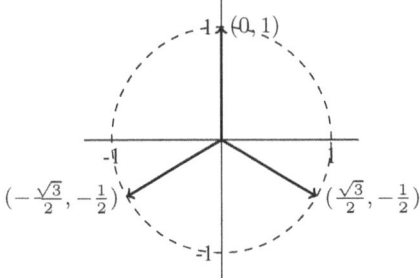

Any sequence satisfying the Plancherel identity will be called a *Parseval frame*. Clearly, the sequence in Example 1.2.1 is not a basis or a union of orthonormal bases. Note that $\|x_n\| = \frac{\sqrt{6}}{3}$ for $n = 1, 2, 3$. If we normalize every vector and denote them by $\{y_n\}_{n=1}^3$, then

$$\sum_{n=1}^{3} |\langle x, y_n \rangle|^2 = \frac{3}{2} \|x\|^2.$$

This system is affectionately referred to as the "Mercedes frame" (See Fig. 1.2).

Parseval frames are exactly the ones such that the ℓ^2-norm of the "Fourier" coefficient sequence of any vector x is the same as the norm of x. A general frame refers to the case when the two norms are equivalent.

Definition 1.3 A frame for a separable Hilbert space $\mathcal{H}$ is a sequence $\{x_n\}_{n=1}^{\infty}$ such that there exist constants $0 < A \le B < \infty$ such that

$$A\|x\|^2 \le \sum_{n=1}^{\infty} |\langle x, x_n \rangle|^2 \le B\|x\|^2, \quad \forall x \in \mathcal{H}. \tag{1.1}$$

The optimal constants are called the upper and lower frame bounds. A frame is called *tight* if $A = B$, and a *Parseval frame* if $A = B = 1$. If we only require the upper bounded condition in Eq. (1.1), then $\{x_n\}$ is called a *Bessel sequence*. We call $\{x_n\}$ an *exact frame* if it ceases to be a frame whenever any single element is deleted from the sequence. A sequence $\{x_k\}_{k=1}^{\infty}$ is called a *frame sequence* if it is a frame for its closed liner span, that is, a frame for $\overline{\text{span}}\{f_k\}_{k=1}^{\infty}$, and a Riesz sequence if it is a Riesz basis for its closed liner span.

Remark 1.1 We make a few remarks some of which can be easily derived from the definitions.

1. Let $\{x_n\}$ be a normalized (i.e., $||x_n|| = 1$) sequence. Then $\{x_n\}$ is a Parseval frame for $\mathcal{H}$ if and only if it is an orthonormal basis.
2. A frame is a complete Bessel sequence and the converse is not true (see Example 1.2.2).
3. Every Riesz basis for a Hilbert space is clearly an exact frame. We will see later that the converse is also true. So inexact frames are always redundant or overcomplete which is an indispensable feature of frames that is crucial in applications.
4. While our definition of the frame is for separable infinite-dimensional Hilbert spaces, frames for finite-dimensional spaces can be defined with the same definition in which we usually consider finite sequences. In this case, frames are exactly the spanning sets, that is, $\text{span}\{x_n\} = \mathcal{H}$. The class of frames in finite-dimensional spaces, called finite frames [54], is very important due to its significant relevance to applications and suitability for computation.
5. If a frame satisfies $||x_n|| = 1$ for all n, then it is called a uniform norm frame or equal norm frame. Benedetto and Fickus [28] gave an elegant characterization of finite uniform norm tight frames (FUNTFs). Moreover, finite frames with good structures, such as group representation frames [210], are of particular interest for both theoretical developments and applications.

The following is a simple example illustrating various terminologies related to the concept of frames.

Example 1.2.2 Let $\mathcal{H}$ be a separable Hilbert space and let $\{e_i\}_{i=1}^{\infty}$ be an orthonormal basis of $\mathcal{H}$.

1. The sequence $\{x_i\}_{i=1}^{\infty} = \{2e_1, e_2, e_3, e_4, \ldots\}$ is an exact frame, that is, a Riesz basis. While $\{y_i\}_{i=1}^{\infty} = \{e_1, e_1, e_2, e_3, e_4, \ldots\}$ is a nontight inexact frame.
2. The sequence

$$\{z_i\}_{i=1}^{\infty} = \{\frac{1}{\sqrt{2}}e_1, \frac{1}{\sqrt{2}}e_1, \frac{1}{\sqrt{2}}e_2, \frac{1}{\sqrt{2}}e_2, \frac{1}{\sqrt{2}}e_3, \frac{1}{\sqrt{2}}e_3, \ldots\}$$

is a Parseval inexact frame.
3. The sequence

$$\{x_i\}_{i=1}^{\infty} = \{e_1, \frac{1}{2}e_2, \frac{1}{3}e_3, \frac{1}{4}e_4, \ldots\}$$

is complete and orthogonal. But it is not a frame since there is no lower frame bound.
4. The sequence

$$\{x_i\}_{i=1}^{\infty} = \{e_1, \frac{e_2}{\sqrt{2}}, \frac{e_2}{\sqrt{2}}, \frac{e_2}{\sqrt{2}}, \frac{e_3}{\sqrt{3}}, \frac{e_3}{\sqrt{3}}, \ldots\}$$

is a Parseval frame that contains an orthogonal basis but not a Riesz basis.

$\square$

Note that the above examples are simple manipulations of an orthonormal basis. Of course, there are many other examples that are not of the above form. Here is a typical example from trigonometric systems [117].

Example 1.2.3 Let $\mathcal{H} = L^2(\mathbf{T})$ be the space of complex-valued functions that are 1-periodic on $\mathbb{R}$ and are square-integrable on the interval $[0, 1]$. Consider the trigonometric system $\{e^{2\pi i bnt}\}_{n\in\mathbb{Z}}$ in $L^2(\mathbf{T})$, where b is a fixed positive real number and $e^{2\pi i bnt}$ is defined on the interval $[0, 1)$ and then extended 1-periodically to $\mathbb{R}$. Then

1. For $b = 1$, $\{e^{2\pi int}\}_{n\in\mathbb{Z}}$ is an orthonormal basis for $L^2(\mathbf{T})$.
2. For $b > 1$, $[0, 1/b]$ is properly contained in $[0, 1)$. Then, for those t such that $t, t + 1/b \in [0, 1)$, we have $e^{2\pi i bnt} = e^{2\pi i bn(t+1/b)}$. Then the extended function on $\mathbb{R}$ in $\overline{\text{span}}\{e^{2\pi i bnt}\}_{n\in\mathbb{Z}}$ are $1/b$-periodic. Not all functions in $L^2(\mathbf{T})$ are of this form, for example $f(t) = t, t \in [0, 1)$. Thus $\{e^{2\pi i bnt}\}_{n\in\mathbb{Z}}$ is incomplete and therefore cannot be a frame for $L^2(\mathbf{T})$
3. For $b < 1$, it can be proved that $\{e^{2\pi int}\}_{n\in\mathbb{Z}}$ is a (redundant) frame for $L^2(\mathbf{T})$. For example, if $b = 1/2$, then we can write $\{e^{2\pi i bnt}\}_{n\in\mathbb{Z}}$ as a union of two orthonormal bases for $L^2(\mathbf{T})$:

$$\{e^{2\pi int/2}\}_{n\in\mathbb{Z}} = \{e^{2\pi int}\}_{n\in\mathbb{Z}} \cup \{e^{2\pi i(n+1/2)t}\}_{n\in\mathbb{Z}}$$
$$= \{e^{2\pi int}\}_{n\in\mathbb{Z}} \cup \{e^{\pi it}e^{2\pi int}\}_{n\in\mathbb{Z}}.$$

Hence, this system is a 2-tight frame. Similarly, if $1/b = M \in \mathbb{N}$ then $\{e^{2\pi i bnt}\}_{n\in\mathbb{Z}}$ is a union of M orthonormal bases hence is an M-tight frame. If a sequence is a union of M orthonormal bases or Riesz bases, it is said to be "M-times overcomplete" or that it has "redundancy M".

$\square$

1.2.2 Frame Operators and Dual Frames

A frame for a Hilbert space behaves exactly like a basis which allows stable reconstructions of every element in the Hilbert space through its dual frames. This is traditionally achieved by using the frame operator of the frame.

Let $\{f_n\}_{n=1}^{\infty}$ be a Bessel sequence with Bessel bound B. Define

$$T : \ell^2(\mathbb{N}) \to \mathcal{H}, \quad T((c_n)) = \sum_{n=1}^{\infty} c_n f_n.$$

Then T is a well-defined bounded linear operator from $\ell^2(\mathbb{N})$ into $\mathcal{H}$ with $\|T\| \leq \sqrt{B}$. We call T the *synthesis operator*. The adjoint operator is given by

$$T^* : \mathcal{H} \to \ell^2(\mathbb{N}), \quad T^*x = (\langle x, x_n \rangle) = \sum_{n=1}^{\infty} \langle x, x_n \rangle e_n,$$

where $\{e_n\}_{n=1}^{\infty}$ is the standard orthonormal basis for $\ell^2(\mathbb{N})$, is called the *analysis operator*. The positive operator $S := TT^* : \mathcal{H} \to \mathcal{H}$ is called the *frame operator* which is given by

$$S : \mathcal{H} \to \mathcal{H}, \quad Sx = TT^*x = \sum_{n=1}^{\infty} \langle x, x_n \rangle x_n, \ \forall\, x \in \mathcal{H}.$$

Following [51], we now list a few crucial properties of the frame operator S.

Proposition 1.1 *Let $\{x_n\}_{n=1}^{\infty}$ be a frame with frame operator S and frame bounds A, B. Then*

1. *S is bounded, invertible, self-adjoint, and positive.*
2. *$\{S^{-1}x_n\}_{n=1}^{\infty}$ is a frame with frame bounds B^{-1}, A^{-1}, and frame operator S^{-1}.*

Proof 1. Then the boundedness of S follows from

$$\|S\| = \|TT^*\| = \|T\|^2 \leq B$$

Since $S^* = (TT^*)^* = TT^* = S$, the operator S is self-adjoint. By definition,

$$A\|x\|^2 \leq \langle Sx, x \rangle \leq B\|x\|^2, \forall x \in \mathcal{H},$$

thus, $AI \leq S \leq BI$ which implies that S is positive and invertible.

2. Note that for $x \in \mathcal{H}$, we have

$$\sum_{n=1}^{\infty} |\langle x, S^{-1}x_k \rangle|^2 = \sum_{n=1}^{\infty} |\langle S^{-1}x, x_k \rangle|^2 \leq B\|S^{-1}x\|^2 \leq B\|S^{-1}\|^2\|x\|^2.$$

That is, $\{S^{-1}x_n\}_{n=1}^{\infty}$ is a Bessel sequence. It follows that the frame operator for $\{S^{-1}f_k\}_{n=1}^{\infty}$ is well-defined. By definition, for $x \in \mathcal{H}$, we have

$$\sum_{n=1}^{\infty} \langle x, S^{-1}x_n \rangle S^{-1}x_n = S^{-1}\left(\sum_{n=1}^{\infty} \langle S^{-1}x, x_n \rangle x_n \right) = S^{-1}SS^{-1}x = S^{-1}x, \quad (1.2)$$

thus, the frame operator for $\{S^{-1}x_n\}_{n=1}^{\infty}$ is S^{-1}. By (1), S^{-1} is also positive then by [117, Theorem 2.18], there is a positive operator $S^{-1/2}$ such that $S^{-1/2}S^{-1/2} = S^{-1}$. Multiplying the inequality $AI \leq S \leq BI$ both sides with $S^{-1/2}$ gives $B^{-1}I \leq S^{-1} \leq A^{-1}I$, that is,

$$B^{-1}\|x\|^2 \leq \langle S^{-1}x, x \rangle \leq A^{-1}\|x\|^2, \forall\, x \in \mathcal{H}.$$

By Eq. (1.2), we have

$$B^{-1}\|x\|^2 \leq \sum_{n=1}^{\infty} |\langle x, S^{-1}x_n \rangle|^2 \leq A^{-1}\|x\|^2, \forall x \in \mathcal{H},$$

thus, $\{S^{-1}x_n\}_{n=1}^{\infty}$ is a frame with frame bounds B^{-1}, A^{-1}.

With the help of the frame operator, every element can be represented as a superposition of the frame elements, which is often referred to as the *frame decomposition*.

Theorem 1.4 *Let $\{x_n\}_{n=1}^{\infty}$ be a frame with frame operator S. Then*

$$x = \sum_{n=1}^{\infty} \langle x, S^{-1}x_n \rangle x_n = \sum_{n=1}^{\infty} \langle x, x_n \rangle S^{-1}x_n, \forall\, x \in \mathcal{H}. \tag{1.3}$$

Proof Let $x \in \mathcal{H}$, by Proposition 1.1, we have

$$x = SS^{-1}x = \sum_{n=1}^{\infty} \langle S^{-1}x, x_n \rangle x_n = \sum_{n=1}^{\infty} \langle x, S^{-1}x_n \rangle x_n$$

and similarly,

$$x = S^{-1}Sx = S^{-1}\left(\sum_{n=1}^{\infty} \langle x, x_n \rangle x_n \right) = \sum_{n=1}^{\infty} \langle x, x_n \rangle S^{-1}x_n.$$

It is natural to view a frame as a generalization of a basis and from Eq. (1.3), the frame $\{S^{-1}x_n\}_{n=1}^{\infty}$ plays the same role in frame theory as the dual of a basis. This leads to the following definition.

Definition 1.4 Let $\{x_n\}_{n=1}^{\infty}$ be a frame with frame operator S. Then the frame $\{S^{-1}x_n\}_{n=1}^{\infty}$ is called the *canonical dual frame* or the *standard dual frame* for $\{x_n\}$.

The sequence $\{\langle x, S^{-1}x_n\rangle\}_{n=1}^{\infty}$ is called the *frame coefficient sequence*, which contains the information of a signal $x \in \mathcal{H}$. In general, it might be very difficult to compute the inverse of a frame operator. However, these technical difficulties can be avoided by using tight frames.

Corollary 1.1 *If a frame* $\{x_n\}_{n=1}^{\infty}$ *is A-tight, then* $S = AI$, $S^{-1} = A^{-1}I$, *and*

$$x = A^{-1} \sum \langle x, x_n \rangle x_n, \quad \forall x \in \mathcal{H}.$$

Definition 1.5 Let $\{x_n\}_{n=1}^{\infty}$ be a frame for a Hilbert space $\mathcal{H}$. If a Bessel sequence $\{y_n\}_{n=1}^{\infty} \subset \mathcal{H}$ satisfies

$$x = \sum_{n=1}^{\infty} \langle x, y_n \rangle x_n, \quad \forall x \in \mathcal{H}$$

where the convergence is unconditional, then $\{y_n\}$ is called an *alternative dual* of $\{x_n\}$.

We remark that an alternative dual is also a frame, and we call $\{x_n, y_n\}_{n=1}^{\infty}$ a *dual frame pair*. Both canonical and alternate duals are often simply referred to as *duals*. It is easy to construct some examples with many dual frames. However, it can be proved that a frame has a unique dual if and only if it is a Riesz basis.

Theorem 1.5 *Let* $\{x_n\}_{n=1}^{\infty}$ *be a frame for a Hilbert space* $\mathcal{H}$ *with frame operator* S. *Set* $\tilde{x}_n = S^{-1}x_n$ *and fix* $x \in \mathcal{H}$. *If* $x = \sum_{n=1}^{\infty} c_n x_n$ *for some scalars* (c_n), *then*

$$\sum_{n=1}^{\infty} |c_n|^2 = \sum_{n=1}^{\infty} |\langle x, \tilde{x}_n \rangle|^2 + \sum_{n=1}^{\infty} |\langle x, \tilde{x}_n \rangle - c_n|^2$$

Proof As $\{\tilde{x}_n\}$ is the canonical dual of $\{x_n\}$ and $x = \sum_{n=1}^{\infty} c_n x_n$ for some $x \in \mathcal{H}$, we get that

$$x = \sum_{n=1}^{\infty} c_n x_n = \sum_{n=1}^{\infty} \langle x, \tilde{x}_n \rangle x_n.$$

Using the self-adjointness of S^{-1}, we have $\langle x_n, S^{-1}x \rangle = \langle \tilde{x}_n, x \rangle$. Taking the inner product of both sides with $S^{-1}x$ gives that

$$\sum_{n=1}^{\infty} c_n \overline{\langle x, \tilde{x}_n \rangle} = \sum_{n=1}^{\infty} |\langle x, \tilde{x}_n \rangle|^2,$$

which is equivalent to the condition that $(c_n - \langle x, \tilde{x}_n \rangle)$ is orthogonal to $(\langle x, \tilde{x}_n \rangle)$ in ℓ_2. Thus

$$\|(c_n)\|_{\ell_2}^2 = \|(c_n - \langle x, \tilde{x}_n \rangle) + (\langle x, \tilde{x}_n \rangle)\|_{\ell_2}^2 = \|(c_n - \langle x, \tilde{x}_n \rangle)\|_{\ell_2}^2 + \|(\langle x, \tilde{x}_n \rangle)\|_{\ell_2}^2,$$

which completes the proof.

Remark 1.2 The above theorem shows that the frame coefficients $(\langle x, \tilde{x}_n \rangle)$ has the minimal ℓ^2-norm among all such sequences $(c_n) \in \ell^2$.

Theorem 1.6 *The removal of a vector x_m from a frame $\{x_n\}_{n=1}^{\infty}$ for $\mathcal{H}$ leaves either a frame or an incomplete set. More precisely:*

1. If $\langle x_m, S^{-1}x_m \rangle \neq 1$, then $\{x_n\}_{n \neq m}$ is a frame for $\mathcal{H}$;
2. If $\langle x_m, S^{-1}x_m \rangle = 1$, then $\{x_n\}_{n \neq m}$ is incomplete.

Proof For simplicity, we use $\tilde{x}_n = S^{-1}x_n$.
1. By assumption, $\langle x_m, \tilde{x}_m \rangle = \langle x_m, S^{-1}x_m \rangle \neq 1$. Since $x_m = \sum_{n=1}^{\infty} \langle x_m, \tilde{x}_n \rangle x_n$, we have $x_m = \frac{1}{1 - \langle x_m, \tilde{x}_m \rangle} \sum_{n \neq m} \langle x_m, \tilde{x}_n \rangle x_n$, and so

$$|\langle x, x_m \rangle|^2 = \left| \frac{1}{1 - \langle x_m, \tilde{x}_n \rangle} \sum_{n \neq m} \langle x_m, \tilde{x}_n \rangle \langle x, x_n \rangle \right|^2 \leq C \sum_{n \neq m} |\langle x, x_n \rangle|^2,$$

where $C = \frac{1}{|1 - \langle x_m, \tilde{x}_m \rangle|^2} \sum_{n \neq m} |\langle x_m, \tilde{x}_n \rangle|^2 > 0$. Therefore,

$$\sum_{n=1}^{\infty} |\langle x, x_n \rangle|^2 = |\langle x, x_m \rangle|^2 + \sum_{n \neq m} |\langle x, x_n \rangle|^2 \leq (1 + C) \sum_{n \neq m} |\langle x, x_n \rangle|^2.$$

Hence, if A, B are the frame bounds for $\{x_n\}$, then

$$\frac{A}{1 + C} \|x\|^2 \leq \frac{1}{1 + C} \sum_{n=1}^{\infty} |\langle x, x_n \rangle|^2 \leq \sum_{n \neq m} |\langle x, x_n \rangle|^2 \leq B\|x\|^2,$$

Thus $\{x_n\}_{n \neq m}$ is a frame with frame bounds $A/(1 + C)$, B.
2. If $\langle x_m, \tilde{x}_m \rangle = 1$. By Theorem 1.1 and 1.5, we have $\langle x_m, \tilde{x}_n \rangle = \langle \tilde{x}_m, x_n \rangle = 0$ for $n \neq m$. That is, $\tilde{x}_m$ is orthogonal to x_n for every $n \neq m$. Thus $\{x_n\}_{n \neq m}$ is incomplete as $\tilde{x}_m \neq 0$.

Similarly, we define the biorthogonal sequence in Hilbert spaces.

Definition 1.6 Let $\{x_n\}_{n=1}^{\infty}$, $\{y_n\}_{n=1}^{\infty}$ be the sequences in the Hilbert space $\mathcal{H}$. $\{x_n\}$ and $\{y_n\}$ are called biorthogonal sequence if $\langle x_n, y_m \rangle = \delta_{n,m}$.

A direct induction gives that a frame is exact if and only if it is biorthogonal to its dual frame, which is one of the equivalent conditions by [51, Theorem 7.1.1] for a frame being a Riesz basis. Therefore, every exact frame for a Hilbert space is equivalent to a Riesz basis, and therefore, both are equivalent to the orthonormal bases. Hence, every exact frame has a unique dual. In contrast, every inexact frame has infinitely many alternative dual frames.

1.2.3   Geometric Characterizations

There is a simple geometric interpretation of Parseval frames and general frames: Frames are exactly orthogonal compressions of Riesz bases for a larger Hilbert space, i.e., is a Parseval frame (resp. frame) for the projection subspace. This was apparently first shown explicitly by Han and Larson in chapter 1 of [122] with an elementary two-line proof. As noted by several authors, it can be alternatively derived by applying the Naimark dilation theorem for operator-valued measures by first passing from a frame sequence to a natural discrete positive operator-valued measure on the power set of the index set. So, this geometric interpretation is often referred to as the Naimark/Han/Larson dilation theorem for frames and is stated as the following proposition.

Proposition 1.2 *Let $\{x_n\}$ be a sequence in a Hilbert space $\mathcal{H}$. Then*

1. *$\{x_n\}$ is a Parseval frame for $\mathcal{H}$ if and only if there exists a Hilbert space $\mathcal{K} \supseteq \mathcal{H}$ and an orthonormal basis $\{u_n\}$ for $\mathcal{K}$ such that $x_n = Pu_n$, where P is the orthogonal projection from $\mathcal{K}$ onto $\mathcal{H}$.*
2. *$\{x_n\}$ is a frame for $\mathcal{H}$ if and only if there exists a Hilbert space $\mathcal{K} \supseteq \mathcal{H}$ and a Riesz basis $\{v_n\}$ for $\mathcal{K}$ such that $x_n = Pv_n$, where P again is the orthogonal projection from $\mathcal{K}$ onto $\mathcal{H}$.*

Furthermore, we also have the following dilation theorem for dual pairs.

Theorem 1.7 *Suppose that $\{x_n\}$ and $\{y_n\}$ are two frames for a Hilbert space $\mathcal{H}$. Then the following are equivalent:*

1. *$\{y_n\}$ is a dual for $\{x_n\}$;*
2. *There exists a Hilbert space $\mathcal{K} \supseteq \mathcal{H}$ and a Riesz basis $\{u_n\}$ for $\mathcal{K}$ such that $x_n = Pu_n$, and $y_n = Pu_n^*$, where $\{u_n^*\}$ is the (unique) dual of the Riesz basis $\{u_n\}$ and P is the orthogonal projection from $\mathcal{K}$ onto $\mathcal{H}$.*

1.2.4 Continuous Frames

The (discrete) frames discussed above represent elements in Hilbert spaces in terms of infinite sums. The notion of continuous frame generalizes this by replacing summing over a discrete set with integrating over a measure space. Following [51], we use the following definition for more general frames.

Definition 1.7 Let $\mathcal{H}$ be a Hilbert space and Ω be a measure spaces with a positive measure μ. A mapping $\mathcal{F} : \Omega \to \mathcal{H}$ is called (Ω, μ)-frame or a continuous frame if

1. for all $x \in \mathcal{H}$, $\omega \mapsto \langle x, \mathcal{F}(\omega) \rangle$ is a measurable function on Ω.
2. there exists constants $A, B > 0$ such that

$$A\|x\|^2 \le \int_\Omega |\langle x, \mathcal{F}(\omega) \rangle|^2 d\mu(\omega) \le B\|x\|^2, \quad \forall x \in \mathcal{H}.$$

As in the discrete frame case, the optimal constants A and B are the upper and lower frame bounds. A tight frame refers to the case when $A = B$, and a Parseval frame or a coherent state refers to the case when $A = B = 1$. The *synthesis operator* for a continuous frame $\mathcal{F}$ is given by

$$V_\mathcal{F} : L^2(\Omega, \mu) \to \mathcal{H}, \quad \langle V_\mathcal{F}(f), x \rangle := \int_\Omega f(\omega) \langle \mathcal{F}(\omega), x \rangle d\mu(\omega).$$

and the *analysis operator* is the operator

$$V_\mathcal{F}^* : \mathcal{H} \to L^2(\Omega, \mu), \quad V_\mathcal{F}^*(x)(\omega) := \langle x, \mathcal{F}(\omega) \rangle.$$

The *frame operator* is the bounded, positive and invertible operator $S_\mathcal{F} := V_\mathcal{F} V_\mathcal{F}^*$ with integral reprsentation of

$$\langle S_\mathcal{F}(x), y \rangle := \int_\Omega \langle x, \mathcal{F}(\omega) \rangle \cdot \langle \mathcal{F}(\omega), y \rangle d\mu(\omega), \forall x, y \in \mathcal{H}.$$

This definition is consistent with the definition of the discrete frames since it follows by setting $X = \mathbb{N}$ and μ being the counting measure. While it seems difficult to verify the frame inequality, many familiar vector-valued functions $\mathcal{F}$ from harmonic analysis are automatically continuous frames. Two such typical examples are associated with short-time Fourier transform (STFT) and continuous wavelet transform (CWT), see [51].

Example 1.2.4 Given a function $g \in L^2(\mathbb{R})\backslash\{0\}$. The *short-time Fourier transform* of a function $f \in L^2(\mathbb{R})$ with respect to the window function g is given by

$$V_g f(a, b) := \int_{\mathbb{R}} f(t)\overline{g(t-a)}e^{-2\pi itb}dt = \langle f(\cdot), e^{2\pi ib\cdot}g(\cdot - a)\rangle_{L^2}.$$

Let T_a be the translation operator defined by $T_a(g)(t) = g(t-a)$, and E_b be the modulation operator defined by $E_b(g)(t) = e^{2\pi itb}g(t)$. Then we have $V_g(f)(a, b) = \langle f, E_b T_a g\rangle$. Let $f_1, f_2, g_1, g_2 \in L^2(\mathbb{R})$. Moyal's formula [102] shows that

$$\int_{-\infty}^{+\infty}\int_{-\infty}^{+\infty} V_{g_1}(f_1)(a, b)\overline{V_{g_2}(f_2)(a, b)}dbda = \langle f_1, f_2\rangle\langle g_2, g_1\rangle.$$

Thus, for $g, f \in L^2(\mathbb{R})$, we get

$$\iint_{\mathbb{R}^2} |\langle f, E_b T_a g\rangle|^2 dbda = \iint_{\mathbb{R}^2} |(F_g f)(t, \omega)|^2 dbda = 2\pi\|g\|_2^2\|f\|_2^2.$$

Therefore $\{E_b T_a g\}_{a,b\in\mathbb{R}} = \{e^{2\pi itb}g(t-a)\}_{a,b\in\mathbb{R}}$ is a continuous (Gabor) frame for $L^2(\mathbb{R})$, where the space we integrate over is $\mathbb{R}^2$, equipped with the Lebesgue measure. In fact, it is a tight frame with $A = B = 2\pi\|g\|_2^2$. □

Example 1.2.5 A function $\psi \in L^2(\mathbb{R})$ is called *admissible* if

$$C_\psi := \int_{-\infty}^{+\infty} \frac{|\hat{\psi}(\gamma)|^2}{|\gamma|}d\gamma < +\infty.$$

Given an admissible function $\psi \in L^2(\mathbb{R})$. The *continuous wavelet transform* of f with respect to ψ is defined by

$$W_\psi(f)(a, b) := \langle f, \psi^{a,b}\rangle = \int_{-\infty}^{+\infty} f(x)\frac{1}{|a|^{\frac{1}{2}}}\overline{\psi(\frac{x-b}{a})}dx.$$

Then, for all functions $f, g \in L^2(\mathbb{R})$, we have

$$\int_{-\infty}^{\infty}\int_{-\infty}^{\infty} W_\psi(f)(a, b)\overline{W_\psi(g)(a, b)}\frac{dadb}{a^2} = C_\psi\langle f, g\rangle.$$

In particular, we get

$$\int_{-\infty}^{\infty}\int_{-\infty}^{\infty} |W_\psi(f)(a, b)|^2\frac{dadb}{a^2} = C_\psi\|f\|^2.$$

Therefore if we set $\psi^{a,b} = \frac{1}{|a|^{\frac{1}{2}}} \psi(\frac{x-b}{a})$, then $\{\psi^{a,b}\}_{a \neq 0, b \in \mathbb{R}}$ is a continuous tight (wavelet) frame for $L^2(\mathbb{R})$ with respect to $(\mathbb{R}\backslash\{0\}) \times \mathbb{R}$ equipped with the measure $\frac{1}{a^2}dadb$. □

Remark 1.3 The continuous Gabor and wavelet frames belong to the category of coherent states which were introduced by Schrödinger and later generalized by Ali et al. [1]. This class of frames appears prominently in quantum mechanics and has been widely used in mathematical physics and harmonic analysis [182].

1.3 Structured Frame Theory

While the structures of both continuous Gabor and wavelet frames make them particularly suitable for applications in mathematical physics, discrete structured frames are crucially important for real-world applications. As mentioned above, Gabor used the time-frequency shifts from a fixed window function as the building blocks to represent signals in his fundamental approach to communication theory [89]. In 1952, Duffin and Schaeffer abstracted Gabor's methods [89] to define *frames* in their paper on nonharmonic Fourier series [69]. Until 1986, frames for $L^2(\mathbb{R})$ based on time-frequency or time-scale translates of functions were constructed by Daubechies, Grossmann, and Meyer in [64]. Using time-frequency shift was also a key element in John von Neumann's fundamental mathematical theory of quantum mechanics, and it led to the theory of rotation C^*-algebras (or twisted C^*-algebras). Discrete frames that have been extensively investigated are the ones that have special structures. These include wavelet frames, Gabor frames and, more generally, group representation frames. Besides, there are several well-known fundamental theorems in Gabor analysis that are naturally connected to the theory of group representation and operator algebras. In this section, we focus on introducing some of our recent work related to the theory of discrete Gabor frames and its generalization to group representation frames.

1.3.1 Gabor Frames

From Example 1.2.3, it follows that $\{e^{i2\pi mx}\}_{m \in \mathbb{Z}}$ is an orthonormal basis for $L^2[0, 1]$, and consquenlty $\{e^{i2\pi mx}\chi_{[n,n+1]}(x)\}_{m,n \in \mathbb{Z}}$ is an orthonormal basis of $L^2(\mathbb{R})$. If we set $g(x) = \chi_{[0,1]}(x)$, then

$$\{E_m T_n g\}_{m,n \in \mathbb{Z}} = \left\{e^{i2\pi mx}\chi_{[n,n+1]}(x)\right\}_{m,n \in \mathbb{Z}}$$

is an orthonormal basis for $L^2(\mathbb{R})$ which is usually referred to as a Gabor basis.

Definition 1.8 Let $g \in L^2(\mathbb{R})$, $a, b \in \mathbb{R}$. Then $\{E_{mb}T_{na}g \mid m, n \in \mathbb{Z}\}$ is called a *Gabor system* or a *Weyl-Heisenberg system*. If $\{E_{mb}T_{na}g(x) \mid m, n \in \mathbb{Z}\}$ is a frame for $L^2(\mathbb{R})$, then it is called a *Gabor (or Weyl-Heisenberg) frame* for $L^2(\mathbb{R})$.

Naturally, one of most fundamental questions in Gabor analysis is to find conditions of g and a, b such that $\{E_{mb}T_{na}g \mid m, n \in \mathbb{Z}\}$ is a frame for $L^2(\mathbb{R})$.

There is a close connection between Gabor systems and the shift-invariant systems considered in [51]. In fact,

$$T_{na}E_{mb}g(x) = e^{-2\pi imnab}e^{2\pi imbx}g(x - na) = e^{-2\pi imnab}E_{mb}T_{na}g(x).$$

This implies that the shift-invariant system $\{T_{na}E_{mb}g\}_{m,n\in\mathbb{Z}}$ is a frame if and only if $\{E_{mb}T_{na}g\}_{m,n\in\mathbb{Z}}$ is a frame. Ron and Shen [184] used this relation to obtain the following characterization of Gabor frames in terms of a matrix inequality.

Given a function $g \in L^2(\mathbb{R})$ and two numbers $a, b > 0$, consider the matrix-valued function

$$M(x) := (g(x - na - m/b))_{m,n\in\mathbb{Z}}, x \in \mathbb{R}$$

which is the bi-infinite matrix with entry in the m-th row and n-th column is $M_{m,n}(x) = g(x - na - m/b)$. Let $M(x)^*$ denote the conjugate transpose of $M(x)$. Then $M(x)M(x)^*$ is a positive operator on $\ell^2(\mathbb{Z})$. The characterization of Gabor frames by Ron and Shen [184] reads as follows:

Theorem 1.8 *Let $A, B > 0$ and the Gabor system $\{E_{mb}T_{na}g\}_{m,n\in\mathbb{Z}}$ be given. Then $\{E_{mb}T_{na}g\}_{m,n\in\mathbb{Z}}$ is a frame for $L^2(\mathbb{R})$ with bounds A, B if and only if*

$$bAI \leq M(x)M(x)^* \leq bBI, \quad a.e.x \in \mathbb{R}$$

where I is the identity operator on $l_2(\mathbb{Z})$.

Another well-known characterization is given in terms of the density of the lattice $a\mathbb{Z} \times b\mathbb{Z}$.

Theorem 1.9 *Let $g \in L^2(\mathbb{R})$ and $a, b > 0$ be given.*

1. *There exists a function g such that $\{E_{mb}T_{na}g\}_{m,n\in\mathbb{Z}}$ is a frame for $L^2(\mathbb{R})$ if and only if $ab \leq 1$.*
2. *There exists a function g such that $\{E_{mb}T_{na}g\}_{m,n\in\mathbb{Z}}$ is a Riesz basis for $L^2(\mathbb{R})$ if and only if $ab = 1$.*

This density theorem, including its high dimensional version, has a long and interesting history of evolution. We refer to the excellent survey of Heil [115] for detailed discussions and various generalizations of this fundamental theorem in Gabor analysis. In the case of $ab \leq 1$, the following results are useful in determining when a window function g produces a Gabor frame.

Theorem 1.10 ([51]) *Let $g \in L^2(\mathbb{R})$, $a, b > 0$ and suppose that*

$$B := \frac{1}{b} \sup_{x \in [0,a]} \sum_{k \in \mathbb{Z}} \left| \sum_{n \in \mathbb{Z}} g(x - na)\overline{g(x - na - k/b)} \right| < \infty.$$

Then $\{E_{mb}T_{na}g\}_{m,n \in \mathbb{Z}}$ is a Bessel sequence with bound B. If in addition the condition

$$A := \frac{1}{b} \inf_{x \in [0,a]} \left[\sum_{n \in \mathbb{Z}} |g(x - na)|^2 - \sum_{k \neq 0} \left| \sum_{n \in \mathbb{Z}} g(x - na)\overline{g(x - na - k/b)} \right| \right] > 0$$

is satisfied, then $\{E_{mb}T_{na}g\}_{m,n \in \mathbb{Z}}$ is a frame for $L^2(\mathbb{R})$ with bounds A, B.

Another basic question in Gabor analysis is to characterize the Gabor frame window functions from a particular class of functions. So far, only very few examples have been fully characterized. For the Gaussian window function $g(x) = e^{-\sigma x^2}$, Daubechies and Grossmann conjectured in 1988 that the Gabor family is a frame for $L^2(\mathbb{R})$ if and only if $|ab| < 1$ in [63], where some supporting evidence was given. This conjecture was later proved by Lyubarskii, Seip and Wallstén [158, 190, 195], and generalized to other functions including Gaussian-related functions [136, 137], totally positive and finite type functions [108].

A function g is said to be *totally positive* if for every pair of sequences $x_1 < x_2 < \cdots < x_N$, and $y_1 < y_2 < \cdots < y_N$, the $N \times N$ matrix $[g(x_i - y_j)]$ has nonnegative determinant, and it has *finite type M* if

$$\widehat{g}(\gamma) = \prod_{\nu=1}^{M}(1 + 2\pi i \delta_\nu \gamma)^{-1}$$

for $\delta_\nu \in \mathbb{R}$, $\delta_\nu \neq 0$, and $M \geq 2$.

Theorem 1.11 *Given $a, b > 0$. Let the window function g be any one of the following cases: Gaussian function $g(t) = e^{-\pi t^2}$, one-sided exponential function $g(t) = e^{-t}\chi_{t>0}(t)$, two-sided exponential function $g(t) = e^{-|t|}$, hyperbolic secant $g(t) = (e^t + e^{-t})^{-1}$, or totally positive functions of finite type $M \geq 2$. Then $\{E_{mb}T_{na}g\}_{m,n \in \mathbb{Z}}$ is a frame if and only if $ab < 1$.*

Another simple and interesting case is the characteristic function $g = \chi_{[0,c]}$. In this case, it asks to find the conditions on a, b, c such that $(E_{mb}T_{na}g\})$ is a frame of $L^2(\mathbb{R})$. This turns out to be a very challenging problem [70, 107, 138].

Finally, we point out that in addition to the density theorem, there are several other fundamental theorems in Gabor analysis including the duality principle which will be discussed in the next subsection. In particular, the following well-known Balian-Low theorem shows that an uncertainty principle phenomenon shows up in the critical $ab = 1$ case, which also indicates that Gabor frames in the overcomplete $ab < 1$ case are more suitable for practical applications.

Theorem 1.12 (Balian-Low [13, 155]) *Let $g \in L^2(\mathbb{R})$ and $\hat{g}$ be its Fourier transform. If $\{e^{2\pi i m x} g(x - n) \mid m, n \in \mathbb{Z}\}$ is an orthonormal basis for $L^2(\mathbb{R})$, then*

$$\int_{\mathbb{R}} x^2 |g(x)|^2 dx = +\infty \quad or \quad \int_{\mathbb{R}} \xi^2 |\hat{g}(\xi)|^2 d\xi = +\infty.$$

1.3.2   Frame Representation

In this subsection we discuss some recent work related to the generalization of Gabor frames. The theory of Gabor analysis is best studied in the time-frequency representations for $L^2(\mathbb{R}^d)$ or even more generally, for the Hilbert space of square-integrable functions on locally compact abelian groups. Similar to the time-frequency analysis on the real line, the time-frequency representation $\pi : \mathbb{R}^{2d} \rightarrow B(L^2(\mathbb{R}^d))$ is defined by

$$\pi(\lambda) = E_{\lambda_1} T_{\lambda_2}$$

where $\lambda = (\lambda_1, \lambda_2) \in \mathbb{R}^d \times \mathbb{R}^d$, and the translation operator T_t and the modulation operator E_s are defined by

$$T_t f(x) = f(x - t), \quad E_s f(x) = e^{2\pi i \langle s, x \rangle} f(x), \forall f \in L^2(\mathbb{R}^d).$$

From the perspective of practical applications, here we are interested in the restrictions of π to a discrete subset in the time-frequency space $\mathbb{R}^d \times \mathbb{R}^d$, and in particular to a full rank lattice. By a *full rank lattice* in $\mathbb{R}^d$, we mean a lattice $\mathcal{L} = M\mathbb{Z}^d$ for some non-singular real $d \times d$ matrix M. The volume of $\mathcal{L}$ is defined to be $v(\mathcal{L}) = |\det(M)|$, which is also the Lebesgue measure of any fundamental domain of $\mathcal{L}$. (A fundamental domain Ω of $\mathcal{L}$ is a measurable subset of $\mathbb{R}^d$ such that $\{\Omega + \lambda : \lambda \in \Lambda\}$ forms a partition, modulo measure-zero sets, of $\mathbb{R}^d$). A full-rank lattice Λ in the time-frequency domain $\mathbb{R}^d \times \mathbb{R}^d$ is often referred to as a time-frequency lattice.

From representation point of view, while the restriction $\pi_\Lambda := .\pi|_\Lambda$ of π to a time-frequency lattice Λ is not necessarily a unitary representation for the abelian group $G =$

$\mathbb{Z}^d \times \mathbb{Z}^d$, it is a projective unitary representation of $\mathbb{Z}^d \times \mathbb{Z}^d$. Recall that a *projective unitary representation* π for a countable group G is a mapping $g \to \pi(g)$ from G into the group $U(\mathcal{H})$ of all the unitary operators on a separable Hilbert space $\mathcal{H}$ such that $\pi(g)\pi(h) = \mu(g, h)\pi(gh)$ for all $g, h \in G$, where $\mu(g, h)$ is a scalar-valued function on $G \times G$ taking values in the circle group $\mathbb{T}$. This function $\mu(g, h)$ is called a *multiplier* or *2-cocycle* of G. In this case, we also say that π is a μ-projective unitary representation. It is clear from the definition that we have

1. $\mu(g_1, g_2 g_3)\mu(g_2, g_3) = \mu(g_1 g_2, g_3)\mu(g_1, g_2)$ for all $g_1, g_2, g_3 \in G$,
2. $\mu(g, e) = \mu(e, g) = 1$ for all $g \in G$, where e denotes the group unit of G.

Any function $\mu : G \times G \to \mathbb{T}$ satisfying (1)–(2) above will be called a multiplier or 2-cocycle for G. It follows from (1) and (2) that we also have

3. $\mu(g, g^{-1}) = \mu(g^{-1}, g)$ for all $g \in G$.

Examples of projective unitary representations include unitary group representations (i.e., $\mu \equiv 1$) and the Gabor representations in time-frequency analysis.

Given a projective unitary representation π of a countable group G on a Hilbert space $\mathcal{H}$. A vector $\xi \in \mathcal{H}$ is called a *frame vector* (resp. Parseval frame vector) for π if $\{\pi(g)\xi\}_{g \in G}$ (here we view this as a sequence indexed by G) is a frame (resp. Parseval frame) for the entire Hilbert space $\mathcal{H}$. A *Bessel vector* for π is a vector $\xi \in \mathcal{H}$ such that $\{\pi(g)\xi\}_{g \in G}$ is Bessel. We will use $\mathcal{B}_\pi$ to denote the set of all the Bessel vectors of π. A projective unitary representation π is called a *frame representation* if it admits a frame vector.

Similar to the group unitary representation case, the left and right regular projective representations with a prescribed multiplier μ for G can be defined by

$$\lambda_g \chi_h = \mu(g, h)\chi_{gh}, \quad \text{and} \quad \rho_g \chi_h = \mu(h, g^{-1})\chi_{hg^{-1}}, \quad h \in G,$$

where $\{\chi_g : g \in G\}$ is the standard orthonormal basis for $\ell^2(G)$. Clearly, λ_g and ρ_g are unitary operators on $\ell^2(G)$. Moreover, λ is a μ-projective unitary representation of G with multiplier $\mu(g, h)$ and ρ is a projective unitary representation of G with multiplier $\overline{\mu(g, h)}$. The representations λ and ρ are called the left regular μ-projective representation and the right regular μ-projective representation of G, respectively.

Recall that a von Neumann algebra $\mathcal{M}$ or w^*-algebra is a $*$-closed algebra of bounded operators on a Hilbert space that is closed in the weak operator topology and contains the identity operator I. A finite von Neumann algebra is a von Neumann algebra in which every isometry is unitary. A von Neumann algebra $\mathcal{M}$ is called a *factor* if $\mathcal{M} \cap \mathcal{M}' = \mathbb{C}I$, where $\mathcal{M}'$ is the commutant of $\mathcal{M}$. We use $\mathcal{M}_\lambda$ and $\mathcal{M}_\rho$ to denote the von Neumann algebras generated by λ and ρ, respectively.

Proposition 1.3 ([92]) *Both $\mathcal{M}_\lambda$ and $\mathcal{M}_\rho$ are finite von Neumann algebras, and $\mathcal{M}_\rho$ is the commutant of $\mathcal{M}_\lambda$. Moreover, if for each $e \neq u \in G$, either $\{vuv^{-1} : v \in G\}$ or $\{\mu(vuv^{-1}, v)\overline{\mu(v, u)} : v \in G\}$ is an infinite set, then both $\mathcal{M}_\lambda$ and $\mathcal{M}_\rho$ are factors.*

The following tells us that frame representations are precisely the subrepresentations of the left regular representation of the group G.

Proposition 1.4 *Let π be a projective unitary representation π of a countable group G on a Hilbert space H. Then π is a frame representation if and only if π is unitarily equivalent to a subrepresentation of the left regular projective unitary representation of G. Consequently, if π is frame representation, then both $\pi(G)'$ and $\pi(G)''$ are finite von Neumann algebras.*

For general group representations, there is a slightly more general result: The set of all the Bessel vectors of π is dense in $\mathcal{H}$ if and only if π is unitarily equivalent to the direct sum of at most countably many subrepresentations of the left regular projective unitary representation of G.

Remark 1.4 Let π be a projective unitary representation π of a countable group G on a Hilbert space $\mathcal{H}$, and ξ be a frame vector for π. Then the frame operator S commutes with $\pi(g)$ for all $g \in G$. Thus, $\{\pi(g)S^{-1/2}\xi\}_{g \in G}$ is a Parseval frame for $\mathcal{H}$. In particular, we have that π admits a Riesz vector ξ (i.e., $\{\pi(g)\xi\}_{g \in G}$ is a Riesz basis for $\mathcal{H}$) if and only if it admits a complete wondering vector η (i.e., $\{\pi(g)\eta\}_{g \in G}$ is an orthonormal basis for $\mathcal{H}$).

1.3.2.1 The Density Property

For time-frequency representation frames for $L^2(\mathbb{R}^d)$, similar to the one dimensional case, one basic question is to characterize the lattice Λ that admits a single function generated frame $\{\pi(\lambda)g : \lambda \in \Lambda\}$ for $L^2(\mathbb{R}^d)$. The following generalizes the one-dimensional density theorem for the existence of Gabor frames.

Theorem 1.13 *Let Λ be a full rank lattice in $\mathbb{R}^d \times \mathbb{R}^d$. The following are equivalent:*

1. *There exists a function $g \in L^2(\mathbb{R}^d)$ such that $\{\pi(\lambda)g : \lambda \in \Lambda\}$ is a Parseval frame for $L^2(\mathbb{R}^d)$.*
2. *There exists a function $g \in L^2(\mathbb{R}^d)$ such that $\{\pi(\lambda)g : \lambda \in \Lambda\}$ is a frame for $L^2(\mathbb{R}^d)$.*
3. *There exists a function $g \in L^2(\mathbb{R}^d)$ such that $\{\pi(\lambda)g : \lambda \in \Lambda\}$ is a complete family for $L^2(\mathbb{R}^d)$.*
4. *$v(\Lambda) \leq 1$.*

Moreover, there exists a Riesz basis (equivalently, an orthonormal basis) if and only if $v(\Lambda) = 1$.

Roughly speaking, the existence of a Gabor frame (even a complete Gabor family) requires that the time-frequency lattice Λ cannot be spread out "too sparsely" throughout the time-frequency plane $\mathbb{R}^{2d}$. This can be exactly measured by the *Beurling density* $\frac{1}{v(\Lambda)}$ for the lattice Λ. Again, for a detailed history of evolution of the above theorem we refer to two excellent survey papers by Chris Heil [115, 116]. We can generalize this result to a more general framework in the context of frame representation theory for more general groups. For this purpose, we will introduce a useful trace formula for the commutant of the von Neumann algebra generated by a projective group representation.

Given a projective representation π of a countable group G on a Hilbert space $\mathcal{H}$ and a vector $x \in \mathcal{H}$. The analysis operator Θ_x is defined by

$$\Theta_x : \mathcal{D}(\Theta_x)(\subseteq \mathcal{H}) \mapsto \ell^2(G), \quad \Theta_x(y) = \sum_{g \in G} \langle y, , \pi(g)x \rangle \chi_g,$$

where $\mathcal{D}(\Theta_x) = \{y \in \mathcal{H} : \sum_{g \in G} |\langle y, \pi(g)x \rangle|^2 < \infty\}$ is the domain space of Θ_x. Clearly, $\mathcal{B}_\pi \subseteq \mathcal{D}(\Theta_x)$ holds for every $x \in \mathcal{H}$. In the case that $\mathcal{B}_\pi$ is dense in $\mathcal{H}$, we have that Θ_x is a densely defined and closable linear operator from $\mathcal{B}_\pi$ to $\ell^2(G)$ (cf. [91]). Moreover, $x \in \mathcal{B}_\pi$ if and only if Θ_x is a bounded linear operator on $\mathcal{H}$, which in turn is equivalent to the condition that $\mathcal{D}(\Theta_x) = \mathcal{H}$.

Lemma 1.1 *Let π be a projective representation of a countable group G on a Hilbert space $\mathcal{H}$ such that $\mathcal{B}_\pi$ is dense in $\mathcal{H}$. Then*

$$\pi(G)' = \overline{\text{span}}^{wot} \left\{ \Theta_\eta^* \Theta_\xi : \xi, \eta \in \mathcal{B}_\pi \right\}$$

where "wot" denotes the closure in the weak operator topology. Moreover, if there exist finitely many vectors $\xi_i (i = 1, \ldots, k)$ such that $\bigcup_{i=1}^k \{\pi(g)\xi_i : g \in G\}$ is a Parseval frame for $\mathcal{H}$, then

$$\text{Tr}_\pi(A) = \sum_{i=1}^k \langle A\xi_i, \xi_i \rangle$$

defines a faithful trace on $\pi(G)'$.

For the reader's benefit, we include a short proof of the first part for the frame representation case and refer to [92] for the rest of the proof.

Proof Assume that π is a frame representation hence $\mathcal{B}_\pi$ is dense in $\mathcal{H}$. First it is easy to directly verify that $\Theta_\eta^* \Theta_\xi \in \pi(G)'$. for every $\xi, \eta \in \mathcal{B}_\pi$. Now let $A \in \pi(G)'$ and assume that φ is a Parseval frame vector for π. Then we have $A\varphi \in \mathcal{B}_\pi$. So we have

$A = A \cdot I = A(\Theta_\varphi^* \Theta_\varphi) = \Theta_{A\varphi}^* \Theta_\varphi$. Thus, we have

$$\pi(G)' = \left\{ \Theta_\eta^* \Theta_\xi : \xi, \eta \in \mathcal{B}_\pi \right\}$$

when π is a frame representation.

Even in the case that $k = \infty$, the quantity $\operatorname{Tr}_\pi(I) = \sum_{i=1}^\infty \|\xi_i\|^2$ is well-defined (could be infinity) and it is independent of the choices of the Parseval frame generators $\{\xi_i\}$.

Theorem 1.14 *Let π be a projective representation of a countable group G on a Hilbert space $\mathcal{H}$ such that $\mathcal{B}_\pi$ is dense in $\mathcal{H}$. If $\operatorname{Tr}_\pi(I) > 1$, then for any $x \in \mathcal{H}$, $\{\pi(g)x : g \in G\}$ is not a complete sequence in $\mathcal{H}$. Moreover, π admits a complete sequence if and only if it admits a Parseval frame vector.*

Proof Suppose that there exists $x \in \mathcal{H}$ such that $\{\pi(g)x : g \in G\}$ is a complete sequence in $\mathcal{H}$. Let $\Theta_x = U|\Theta_x|$ be the polar decomposition of the (possibly unbounded) operator Θ_x. Then U induces an equivalence between π and the subrepresentation of the left regular representation of G restricted to the closure of the range space of Θ_x. Thus π admits a complete Parseval frame vector ξ and so $\operatorname{Tr}_\pi(I) = \|\xi\|^2 \leq 1$, which leads to a contradiction. The above argument also proves the "moreover" part of the theorem.

Returning to the density theorem for the Gabor family, we can prove that $Tr_\pi(I) = v(\Lambda)$ and hence we get (3) $\implies$ (4). It is clear that (1) $\implies$ (2) $\implies$ (3). The implication of (4) $\implies$ (1) turns out to be the most difficult part except for the one-dimensional separable lattices (i.e., $\Lambda = a\mathbb{Z} \times b\mathbb{Z}$). The general case was settled by Bekka [26]. However, his proof is not constructive and relies on techniques from the theory of von Neumann algebras. Prior to that, Han and Wang settled the separable lattice case by solving a conjecture on common lattice tilings [133, 134].

Let Ω be a measurable set in $\mathbb{R}^d$ (not necessarily bounded), and let $\mathcal{L}$ be a full rank lattice in $\mathbb{R}^d$. We say Ω *tiles* $\mathbb{R}^d$ by $\mathcal{L}$, or Ω is a *fundamental domain* of $\mathcal{L}$, if

(i) $\bigcup_{\ell \in \mathcal{L}}(\Omega + \ell) - \mathbb{R}^d$ a.e.;
(ii) $(\Omega + \ell) \cap (\Omega + \ell')$ has Lebesgue measure 0 for any $\ell \neq \ell'$ in $\mathcal{L}$.

We say that Ω *packs* $\mathbb{R}^d$ by $\mathcal{L}$ if only (ii) holds. Equivalently, Ω tiles $\mathbb{R}^d$ by $\mathcal{L}$ if and only if

$$\sum_{\ell \in \mathcal{L}} \chi_\Omega(x - \ell) = 1 \quad \text{for a.e. } x \in \mathbb{R}^d,$$

and Ω packs $\mathbb{R}^d$ by $\mathcal{L}$ if and only if

$$\sum_{\ell \in \mathcal{L}} \chi_\Omega(x - \ell) \leq 1 \text{ for a.e. } x \in \mathbb{R}^d.$$

One of the questions in lattice tiling is: Let $\mathcal{L}_1, \mathcal{L}_2, \ldots, \mathcal{L}_m$ be full rank lattices in $\mathbb{R}^d$ such that $v(\mathcal{L}_1) = v(\mathcal{L}_2) = \cdots = v(\mathcal{L}_m)$. Does there exist a measurable set Ω in $\mathbb{R}^d$ such that Ω tiles $\mathbb{R}^d$ by $\mathcal{L}_j$ for each $1 \leq j \leq m$? This question was confirmed in [133] to be positive for $m = 2$ but negative for $m \geq 3$.

Theorem 1.15 *Let $\mathcal{L}, \mathcal{K}$ be two full rank lattices in $\mathbb{R}^d$ such that $v(\mathcal{L}) \leq v(\mathcal{K})$. Then there exists a measurable set Ω in $\mathbb{R}^d$ such that Ω tiles $\mathbb{R}^d$ by $\mathcal{L}$ and packs $\mathbb{R}^d$ by $\mathcal{K}$. In particular, if $v(\mathcal{L}) = v(\mathcal{K})$, then they have a common fundamental domain.*

Now we can give the proof of "(4) $\Rightarrow$ (1)" of Theorem 1.13 for the separable lattice $\Lambda = A\mathbb{Z}^d \times B\mathbb{Z}^d$ case: Assume that $v(\Lambda) = |det(AB)| \leq 1$. Then $|\det B| \leq |\det(A^T)^{-1}|$. Let Ω in $\mathbb{R}^d$ be such that it tiles $\mathbb{R}^d$ by $B\mathbb{Z}^d$ and packs $\mathbb{R}^d$ by $(A^T)^{-1}\mathbb{Z}^d$. Define $g = \frac{1}{\sqrt{|\det A|}} \chi_\Omega$. Then it is easy to verify that the Gabor family $\{e^{2\pi i \langle Am, x \rangle} g(x - Bn) : m, n \in \mathbb{Z}^d\}$ is a Parseval frame for $L^2(\Omega)$. □

We remark that the condition $Tr_{\pi(G)'}(I) \leq 1$ is also sufficient for a Gabor representation $\pi := \pi_\Lambda$ to admit a complete frame vector for $L^2(\mathbb{R}^d)$, but this condition is not sufficient in general for π to be a frame representation of a group G. In fact, this condition is no longer sufficient even for the Gabor representation restricted to a proper π_Λ-invariant subspace (see [91]). Another simple example is the following: Let $G = \mathbb{Z}$ and $\mathcal{H} = L^2[0, 1/2] \oplus L^2[0, 1/2]$. Define $\pi(n) = diag(E_n, E_n)$ with $E_n f(x) = e^{2\pi i n x} f(x)$ for $f \in L^2[0, 1/2]$. Then $Tr_{\pi(G)'}(I) = 1$. But π does not admit a cyclic vector for $\mathcal{H}$. This leads to the following problem:

Problem 1.1 Identify classes of group and 2-cocycle pairs (G, μ) such that the trace condition $Tr_{\pi(G)'}(I) \leq 1$ is sufficient for the existence of a frame generator for every μ-projective unitary representation π.

One such class consists of the pairs that are associated with factor von Neumann algebras.

Theorem 1.16 ([112]) *Let G be a countable group and μ be a 2-cocycle of G. Assume that the von Neumann algebra generated by the μ-projective left regular representation λ is a factor. Then for any μ-projective unitary representation π of G on a Hilbert space $\mathcal{H}$ such that $\mathcal{B}_\pi$ is dense in $\mathcal{H}$, the following statements are equivalent:*

1. *π is a cyclic representation, i.e., there exists a vector ξ such that $\{\pi(g)\xi\}_{g \in G}$ is a complete sequence.*
2. *π is a frame representation.*
3. *$Tr_{\pi(G)'}(I) \leq 1$.*

1.3.2.2 The Duality Principle

The well-known duality principle is another fundamental theorem in time-frequency analysis, and it is closely related to the fundamental identity (Janssen representation) and the Wexler-Raz biorthogonal relations for Gabor representation. For a full rank time-frequency lattice Λ, the *adjoint (or dual) lattice* is the lattice defined by

$$\Lambda^\circ = \left\{ x \in \mathbb{R}^d \times \mathbb{R}^d : \langle x, \lambda \rangle \in \mathbb{Z}, \forall \lambda \in \Lambda \right\}.$$

In the separable lattice $\Lambda = A\mathbb{Z}^d \times B\mathbb{Z}^d$ case, we have $\Lambda^\circ = A^{-1}\mathbb{Z}^d \times B^{-1}\mathbb{Z}^d$.

The fundamental identity of Gabor analysis states that if f, g, h, k are Bessel vectors for π_Λ, then

$$\sum_{m,n \in \mathbb{Z}^d} \langle f, \pi_\Lambda(m,n)g \rangle \langle \pi_\Lambda(m,n)h, k \rangle$$

$$= \frac{1}{\mathrm{v}(\Lambda)} \sum_{m,n \in \mathbb{Z}^d} \langle f, \pi_{\Lambda^\circ}(m,n)k \rangle \langle \pi_{\Lambda^\circ}(m,n)h, g \rangle,$$

and the Wexler-Raz biorthogonality presents another fundamental identity connecting π_Λ with π_{Λ°:

$$\langle \pi_{\Lambda^\circ}(m,n)g, S^{-1}g \rangle = \frac{1}{\mathrm{v}(\Lambda)} \delta_{(m,n),(0,0)},$$

where $\{\pi_\Lambda(m,n)g\}$ is a frame for $L^2(\mathbb{R}^d)$ and S is its frame operator. Note that the Wexler-Raz biorthogonality implies that if $\{\pi_\Lambda(m,n)g\}$ is a frame for $L^2(\mathbb{R}^d)$, then $\{\pi_{\Lambda^\circ}(m,n)g\}$ and $\{\pi_{\Lambda^\circ}(m,n)S^{-1}g\}$ are biorthogonal sequences, and in particular $\{\pi_{\Lambda^\circ}(m,n)g\}$ is a Riesz sequence. The duality principle tells us that the converse is also true.

Theorem 1.17 *Let Λ be a lattice and $g \in L^2(\mathbb{R}^d)$. Then the following are true:*

(i) *The Gabor family $\{\pi_\Lambda(m,n)g\}$ is a frame for $L^2(\mathbb{R}^d)$ if and only if $\{\pi_{\Lambda^\circ}(m,n)g\}$ is a Riesz sequence;*

(ii) *The Gabor family $\{\pi_\Lambda(m,n)g\}$ is a Parseval frame for $L^2(\mathbb{R}^d)$ if and only if*

$$\left\{ \frac{1}{\mathrm{v}(\Lambda)} \pi_{\Lambda^\circ}(m,n)g \right\}$$

is an orthonormal sequence.

This well-known duality principle for Gabor analysis was independently and simultaneously discovered by Daubechies, Landau, and Landau [67], Janssen [136], and Ron and Shen [184]. The techniques used in these three articles to prove the duality principle are quite different. In particular, Daubechies/Landau/Landau's approach utilized von Neumann algebra techniques in their arguments.

It turns out that this duality principle is not accidental, and in fact, it represents a general phenomenon that belongs to the general representation theory. A few key facts are important in order to explain the generality of this duality phenomenon. First, the density theorem (c.f. [133]) implies that one of two projective unitary representations π_Λ and π_{Λ° for the group $G = \mathbb{Z}^d \times \mathbb{Z}^d$ must be a frame representation and the other admits a Riesz sequence vector. So we can always assume that π_Λ is a frame representation of $\mathbb{Z}^d \times \mathbb{Z}^d$ and hence π_{Λ° admits a Riesz sequence vector. Secondly, the von Neumann algebras generated by the two representations are commutants to each other, i.e., $\pi_\Lambda(G)' = \pi_{\Lambda^\circ}(G)''$. Moreover, both representations share the same Bessel vectors. This motivated us to propose the following definition:

Definition 1.9 Let π be a μ-projective frame representation and σ be a $\bar{\mu}$-projective unitary representation of a countable group G on the same Hilbert space $\mathcal{H}$. We say that π and σ form a commutant pair if the following conditions are satisfied:

1. $\pi(G)' = \sigma(G)''$,
2. $\mathcal{B}_\pi = \mathcal{B}_\sigma$.
3. There exists a vector $\xi \in \mathcal{H}$ such that $\{\sigma(g)\xi : g \in G\}$ is a Riesz sequence.

The general duality principle shows that any commutant pair shares the same duality principle as the one a time-frequency representation pair $(\pi_\Lambda, \pi_{\Lambda^\circ})$ presented to us, where Λ° is the dual lattice of Λ.

Theorem 1.18 *Let (π, σ) be a commutant pair of projective unitary representations of G on $\mathcal{H}$. Then we have*

1. *$\{\pi(g)\xi\}_{g \in G}$ is a frame sequence if and only if $\{\sigma(g)\xi\}_{g \in G}$ is a frame sequence.*
2. *$\{\pi(g)\xi\}_{g \in G}$ is a frame for $\mathcal{H}$ if and only if $\{\sigma(g)\xi\}_{g \in G}$ is a Riesz sequence.*
3. *$\{\pi(g)\xi\}_{g \in G}$ a Parseval frame for $\mathcal{H}$ if and only if $\{\sigma(g)\xi\}_{g \in G}$ is an orthogonal sequence.*

The proof of the above result relies on several key ingredients developed by the author and his collaborators for frame theory and frame representations. In particular, the concepts of super-frames (i.e., orthogonal frames) and the theory of dilations/parameterizations of frame generators played essential roles in the proof.

Definition 1.10 Suppose π is a projective unitary representation of a countable group G on a Hilbert space $\mathcal{H}$ such that the set $\mathcal{B}_\pi$ of Bessel vectors for π is dense in $\mathcal{H}$. We say that two vectors x and y in $\mathcal{H}$ are π-*orthogonal* if the ranges of Θ_x and Θ_y are orthogonal, and that they are π-*weakly equivalent* if the closures of the ranges of Θ_x and Θ_y are the same.

The following result characterizes the π-orthogonality and π-weakly equivalence in terms of the commutant of $\pi(G)$.

Proposition 1.5 *Let π be a projective representation of a countable group G on a Hilbert space $\mathcal{H}$ such that $\mathcal{B}_\pi$ is dense in $\mathcal{H}$, and let $x, y \in \mathcal{H}$. Then*

(i) x and y are π-orthogonal if and only if $[\pi(G)'x] \perp [\pi(G)'y]$ (or equivalently, $x \perp \pi(G)'y$);
(ii) x and y are π-weakly equivalent if and only if $[\pi(G)'x] = [\pi(G)'y]$.

For reader's convenience, we include the proof of (i): We first verify that

$$\langle \Theta_x(\xi), \Theta_y(\eta) \rangle = \langle \Theta_\eta(y), \Theta_\xi(x) \rangle$$

holds for any $\xi, \eta \in \mathcal{B}_\pi$ and $x, y \in \mathcal{H}$. Indeed, it follows from the following direct calculation:

$$\langle \Theta_x(\xi), \Theta_y(\eta) \rangle = \sum_{g \in G} \langle \xi, \pi(g)x \rangle \langle \pi(g)y, \eta \rangle$$

$$= \sum_{g \in G} \langle \pi(g)^*\xi, x \rangle \langle y, \pi(g)^*\eta \rangle$$

$$= \sum_{g \in G} \overline{\mu(g, g^{-1})}\mu(g, g^{-1})\langle \pi(g^{-1})\xi, x \rangle \langle y, \pi(g^{-1})\eta \rangle$$

$$= \sum_{g \in G} \langle y, \pi(g^{-1})\eta \rangle \langle \pi(g^{-1})\xi, x \rangle$$

$$= \langle \Theta_\eta(y), \Theta_\xi(x) \rangle.$$

By Lemma 1.1 $\pi(G)'$ is the weak operator closure of the linear span of $\{\Theta_\eta^*\Theta_\xi : \eta, \xi \in \mathcal{B}_\pi\}$. Thus, we have that $[\pi(G)'x] \perp [\pi(G)'y]$ if and only if $\langle y, \Theta_\eta^*\Theta_\xi(x) \rangle = 0$ for all $\xi, \eta \in \mathcal{B}_\pi$. However, from the above identity, we obtain that

$$\langle \Theta_x(\xi), \Theta_y(\eta) \rangle = \langle \Theta_\eta(y), \Theta_\xi(x) \rangle = \langle y, \Theta_\eta^*\Theta_\xi(x) \rangle, \forall \xi, \eta \in \mathcal{B}_\pi.$$

This implies that $[\pi(G)'x] \perp [\pi(G)'y]$ if and only if $\Theta_x(\mathcal{B}_\pi) \perp \Theta_y(\mathcal{B}_\pi)$.

One natural question: other than the example of dual pairs for time-frequency representations, do commutant pairs exist for other types of groups and representations? More specifically, we ask:

Problem 1.2 (i) Does every frame representation π admit a commutant pair (π, σ)?

(ii) If the answer to (i) is negative in general, then identify the group G and the 2-cocycle μ that have the property that every μ-projective frame representation π of G admits a commutant pair (π, σ).

(iii) Give a group G and a 2-cocycle μ. Is it possible that some of the μ-projective frame representations of G admit commutant pairs and some others do not admit any commutant pairs?

Although partial results are known for the case when the von Neumann algebra generated by the μ-projective left regular representation is a factor, the problem is open in most cases. In fact, even for the free group case, question (i) is equivalent to the classification problem of von Neumann algebras, one of the most famous open problems in the theory of operator algebras: Are all the free groups (with more than one generator) von Neumann algebras isomorphic? This problem was one of the inspirations for Voiculescu's theory of free probability.

Recall that the fundamental group $F(\mathcal{M})$ of a II_1 factor $\mathcal{M}$ is an invariant that was considered by Murray and von Neumann in connection with their notion of continuous dimension in [164], where they proved that $F(\mathcal{M}) = \mathbb{R}_+^*$ when $\mathcal{M}$ is isomorphic to a hyperfinite II_1 factor, and more generally when it splits off such a factor. For free groups $\mathcal{F}_n$ of n-generators, by using Voiculescu's free probability theory [166], Radulescu [180, 181] showed that the fundamental group $F(\mathcal{M}) = \mathbb{R}_+^*$ for $\mathcal{M} = \lambda(\mathcal{F}_\infty)'$. But the problem of calculating $F(\mathcal{M})$ for $\mathcal{M} = \lambda(\mathcal{F}_n)'$ with $2 \le n < \infty$ remains a central open problem in the classification of II_1 factors, and it can be rephrased as:

Problem 1.3 Let $\mathcal{F}_n(n > 1)$ be the free group of n-generators and $P \in \lambda(\mathcal{F}_n)'$ is a nontrivial projection. Is $\lambda(\mathcal{F}_n)'$ $*$-isomorphic to $P\lambda(\mathcal{F}_n)'P$?

It was proved independently by Dykema [73] and Radulescu [181] that either all the von Neumann algebras $P\lambda(\mathcal{F}_n)'P(0 \ne P \in \lambda(\mathcal{F}_n)')$ are $*$-isomorphic, or no two of them are $*$-isomorphic. How does this connect with the existence problem of commutant pairs? The following result builds such a connection at least for the free infinite-conjugacy-classes (ICC) group case.

Theorem 1.19 *Suppose that the von Neumann algebra generated by the μ-projective left regular representation λ is a factor. Let $\pi = \lambda|_P$ for some projection $P \in \lambda(G)'$. Then the following are equivalent:*

1. *$\lambda(G)'$ and $P\lambda(G)'P$ are isomorphic von Neumann algebras;*
2. *there exists a $\bar{\mu}$-projection representation σ such that (π, σ) form a commutant pair.*

Thus, we immediately establish the following equivalence for free groups with finitely many generators.

Corollary 1.2 *Let $n \in \mathbb{N}$ with $2 \leq n < \infty$. The following statements are equivalent:*

1. *All the free group von Neumann algebras with n-generators are $*$-isomorphic.*
2. *There is a free group $\mathcal{F}_n$ and a non-trivial frame representation π (i.e. π is a proper subrepresentation of λ) such that π admits a commutant pair.*
3. *There is a free group $\mathcal{F}_n$ such that every frame representation admits a commutant pair.*
4. *For every $n \geq 2$, every frame representation of $\mathcal{F}_n$ admits a commutant pair.*

Example 1.3.1 Suppose that G is either an abelian group or an amenable ICC group or $\mathcal{F}_\infty$, and assume that $\mu \equiv 1$. Then, every frame representation of G admits a commutant pair.   □

Proof The abelian group case is proved in [65]. If G is an amenable ICC group, then the statement follows immediately from Theorem 1.19 and the famous result of A. Connes [57]: If G is an amenable ICC group, then $\lambda(G)'$ is the hyperfinite II_1 factor, and so $\lambda(G)'$ and $P\lambda(G)'P$ are isomorphic for any nonzero projection $P \in \lambda(G)'$.

For the free group $\mathcal{F}_\infty$ case, Radulescu [180, 181] proved that the fundamental group $F(\mathcal{M}) = \mathbb{R}_+^*$ for $\mathcal{M} = \lambda(\mathcal{F}_\infty)$. Therefore for $\lambda(\mathcal{F}_\infty)' \simeq P\lambda(\mathcal{F}_\infty)'P$ for any nonzero projection $P \in \lambda(\mathcal{F}_\infty)'$, and thus, $\lambda|_P$ admits a commutant pair for free group $\mathcal{F}_\infty$.

Another example is due to S. Popa.

Example 1.3.2 If $G = \mathbb{Z}^2 \rtimes SL(2, \mathbb{Z})$), then none of the nontrivial subrepresentations $\lambda|_P$ admits a commutant pair.   □

Proof Recall that the fundamental group of a type II_1 factor $\mathcal{M}$ is the set of numbers $t > 0$ for which the amplification of $\mathcal{M}$ by t is isomorphic to $\mathcal{M}$, $F(\mathcal{M}) = \{t > 0 : \mathcal{M} \simeq \mathcal{M}^t\}$. Let $G = \mathbb{Z}^2 \rtimes SL(2, \mathbb{Z})$. Then, by [171,172], the fundamental group of $\lambda(G)'$ is $\{1\}$, which implies that von Neumann algebras $P\lambda(G)'P$ is not $*$-isomorphic to $\lambda(G)'$ for any nontrivial projection $P \in \lambda(G)'$. Thus, by Theorem 1.19, none of the nontrivial subrepresentations $\lambda|_P$ admits a commutant pair.

Remark 1.5 By Corollary 1.2, we know that for the free group $G = \mathcal{F}_n$, either all the frame representations of G admit commutant pairs, or none of them admit a commutant pair. While we don't have a concrete example at this moment, we conjecture this is not true in general. However, it would be interesting to identify some other types of groups for which this remains to be true.

1.3.2.3 Frame Vector Classifications

Many of the results discussed in the previous sections are closely connected to one of the very basic questions: How to classify frame generators for projective unitary representations? This question was examined in [122]. We first point out the following observation:

Lemma 1.2 *Let ξ and η be two frame vectors for a projective unitary representation π of G on a Hilbert space $\mathcal{H}$. Then they are π-weakly equivalent if and only if they are π-equivalent, meaning that $\pi(g)\eta = T\pi(g)\xi$ for some invertible operator $T \in B(\mathcal{H})$. In this case, $T \in \pi(G)'$.*

Proof Without loss of generality, we can assume that both are Parseval frame vectors. Note that both $\Theta_\xi(\mathcal{H})$ and $\Theta_\eta(\mathcal{H})$ are closed subspace of $\ell^2(G)$ that are invariant under the projective left regular representation λ. So we have $\Theta_\xi(\mathcal{H}) = \Theta_\eta(\mathcal{H})$ if ξ and η are π-weakly equivalent, and hence $\Theta_\xi\Theta_\xi^* = \Theta_\eta\Theta_\eta^*$. Let $T = \Theta_\eta^*\Theta_\xi$. Then $T \in \pi(G)'$ is invertible and

$$T\pi(g)\xi = T\Theta_\xi^*(\chi_g) = \Theta_\eta^*\Theta_\xi\Theta_\xi^*(\chi_g) = \Theta_\eta^*\Theta_\eta\Theta_\eta^*(\chi_g) = \Theta_\eta^*(\chi_g) = \pi(g)\eta.$$

Therefore, we immediately get the following.

Proposition 1.6 *Given a projective unitary representation π and let P be the orthogonal projection on the range space of a Parseval frame $\{\pi(g)\xi\}$. Then, the equivalent class of frame vectors for π is in one-to-one correspondence with an orthogonal projection $Q \in \lambda(G)'$ that is equivalent to P.*

So if there exists more than one equivalent class of frame generators, then they can not be parameterized by the group of invertible operators in the commutant of $\pi(G)$. Interestingly enough is the fact that they can be parametrized by the invertible operator group in the von Neumann algebra generated by $\pi(G)$. This is often referred to as the "Parametrization Theorem".

Theorem 1.20 *Let G be a countable group and let π be a projective unitary representation of the group G on a Hilbert space $\mathcal{H}$. Suppose $\xi \in \mathcal{H}$ is a Parseval frame vector for $\mathcal{H}$. Then*

1. *$\eta \in \mathcal{H}$ is a Parseval frame vector for $\mathcal{H}$ if and only if there exists a unitary $U \in \pi(G)''$ such that $U\xi = \eta$.*
2. *$\eta \in \mathcal{H}$ is a frame vector for $\mathcal{H}$ if and only if there exists an invertible operator $A \in \pi(G)''$ such that $A\xi = \eta$.*
3. *$\eta \in \mathcal{B}_\pi$ if and only if there exists an operator $A \in \pi(G)''$ such that $A\xi = \eta$.*

Then the question that follows is: Can we still use the commutant $\pi(G)'$ to parametrize the frame vectors by using possibly several operators? The answer to this question is affirmative, and the number of operators involved depends on the so-called "π-orthogonality index" for the frame representation.

Definition 1.11 For a projective representation π of a countable group G on a Hilbert space $\mathcal{H}$, we define the *decomposition space* of π to be the subspace $\mathcal{D}_\pi = \overline{span}\{\Theta_\xi(\mathcal{H}) : \xi \in \mathcal{B}_\pi\}$ of $\ell^2(G)$. Let N be a positive integer. We call N the *orthogonality index* of π if N is the smallest natural number such there exist N π-orthogonal vectors $\xi_i \in \mathcal{B}_\pi$ such $\{\Theta_{\xi_i}(\mathcal{B}_\pi) : i = 1, \ldots, N\}$ generates $\mathcal{D}_\pi$. We say that π has the orthogonality index ∞ if such a finite integer N does not exist.

Theorem 1.21 *Let π be a projective representation of a countable group G on a Hilbert space $\mathcal{H}$ such that $\mathcal{B}_\pi$ is dense in H. Then π has the orthogonality index N if and only if $\pi(G)'$ has the cyclic multiplicity N.*

Now assume that a frame representation π has the orthogonality index $N + 1$ and ξ_1 is a Parseval frame vector for $\mathcal{H}$. Then it can be shown that there exist $\xi_i \in \mathcal{H}$ for $i = 2, \ldots, N + 1$ satisfying the properties:

1. $\{\pi(g)\xi_i : g \in G\}$ is a Parseval frame for $M_i := \overline{span}\{\pi(g)\xi_i : g \in G\}$ ($i = 2, \ldots, N + 1$).
2. ξ_i ($i = 1, \ldots, N + 1$) are mutually π-orthogonal.
3. There is no nonzero Bessel vector which is π-orthogonal to all ξ_i.

With this setup, we have the following parametrization theorem by operators in $\pi(G)'$ (For simplicity, we only state the result for the Parseval frame generator case).

Theorem 1.22 *Let G be a countable group and let π be a unitary representation of the group G on the Hilbert space $\mathcal{H}$. Suppose ξ_i, $i = 1, \ldots, N + 1$ are as above.*

(i) Let $\eta \in \mathcal{H}$. Then η is a Parseval frame vector for $\mathcal{H}$ if and only if there exist $U_i \in \pi(G)'$, $i = 1, \ldots, N + 1$, with $U_i P_i = U_i$, $i = 1, \ldots, N + 1$ (where P_i is the projection onto the closed linear span of $\{\pi(g)\xi_i : g \in G\}$) such that

$$\eta = U_1\xi_1 + \cdots + U_{N+1}\xi_{N+1}$$

and $\sum_{i=1}^{N+1} U_i U_i^* = I$. Moreover, these U_is are unique.
(ii) Let $\eta, \zeta \in \mathcal{H}$, be two Parseval frame vectors for $\mathcal{H}$ such that

$$\eta = U_1\xi_1 + \cdots + U_{N+1}\xi_{N+1}, \quad \zeta = U_1'\xi_1 + \cdots + U_{N+1}'\xi_{N+1},$$

with U_i and U_i' satisfying the requirement as in (i). Then

(a) η and ζ are π-orthogonal if and only if

$$U_1' U_1^* + \ldots U_{N+1}' U_{N+1}^* = 0,$$

(b) η and ζ are equivalent if and only if

$$[U_1^*, \ldots, U_{N+1}^*]^t [U_1, \ldots, U_{N+1}] = [U_1'^*, \ldots, U_{N+1}'^*]^t [U_1', \ldots, U_{N+1}'].$$

Finally, as we have discussed before, there is a simple geometric interpretation of Parseval frames and general frames: Frames are exactly orthogonal compressions of Riesz bases for a larger Hilbert space, i.e., is a Parseval frame (resp. frame) for the projection subspace. More generally, dual frame pairs are orthogonal compressions of dual Riesz basis pairs. This simple observational result later led to some work that is crucial in the study of the representation theory for various types of frames, dilation theory of wavelet frames, dilations of operator-valued measures and their applications to the noncommutative measures theory [125, 127]. Here we point out the following useful result.

Theorem 1.23 *Let π be a frame representation of a countable group G on a Hilbert space $\mathcal{H}$. If $\{\pi(g)\xi\}$ and $\{\pi(g)\eta\}$ form a dual frame pair for a π-invariant subspace M, then there exists $\varphi \in \mathcal{H}$ such that*

(i) *$\{\pi(g)\varphi : g \in G\}$ is a frame for $\mathcal{H}$, and*
(ii) *$P\varphi = \xi$ and $P\varphi^* = \eta$, where P is the orthogonal projection from H onto M and $\{\pi(g)\varphi^* : g \in G\}$ is the canonical dual of $\{\pi(g)\varphi : g \in G\}$.*

In particular, every frame sequence $\{\pi(g)\xi\}$ is the orthogonal compression of a frame $\{\pi(g)\varphi\}$ for $\mathcal{H}$ with $\xi = P\varphi$, where $P \in \pi(G)'$ is the orthogonal projection onto the closed subspace generated by $\{\pi(g)\xi\}$. We remark that unlike the proof of Proposition 1.2, the techniques from the theory of von Neumann algebras are heavily involved in the proof of Theorem 1.23.

Frames in Banach Spaces

2

Banach frames emerged in the theory of frames related to Gabor and Wavelet analysis. Before the concept of Banach frames was formalized, in 1991, Gröchenig [101] introduced atomic decompositions as an extension of the notion of frames from Hilbert space theory to a Banach space setting, Feichtinger and Gröchenig [81–83] then developed a general theory for a large class of function spaces and group representations. In 1999, Casazza, Han and Larson [52] introduced the concept of Schauder frames for Banach spaces, which is a natural generalization of frames for Hilbert spaces and Schauder bases for Banach spaces. In some subsequent work, Casazza et al. in 2008 tackled the problem of coefficient quantization with respect to frames [45]; Carando, Lassalle [55, 56], Liu [153] and [75] related duality and reflexivity with atomic decompositions and framings. In [144], Junge, Nielsen, Ruan and Xu introduced the notion of cb-basis for operator spaces. Recently, Liu and Ruan [156] constructed the complete bounded frame for the reduced group C^*-algebra $C_r^*(\mathbb{F}_2)$ over the free group of two generators $\mathbb{F}_2$. In addition, Liu et al. [157] defined the Lipschitz frame for Banach spaces and showed that a Banach space has an (unconditional) Lipschitz frame if and only if it has an (unconditional) Schauder frame.

2.1 Schauder Frames in Banach Spaces

2.1.1 Atomic Decompositions

We start by recalling the definition of atomic decomposition [101].

Definition 2.1 For a Banach space X, let X_d be an associated Banach space of scalar-valued sequences indexed by $\mathbb{N}$. Let $\{y_j\}_{j=1}^\infty$ be a sequence of elements from X^* and $\{x_j\}_{j=1}^\infty$ a sequence of elements of X. We say that $\{x_j, y_j\}_{j=1}^\infty$ is an *atomic decomposition*

© The Author(s), under exclusive license to Springer Nature Switzerland AG 2026
D. Han et al., *Banach Space Aspects of Frame Theory with Applications*, Synthesis
Lectures on Mathematics & Statistics, https://doi.org/10.1007/978-3-032-10174-7_2

of X with respect to X_d if the following conditions are satisfied:

1. $(f_j(x)) \in X_d$, for each $x \in X$,
2. there exist constants $A, B > 0$ such that

$$A\|x\|_X \le \|(f_j(x))\|_{X_d} \le B\|x\|_X,$$

3. $x = \sum_{j=1}^{\infty} f_j(x)x_j, \forall x \in X$.

In this case we also simply write $(\{x_j, y_j\}, X_d)$ as the atomic decomposition of X. The property (3) in the above definition is often referred to as the reconstruction formula associated with atomic decomposition.

For a more general setting, Gröchenig [101] defined:

Definition 2.2 Let X be a Banach space and X_d be an associated Banach space of scalar-valued sequences indexed by $\mathbb{N}$. Let $\{f_i\}_{i=1}^{\infty}$ be a sequence of elements from X^* and $S : X_d \to X$ be given. We say that $(\{f_i\}, S)$ is a *Banach frame* for X with respect to X_d if

1. $(f_i(x)) \in X_d$, for each $x \in X$,
2. The norms $\|x\|_X$ and $\|\{f_i(x)\}\|_{X_d}$ are equivalent, that is

$$A\|x\|_X \le \|(f_i(x))\|_{X_d} \le B\|x\|_X,$$

for some positive constants $A, B > 0$,
3. S is bounded and linear, and $S((f_i(x))) = x$ for each $x \in \overset{,}{X}$,

The map S is called the *reconstruction operator*.

There is a natural relation between the two definitions, which can be stated formally in the following proposition.

Proposition 2.1 *Let X be a Banach space and X_d be an associated Banach space of scalar-valued sequences indexed by $\mathbb{N}$. Let $\{f_i\}_{i=1}^{\infty}$ be a sequence of elements from X^* and $S : X_d \to X$ be given. Let $\{e_i\}_{i=1}^{\infty}$ be the unit vectors in X_d. Then the following are equivalent:*

1. $(\{f_i\}, S)$ is a Banach frame for X with respect to X_d and $\{e_i\}$ is a Schauder basis for X_d.
2. $\{Se_i, f_i\}$ is an atomic decomposition for X with respect to X_d.

Proof Let $\{e_i\}$ be a basis of unit vectors of X_d in Definition 2.2. Set $x_i = Se_i \in X$, for all $i \in \mathbb{N}$. Then the conditions (1), (2), and (3) of Definition 2.2 are now equivalent to the corresponding conditions in Definition 2.1.   □

The following result tells us that every separable Banach space has a Banach frame.

Proposition 2.2 *Every separable Banach space has a Banach frame with frame bounds* $A = B = 1$.

Proof Let X be a separable Banach space. Choose a sequence $\{f_i\}_{i=1}^{\infty} \subset X^*$ of unit vectors such that

$$\|x\| = \sup_{i \in \mathbb{N}} |f_i(x)| \tag{2.1}$$

holds for every $x \in X$. Let X_d be the subspace of ℓ^{∞} given by:

$$X_d = \left\{ (f_i(x)) : x \in X \right\}.$$

Let $S : X_d \to X$ be given by $S((f_i(x)) = x$. Then, by Eq. (2.1), S is an isometry of X onto X_d and $(\{f_i\}, S)$ is a Banach frame for X with respect to X_d.   □

In 1999, Casazza et al. [52] used geometry interpretation or dilation characterization of Hilbert space frames (see Proposition 1.7) to introduce the following generalization of frames for general Banach spaces.

Definition 2.3 A sequence $\{x_i\}_{i=1}^{\infty}$ in a Banach space X is a *frame* for X if there is a Banach space Z with an unconditional basis $\{z_i\}_{i=1}^{\infty}$ with $X \subset Z$ and a (onto) projection $P : Z \to X$ so that $Pz_i = x_i$ for all $i \in \mathbb{N}$. If $\{z_i\}$ is a 1-unconditional basis for Z and $\|P\| = 1$, we will call $\{x_i\}$ a *normalized tight frame* for X.

Let $\{z_i^*\}$ be the *biorthogonal functionals* associated to $\{z_i\}$. In this case, we have for all $x \in X$ that

$$x = \sum_{i=1}^{\infty} \langle x, z_i^* \rangle z_i = Px = \sum_{i=1}^{\infty} \langle x, z_i^* \rangle Pz_i = \sum_{i=1}^{\infty} \langle x, z_i^* \rangle x_i,$$

and this series converges unconditionally in X, which is also called *projective frame*.

Note that the definition recaptures the unconditional convergence of the frame representation. However, there exists a projective frame in the sense of Definition 2.3 for an infinite-dimensional Hilbert space that fails to be a frame. Such a concrete example was constructed by [52, Example 3.9] or in chapter 5 of [125].

Recall that if $\{x_i\}_{i=1}^\infty$ is a frame for a Hilbert space $\mathcal{H}$ and $\{y_i\}_{i=1}^\infty$ is a dual frame for $\{x_i\}$, then we have the reconstruction formula

$$x = \sum_{i=1}^\infty \langle x, y_i \rangle x_i, \quad \forall x \in \mathcal{H}.$$

and the series convergence unconditionally for all $x \in \mathcal{H}$. The next definition of the Banach space frame comes naturally from this representation.

Definition 2.4 A *framing* for a Banach space X is a pair of sequences $\{x_i, y_i\}_{i=1}^\infty$ with $x_i \in X$ and $y_i \in X^*$ satisfying for all $x \in X$,

$$x = \sum_{i=1}^\infty \langle x, y_i \rangle x_i, \forall x \in X,$$

and this series converges unconditionally.

By unconditionalizing the definition of atomic decompositions for Banach spaces, Casazza et al. [52] also gave the following definition.

Definition 2.5 A *framing model* is a Banach space Z with a fixed unconditional basis $\{e_i\}_{i=1}^\infty$ for Z. A *framing modeled on* $(Z, \{e_i\}_{i=1}^\infty)$ for a Banach space X is a pair of sequences $\{y_i\}_{i=1}^\infty$ in X^* and $\{x_i\}_{i=1}^\infty$ in X so that the operator $\theta : X \to Z$ defined by

$$\theta x = \sum_{i=1}^\infty \langle x, y_i \rangle e_i,$$

is an into isomorphism (that is, an isomorphism onto the range of θ) and $\Gamma : Z \to X$ given by

$$\Gamma \left(\sum_{i=1}^\infty a_i e_i \right) = \sum_{i=1}^\infty a_i x_i$$

is bounded and $\Gamma \theta = I_X$.

In this setting, Γ is the reconstruction operator for the frame. The next theorem, due to Casazza, Han, Larson [52], shows three methods to define a frame for a Banach space are really all the same.

Theorem 2.1 *Let X be a Banach space and $\{x_i\}_{i=1}^\infty$ be a sequence of elements of X. The following are equivalent:*

1. $\{x_i\}_{i=1}^{\infty}$ is a frame for X.
2. There is a sequence $\{y_i\}_{i=1}^{\infty} \subset X^*$ so that $\{x_i, y_i\}_{i=1}^{\infty}$ is a framing for X.
3. There is a sequence $\{y_i\}_{i=1}^{\infty} \subset X^*$ and a framing model $(Z, \{e_i\}_{i=1}^{\infty})$ so that $\{x_i, y_i\}_{i=1}^{\infty}$ is a framing modeled on $(Z, \{e_i\}_{i=1}^{\infty})$.

Proof (1) $\Longrightarrow$ (3): Since $\{x_i\}_{i=1}^{\infty}$ is a frame for X, there is a Banach space Z with a unconditional basis $\{e_i, e_i^*\}_{i=1}^{\infty}$ and a bounded linear projection $P : Z \to X$ so that $Pe_i = x_i$. Now, letting $y_i = P^* e_i^*$ we have for all $u \in X$,

$$\sum_{i=1}^{\infty} \langle u, y_i \rangle x_i = \sum_{i=1}^{\infty} \langle u, P^* e_i^* \rangle Pe_i = \sum_{i=1}^{\infty} \langle u, e_i^* \rangle Pe_i = \sum_{i=1}^{\infty} \langle u, e_i^* \rangle e_i = u.$$

So if we let θ be the injection $I : X \to Z$ and let $\Gamma = P$, then we see that $\{x_i, y_i\}_{i=1}^{\infty}$ is a framing modeled on $(Z, \{e_i\}_{i=1}^{\infty})$ for X.

(3) $\Longrightarrow$ (2): This is obvious by the definitions.

(2) $\Longrightarrow$ (1): We first construct our framing model. To do this, define a new Banach space C_X by:

$$C_X = \left\{ (a_i) : \sup_{\varepsilon_i = \pm 1} \left\| \sum_{i=1}^{\infty} \varepsilon_i a_i x_i \right\|_X < \infty \right\},$$

with the norm on C_X being given by:

$$\|(a_i)\| = \sup_{\varepsilon_i = \pm 1} \left\| \sum_{i=1}^{\infty} \varepsilon_i a_i x_i \right\|_X.$$

Let $\{e_i\}_{i=1}^{\infty}$ be the natural unit vectors in C_X and let Z denote the Banach space spanned by the $\{e_i\}_{i=1}^{\infty}$. Then Z is the sequence of scalars for which the sum $\sum a_i x_i$ converges unconditionally in X. Without losing any generality, we assume that each x_i is nonzero. It follows immediately that $\|e_i\| = \|x_i\|$ and $\{e_i\}$ is an unconditional basis for Z.

Now, define: $\theta : X \to Z$ by

$$\theta u = (\langle u, y_i \rangle) = \sum_{i=1}^{\infty} \langle u, y_i \rangle e_i.$$

Also, define a map $\Gamma : Z \to X$ by

$$\Gamma(a_i) = \sum_{i=1}^{\infty} a_i x_i.$$

This series clearly converges unconditionally in X. It is easy to see that $\Gamma\theta = I_X$, and so $P = \theta\Gamma$ is a projection from Z onto θX. Immediately, we have

$$Pe_i = \theta x_i.$$

Since θ is injective, we can identify X with θX, and thus we have $\{x_i\}$ is a frame for X. □

In connection with the Schauder bases, we give the definition of the Schauder frames for Banach spaces, introduced by Casazza, Han and Larson [52, 122] as a unified generalization of frames for Hilbert spaces and Schauder bases for Banach spaces.

Definition 2.6 Let X be a (finite or infinite-dimensional) separable Banach space. A pair of sequences $\{x_n, f_n\}_{n=1}^{\infty} \subseteq X \times X^*$ with $\{x_n\}_{n=1}^{\infty} \subset X$ and $\{f_n\}_{n=1}^{\infty} \subset X^*$, is called a *Schauder frame* of X^1 if

$$x = \sum_{i=1}^{\infty} f_i(x)x_i \quad \forall x \in X,$$

where the series converges in norm.

Remark 2.1 A Schauder frame essentially takes, as its definition, an extension of the reconstruction formula in Hilbert spaces to Banach spaces so that it can represent general vectors by using quantized coefficients. Moreover, if $\{x_i\}_{i=1}^{\infty}$ and $\{f_i\}_{i=1}^{\infty}$ both are standard frames for a Hilbert space $\mathcal{H}$, then $\{f_i\}$ is an alternate dual frame for $\{x_i\}$ if and only if $\{x_i, f_i\}$ is a Schauder frame for $\mathcal{H}$.

Similar to the framing model for framing in Definition 2.5, the associated sequence space of a Schauder frame was introduced in [52] and the minimal bases was defined in [159, Definition 3.4 and Theorem 3.5], which will play an important role in the study of reflexivity in the following sections and approximation property in the next chapter. So we introduce the notion of minimal associated bases of the associated sequence space for Schauder frames.

Definition 2.7 Let $\{x_i\}_{i=1}^{\infty}$ be a non-zero sequence in a Banach space X. Define a norm on c_{00} as follows

$$\left\| \sum a_i e_i \right\|_{Min} = \max_{m \leq n} \left\| \sum_{i=m}^{n} a_i x_i \right\|_X, \quad \forall (a_i) \in c_{00}.$$

---

[1] It will be our convention that we only consider nonzero frames or framings $\{x_i, f_i\}$, that is, $x_i \neq 0$, and $f_i \neq 0$ for all $i \in \mathbb{N}$.

Denote by Z_{Min} the completion of c_{00} endowed with the norm $\|\cdot\|_{Min}$. It is easy to prove that $\{e_i\}$, denoted by $\{e_i^{Min}\}$, is a basis of Z_{Min}. We call Z_{min} and $\{e_i^{Min}\}$ the minimal sequence space and the minimal associated basis with respect to $\{x_i\}$, respectively.

2.1.2   Duality, Reflexivity and Schauder Frames

As discussed in the last subsection, the concepts of atomic decompositions and Banach frames serve as natural generalizations of the concept of Hilbert space frames (and dual frame pairs) that capture the conditional and unconditional convergence aspects of the reconstruction theory. In what follows, we will generalize the shrinking and boundedly complete properties to the atomic decompositions, or Schauder frames, or framing setting and present some results similar to James' theorem. Before proceeding, let us recall James' results [135] on bases and duality.

Let $\{x_n\}_{n=1}^{\infty}$ be a basis for X and $\{x_n^*\}_{n=1}^{\infty}$ be the coefficient functionals for the basis $\{x_n\}$. A basis $\{x_n\}$ is called *shrinking* if

$$\lim_{n \to \infty} \|x^*\|_n = 0, \ \forall x^* \in X^*,$$

where $\|x^*\|_n$ is the norm of x^* restricted to the "tail space" $\mathrm{span}\{x_i : i > n\}$. It can be verified that $\{x_n\}$ is shrinking if $\{x_n^*\}$ is a basis for X^*. The natural basis $\{e_n\}_{n=1}^{\infty}$ is shrinking in c_0 and $\ell_p(1 < p < \infty)$, but not in ℓ_1.

While, a basis $\{x_n\}_{n=1}^{\infty}$ for a Banach space X is called *boundedly complete* if $\sum_n a_n x_n$ converges whenever $\sup_n \|\sum_{i=1}^n a_i x_i\| < \infty$. The usual basis $\{e_n\}_{n=1}^{\infty}$ is a boundedly complete basis for $\ell_p(1 \le p < \infty)$ but not for c_0.

James [135] took the shrinking and boundedly complete notions together to characterize reflexivity for spaces possessing a basis.

Theorem 2.2 *Let X be a Banach space with a basis $\{x_n\}_{n=1}^{\infty}$. Then, X is reflexive if and only if $\{x_n\}$ is both shrinking and boundedly complete. Moreover, if X is a Banach space with an unconditional basis, then X is reflexive if and only if X contains no isomorphic copy of ℓ_1 or c_0.*

In 2010, Liu introduced in [153] the concepts of minimal and maximal associated spaces and bases with respect to Schauder frames as well as shrinking and boundedly complete Schauder frames.

Definition 2.8 Let $\{x_i, f_i\}_{i=1}^{\infty}$ be a Schauder frame of a Banach space X. We say that $\{x_i, f_i\}$ is called *locally shrinking* if for all $m \in \mathbb{N}$

$$\|f_m|_{\mathrm{span}\{x_i : i \ge n\}}\| \to 0, \quad n \to \infty.$$

And $\{x_i, f_i\}$ is called *locally boundedly complete* if for all $m \in \mathbb{N}$,

$$\|x_m|_{\text{span}\{f_i : i \geq n\}}\| \to 0, \quad n \to \infty.$$

A Schauder frame $\{x_i, f_i\}$ is called *weakly localized* if it is locally shrinking and locally boundedly complete. A frame $\{x_i, f_i\}$ is called *pre-shrinking* if $\{f_i, x_i\}$ is a frame of X^*. It is called *pre-boundedly complete* if for all $x^{**} \in X^{**}$, $\sum_{i=1}^{\infty} x^{**}(f_i)x_i$ converges. We call a Schauder frame $\{x_i, f_i\}$ *shrinking* if it is locally shrinking and pre-shrinking, and say that $\{x_i, f_i\}$ *boundedly complete* if it is weakly localized and pre-boundedly complete.

We have the following two characterizations for shrinking and boundedly complete Schauder frames.

Theorem 2.3 ([153, Theorem A.]) *Let* $\{x_i, f_i\}_{i=1}^{\infty} \subset X \times X^*$ *be a Schauder frame of a Banach space X and assume that for all $m \in \mathbb{N}$,*

$$\lim_{n \to \infty} \|f_m|_{\text{span}\{x_i : i \geq n\}}\| = 0.$$

Then the following are equivalent.

1. $\{x_i, f_i\}$ *is shrinking.*
2. *Every normalized block of $\{x_i\}$ is weakly null.*
3. $X^* = \overline{\text{span}\{f_i : i \in \mathbb{N}\}}$.
4. *The minimal associated basis is shrinking.*

Theorem 2.4 ([153, Theorem B.]) *Let* $\{x_i, f_i\}_{i=1}^{\infty} \subset X \times X^*$ *be a Schauder frame of a Banach space X and assume that for all $m \in \mathbb{N}$*

$$\lim_{n \to \infty} \|f_m|_{\text{span}\{x_i : i \geq n\}}\| = 0 \quad and \quad \lim_{n \to \infty} \|x_m|_{\text{span}\{f_i : i \geq n\}}\| = 0$$

Then the following are equivalent:

1. $\{x_i, f_i\}$ *is boundedly complete.*
2. X *is isomorphic to* $\text{span}\{f_i : i \in \mathbb{N}\}^*$ *under the natural canonical map.*
3. *The maximal associated basis is boundedly complete.*

By the above characterizations, Liu in [153] extended the well-known Theorem 2.2 of James' reflexivity from unconditional bases to framings.

Theorem 2.5 ([153, Theorem C.]) *Let* $\{x_i, f_i\}_{i=1}^{\infty} \subset X \times X^*$ *be a unconditional Schauder frame of a Banach space X and assume that for all $m \in \mathbb{N}$,*

$$\lim_{n \to \infty} \| f_m |_{\text{span}\{x_i : i \geq n\}} \| = 0. \tag{2.2}$$

Then X is reflexive if an only if X does not contain isomorphic copies of c_0 and ℓ_1.

In 2009, D. Carando and S. Lassalle studied atomic decompositions and their relationship with duality and reflexivity of Banach spaces [55]. We have proved the equivalence between atomic decomposition and Banach frames in Proposition 2.1 and furthermore, the Schauder frame in Theorem 3.7 and Proposition 3.4 and all the results stated below for atomic decompositions can be reformulated for Banach frames with respect to the sequence spaces. In 2011, Carando, Lassalle and Schmidberg [56] proved Theorem 2.5 without the local condition in Eq. (2.2). They first extended the notion of shrinking and boundedly complete to the atomic decompositions.

Let $\{x_i, x_i^*\}_{i=1}^{\infty}$ be an atomic decomposition of X with respect to a Banach sequence space Z. For each $n \in \mathbb{N}$, define the linear operator $T_n : X \to X$ by

$$T_n(x) = \sum_{i \geq n} \langle x_i', x \rangle x_i.$$

Then $\{x_i, x_i'\}_{i \in \mathbb{N}}$ is called *shrinking* if for all $x^* \in X^*$,

$$\| x^* \circ T_n \| \longrightarrow 0,$$

and *boundedly complete* if for each $x^{**} \in X^{**}$, the series $\sum_{i=1}^{\infty} \langle x^*, x_i' \rangle x_i$ converges in X. The following is a James-type result for atomic decompositions.

Theorem 2.6 ([56]) *Let X be a Banach space with an unconditional atomic decomposition $\{x_i, x_i^*\}_{i=1}^{\infty}$. Then $\{x_i, x^*\}$ is shrinking if and only if X does not contain a copy of ℓ^1 and is boundedly complete if and only if X does not contain a copy of c_0.*

Remark 2.2 Let X be a Banach space with an unconditional atomic decomposition $\{x_i, x_i^*\}_{i=1}^{\infty}$. Then the following are equivalent:

1. $\{x_i, x^*\}$ is shrinking and boundedly complete.
2. X does not contain a copy of c_0 or ℓ_1.
3. X is reflexive.

Beanland, Freeman and Liu in 2015 proved that the upper and lower estimates theorems for finite-dimensional decompositions of Banach spaces can be extended to Schauder frames and gave a complete characterization of the duality theorem for Schauder frames.

Theorem 2.7 ([31]) *Let X be a Banach space and Z be a Banach sequence space whose standard unit vectors $\{e_i\}$ form a basis for Z. Let $\{x_i, f_i\}_{i=1}^{\infty}$ be an atomic decomposition of*

X with respect to Z. Then $\{x_i, f_i\}$ is shrinking if and only if there exists a Banach sequence space Z' whose standard unit vectors $\{e_i\}$ form a shrinking basis for Z' such that $\{x_i, f_i\}$ is an atomic decomposition of X with respect to Z'. Furthermore, $\{x_i, f_i\}$ is shrinking and boundedly complete if and only if there exists a reflexive Banach sequence space Z' whose standard unit vectors $\{e_i\}$ form a basis for Z' such that $\{x_i, f_i\}$ is an atomic decomposition of X with respect to Z'.

Recently, Eisner and Freeman [75] introduced the continuous Schauder frames for Banach spaces, which generalize both continuous frames for Hilbert spaces and unconditional Schauder frames for Banach spaces and defined the shrinking and boundedly complete properties of continuous Schauder frames, and showed that many of the fundamental James' theorems still hold in this general context.

The definition of continuous Schauder frame is based on the Pettis integral to integrate vector-valued functions, and we recommend [201] for reference.

Definition 2.9 ([75]) Given a Banach space X with dual X^* and a measure space (M, Σ, μ), a (w, w^*)-measurable function $t \mapsto \{x_t, f_t\} \in X \times X^*$ is called a *continuous Schauder frame* of X with respect to (M, Σ, μ) if

$$x = \int_M f_t(x) x_t d\mu(t) \quad \text{for all } x \in X.$$

The dual map $t \mapsto \{f_t, x_t\} \in X^* \times X$ is called a continuous* Schauder frame of X^* with respect to (M, Σ, μ) if

$$f = \int_M^* f(x_t) f_t d\mu(t) \quad \forall f \in X^*.$$

And a continuous Schauder frame $\{x_t, f_t\}_{t \in M}$ is called *bounded* if $\sup_{t \in M} \|x_t\| \|f_t\| < \infty$.

Unlike series, there is no order for integration and so all continuous Schauder frames are by necessity unconditional. By [75, Lemma 2.4], any continuous frame can be converted into a bounded continuous Schauder frame by changing a measure.

Recall that if $\{x_n\}_{n=1}^{\infty}$ is a Schauder basis, then the basis projections are the operators $P_n : X \rightarrow X$ are defined by $P_n(\sum_{j=1}^{\infty} a_j x_j) = \sum_{j=1}^{n} a_j x_j$. Let $\{x_t, f_t\}_{t \in M}$ be a continuous Schauder frame of a Banach space X with respect to the measure space (M, Σ, μ). For $E \in \Sigma$, the operator $P_E : X \rightarrow X$ defined by

$$P_E(x) = \int_E f_t(x) x_t d\mu(t) \quad \text{for all } x \in X \tag{2.3}$$

is referred to as a restriction operator and it will play a similar role for continuous Schauder frames as the basis projections for Schauder bases.

Definition 2.10 Let (M, Σ, μ) be a measure space and X be a Banach space. Suppose that $t \mapsto x_t$ is a w-measurable map from M to X and that $t \mapsto f_t$ is a w^*-measurable map from M to X^*. Set

$$\mathcal{D} = \left\{ E \in \Sigma : \int_E \|f_t\| \|x_t\| d\mu(t) < \infty \right\}$$

and define $E \preceq F$ whenever $E \subseteq F$ to make $\mathcal{D}$ a directed set. The elements of $\mathcal{D}$ are referred to as *absolutely finite sets*.

Now, the shrinking and boundedly complete definitions in [55] can be extended by using the limits over the net $\mathcal{D}$.

Definition 2.11 A continuous Schauder frame $\{x_t, f_t\}_{t \in M}$ for a Banach space X is called *shrinking* if

$$\lim_{E \in \mathcal{D}} \|P_{E^c}^* f\| = 0, \forall f \in X^*,$$

where $P_{E^c}^*$ is adjoint of restriction operator defined in Eq. (2.3). And $\{x_t, f_t\}_{t \in M}$ is called boundedly complete if for all $x^{**} \in X^{**}$, we have that

$$P_E^{**} x^{**} = \int_E x^{**}(f_t) x_t d\mu, \ \forall \ E \in \mathcal{D}$$

and $\lim_{E \in \mathcal{D}} P_E^{**} x^{**} \in X$.

The following theorem extends the characterizations of shrinking property for unconditional bases in [135] or Schauder frames in [55, 56, 153] to continuous Schauder frames for separable Banach spaces.

Theorem 2.8 ([75]) *Let* $\{x_t, f_t\}_{t \in M}$ *be a continuous Schauder frame of a separable Banach space X with respect to a measure space* (M, Σ, μ) *such that* $t \mapsto f_t$ *is w-measurable. Then the following are equivalent:*

1. *The continuous Schauder frame* $\{x_t, f_t\}_{t \in M}$ *is shrinking.*
2. *The dual frame* $\{f_t, x_t\}_{t \in M} \subseteq X^* \times X^{**}$ *is a continuous Schauder frame for X^*.*
3. ℓ_1 *does not embed isomorphically into X.*
4. *Every* $x^{**} \in X^{**}$ *is the w^*-limit of a sequence in X.*

Recall that a map $\Psi : M \to X$ is semi-discrete with respect to (M, Σ, μ) if for all $\varepsilon > 0$ and all $F \in M$ with $0 < \mu(F)$ there exists $E \in \Sigma$ with $E \subseteq F$ and $0 < \mu(E)$ so that $\|\Psi(s) - \Psi(t)\| < \varepsilon$ for all $s, t \in E$. When working on the non-separable Banach, the theorem of Odell and Rosenthal that states that if X is a separable Banach space then ℓ_1 does not embed into X if and only if every $x^{**} \in X^{**}$ is the w^*-limit of a sequence in X [169], is not applicable to this case. However, if additional structures are imposed, it gives the following results:

Theorem 2.9 ([75, Theorem 3.10]) *Let* $\{x_t, f_t\}_{t \in M}$ *be a continuous Schauder frame for* X *such that either* $\{f_t\}_{t \in M}$ *is semi-discrete or every* x^{**} *is the* w^*-*limit of a sequence in* X. *Then* $\{x_t, f_t\}_{t \in M}$ *is boundedly complete if and only if* c_0 *does not embed into* X.

Theorem 2.10 ([75, Theorem 3.11]) *Let* $\{x_t, f_t\}_{t \in M}$ *be a continuous Schauder frame for a Banach space* X *such that either:* X *is separable,* $\{f_t\}_{t \in M}$ *is semi-discrete, or every* $x^{**} \in X^{**}$ *is the* w^*-*limit of a sequence in* X. *Then the following are equivalent:*

1. $\{x_t, f_t\}_{t \in M}$ *is shrinking and boundedly complete.*
2. X *does not contain an isomorphic copy of* c_0 *or* ℓ_1.
3. X *is reflexive.*

In [159], Liu and Zheng obtained a characterization of Schauder frames that are near-bases, which extended Holub's results [130] from Hilbert frames to Schauder frames.

Definition 2.12 Let X a Banach space and $\{x_i\}_{i=1}^{\infty} \subset X$. We call $\{x_i\}$ a *near-Schauder basis* of X if there is a finite set $\sigma \subset \mathbb{N}$ such that $\{x_i\}_{i \notin \sigma}$ is a Schauder basis of X. We call a Schauder frame $\{x_i, f_i\}$ a *near-Schauder basis* if $\{x_i\}$ is a near-Schauder basis. The *excess* of a near-Schauder basis $\{x_i\}$ of X is defined by

$$\text{exc}(x_i) = \left\{ \text{card } \sigma : \exists \text{ a finite subset } \sigma \subset \mathbb{N} \text{ s.t. } \{x_i\}_{i \notin \sigma} \text{ is a basis of } X \right\}.$$

where card σ denotes the cardinal number of the set σ.

Theorem 2.11 *Let* $\{x_i, f_i\}$ *be a Schauder frame of a Banach space* X *and let* $(E_{\min}, (\hat{e}_i))$ *be the minimal-associated sequence space to* $\{x_i, f_i\}$ *with* $S_{\min}$ *the minimal-associated reconstruction operator. Then the following conditions are equivalent:*

1. *The kernel of* $S_{\min}$ *contains no copy of* c_0.
2. $S_{\min}$ *has a finite-dimensional kernel.*
3. $\{x_i\}$ *is a near-Schauder basis of* X.

Furthermore, in this case, we have

$$\text{exc}(x_i) = \dim(\ker S_{\min})$$

2.2 Completely Bounded Frames for Operator Spaces

Operator spaces belong to a special type of Banach spaces which are the closed linear subspaces of general C^*-algebras. They are particularly appropriate for studying spaces or algebras of operators on Hilbert spaces. Thus, the field of operator spaces provides a bridge from the world of Banach and function spaces to the world of spaces of operators on a Hilbert space, and of "noncommutative mathematics". This section is focused on presenting some recent results on complete bounded frames for operator spaces.

2.2.1 Operator Spaces and Completely Bounded Maps

There exists an abstract definition or characterization of operator spaces by Ruan [186]. In this section, we use the following concrete and well-adapted definition.

Definition 2.13 An operator space E is a closed subspace of $B(\mathcal{H})$ of all bounded linear operators on a Hilbert space $\mathcal{H}$.

Due to the requirement of "closedness", operator spaces are viewed as "quantized (or non-commutative)" Banach spaces. On the one hand, operator spaces are indeed Banach spaces with respect to the following canonical operator matrix norm $\| \cdot \|_n$ on each matrix space $M_n(E)$. On the other hand, every Banach space X can be embedded isometrically into some $C(\Omega)$, as the unit ball of X^* equipped with the weak* topology is a compact Hausdorff space (by Alaoglu's Theorem), denoted by Ω, we then use an isometric representation of the C^*-algebra $C(\Omega)$ on some Hilbert space $\mathcal{H}$ and have isometric embedding of X into $B(\mathcal{H})$. In particular, C^*-algebras and von Neumann algebras are operator spaces.

Note that, for each Hilbert space $\mathcal{H}$, there is a natural identification between $M_n(B(\mathcal{H}))$ and $B(\mathcal{H}^n)$, where

$$\mathcal{H}^n := \underbrace{\mathcal{H} \oplus \cdots \oplus \mathcal{H}}_{n \text{ copies}}.$$

For each operator space $E \subset B(\mathcal{H})$, we use $M_n(E)$ to denote the set of all $n \times n$ matrices with entries from E. That is,

$$M_n(E) = \left\{ (a_{ij}) : a_{ij} \in E, 1 \leq i, j \leq n \right\} \subset M_n(B(\mathcal{H})).$$

Then, each $M_n(E)$ inherits a norm, given by

$$\|a\|_n := \|a\|_{B(\mathcal{H}^n)} = \sup \left\{ \left(\sum_{i=1}^{n} \left\| \sum_{j=1}^{n} a_{ij}(h_j) \right\|^2 \right)^{1/2} : h_j \in \mathcal{H}, \sum_{j=1}^{n} \left\| h_j \right\|^2 \leq 1 \right\}$$

for all $a = [a_{ij}] \in M_n(E)$.

In Banach spaces, we consider norms and bounded linear maps, while in operator spaces, we are interested in matrix norms and completely bounded maps.

Definition 2.14 Given two operator spaces E and F and a linear map $\phi : E \to F$. For each $n \geq 1$, let $\phi_n : M_n(E) \to M_n(F)$ be defined by

$$\phi_n\big((a_{i,j})\big) = \big(\phi(a_{i,j})\big).$$

The map ϕ is called *completely bounded* (CB) if ϕ is bounded and $\|\phi\|_{cb} = \sup_{n \geq 1} \|\phi_n\|_{M_n(E) \to M_n(F)}$ is finite.

Let $CB(E, F)$ be the Banach space of all completely bounded maps from E into F equipped with the completely bounded norm. Complete boundedness has many properties similar to boundedness, for example: $\|vu\|_{cb} \leq \|v\|_{cb}\|u\|_{cb}$. We will see later that $CB(E, F)$ also inherits an appropriate operator space structure.

Let $E \subseteq B(\mathcal{H})$ be an operator space. Then there is a natural operator norm $\| \cdot \|_n$ on each matrix space $M_n(E) \subseteq M_n(B(\mathcal{H}))$ such that

M1. $\left\| \begin{bmatrix} x & 0 \\ 0 & y \end{bmatrix} \right\|_{n+m} = \max \big\{ \|x\|_n, \|y\|_m \big\}$

M2. $\|\alpha x \beta\|_n \leq \|\alpha\| \|x\|_n \|\beta\|$ for all $x \in M_n(E)$, $y \in M_m(E)$ and $\alpha, \beta \in M_n(C)$.

Conversely, the following Ruan's fundamental theorem [186] gives an abstract matrix norm characterization.

Theorem 2.12 *Let E be a Banach space with a norm $\| \cdot \|_n$ on each matrix space $M_n(E)$. Then E is completely isometrically isometric to a concrete operator space*

$$E \to \pi(E) \subseteq B(\mathcal{H})$$

if and only if it satisfies M1 and M2.

Remark 2.3 The axioms M1 and M2 are called *Ruan's axioms*. We call a complex linear space X an *abstract operator space* if it comes equipped with a family of norms $\| \cdot \|_n$ on $M_n(X)$ for each $n \in \mathbb{N}$, which satisfies M1 and M2.

2.2.2 Completely Bounded Frames (Cb-Frames) for $C_r^*(\mathbb{F}_2)$

In [144], Junge, Nielsen, Ruan and Xu introduced the notion of the completely bounded bases for operator spaces.

Definition 2.15 An operator space X has a *completely bounded basis* (in short, cb-basis) if X has a Schauder basis $\{e_j\}_{j=1}^{\infty}$ and the natural initial sum projections

$$P_n\left(\sum_{j=1}^{\infty}\alpha_j e_j\right) = \sum_{j=1}^{n}\alpha_j e_j$$

satisfy $\sup_n \|P_n\|_{cb} < \infty$. In this case, we call $\{e_j\}$ a *cb-basis* of X.

For the free group $\mathbb{F}_n$ of n generators, let λ be the left regular representation of $\mathbb{F}_n$. The reduced group C^*-algebra $C_r^*(\mathbb{F}_n)$ is defined to be the norm-closure of $\operatorname{span}\{\lambda_s \,|\, s \in \mathbb{F}_n\}$ in $B(\ell_2(\mathbb{F}_n))$. It is shown in [144] that every separable nuclear C^*-algebra has a cb-basis, and the noncommutative L^p space $L^p(\operatorname{vN}(\mathbb{F}_n))(1 < p < \infty)$ [179] over the free group von Neumann algebra $\operatorname{vN}(\mathbb{F}_n)$ has a cb-basis, where $\mathbb{F}_n$ is the free group of n generators. In particular, the reduced group C^*-algebra $C_r^*(\Gamma)$ of an amenable group Γ has a cb-basis. However, the existence of Schauder bases, even cb-bases for some operator spaces, is still unknown and remains an open problem. In particular, we ask the following question:

Problem 2.1 Is there a Schauder basis for $L^1(\operatorname{vN}(\mathbb{F}_n))$, the noncommutative L^1 space over the free group von Neumann algebra $\operatorname{vN}(\mathbb{F}_n)$, or for the reduced free group C^*-algebra $C_r^*(\mathbb{F}_n)$?

Motivated by this problem, Liu and Ruan [156] introduced the following concept for operator spaces.

Definition 2.16 Let X be an operator space and X^* be its canonical operator dual. A sequence $\{x_n, f_n\}_{n=1}^{\infty} \subset X \times X^*$ is a *completely bounded frame (in short, cb-frame)* for X if $\{x_n, f_n\}$ is a frame, i.e.

$$x = \sum_{n=1}^{\infty} f_n(x)x_n = \sum_{n=1}^{\infty}(x_n \otimes f_n)(x)$$

for all $x \in X$, and the initial sums

$$S_m(x) = \sum_{n=1}^{m} x_n \otimes f_n$$

define completely bounded maps on X with $\sup_m \|S_m\|_{cb} < \infty$.

Liu and Ruan [156] showed that there is a concrete cb-frame for $C_r^*(\mathbb{F}_2)$, which is derived from the infinite convex decomposition of the biorthogonal system $\{\lambda_s, \delta_s\}_{s \in \mathbb{F}_2}$.

Here λ is the left regular representation of $\mathbb{F}_2$ and $\delta_s \in B_r(\mathbb{F}_2) = C_r^*(\mathbb{F}_2)^*$ is the characteristic function at $s \in \mathbb{F}_2$.

Theorem 2.13 *There exists a cb-frame (x_n, f_n) for $C_r^*(\mathbb{F}_2)$, which is derived from the infinite convex decomposition of the biorthogonal system $(\lambda_s, \delta_s)_{s \in \mathbb{F}_2}$. More precisely, there is a surjective map $\varphi : \mathbb{N} \to \mathbb{F}_2$ and a sequence of positive scalars (a_n) satisfying*

(i) $x_n = \lambda_{\varphi(n)}$, $f_n = a_n \delta_{\varphi(n)}$ *for all $n \in \mathbb{N}$;*
(ii) $\sum_{n \in \varphi^{-1}(s)} a_n = 1$ *for all $s \in \mathbb{F}_2$.*

Proof We divide our proof into three steps: Let us first recall from [109] that the word length function $s \in \mathbb{F}_2 \mapsto |s| \in [0, \infty)$ is conditionally negative definite on $\mathbb{F}_2$, and thus by Schoenburg's theorem the map $\varphi_t(s) = e^{-t|s|}$ is a positive definite function (with $\varphi_t(e) = 1$) on $\mathbb{F}_2$ for any $t > 0$. This gives us a family of unital completely positive maps $(\Phi_t)_{t>0}$ on $C_r^*(\mathbb{F}_2)$ such that

$$\Phi_t(\lambda_s) = \varphi_t(s)\lambda_s = e^{-t|s|}\lambda_s, \quad \forall s \in \mathbb{F}_2. \tag{2.4}$$

It is clear from Eq. (2.4) that $\lim_{t \to 0^+} \|\Phi_t(\lambda_s) - \lambda_s\| = 0$ for all $s \in \mathbb{F}_2$. Then $(\Phi_t)_{t>0}$ is a family of unital completely positive maps converging to the identity on $C_r^*(\mathbb{F}_2)$ in the point-norm topology.

Let $W_d = \{s \in \mathbb{F}_2 : |s| = d\}$ be the set of all words in $\mathbb{F}_2$ with length d, and let χ_{W_d} be the characteristic function on W_d. Then

$$P_d : \lambda_s \in C_r^*(\mathbb{F}_2) \mapsto \lambda_s \chi_{W_d}(s) \in C_r^*(\mathbb{F}_2)$$

is the finite-rank projection onto the subspace $E_d = \text{span}\{\lambda_s : |s| = d\}$. It is known (see [39, 177] that this projection is completely bounded with $\|P_d\|_{cb} \leq 2d$. For each $t > 0$ and $m \in \mathbb{N}$, we get a completely bounded finite-rank map

$$\Phi_{t,m} = \Phi_t(P_0 + \cdots + P_m)$$

from $C_r^*(\mathbb{F}_2)$ onto $\sum_{d=0}^m E_d = \text{span}\{\lambda_s : |s| \leq m\}$.

Step 1.   For each $t > 0$ and $m \in \mathbb{N}$, we have

$$\|\Phi_t - \Phi_{t,m}\|_{cb} \leq 2 \sum_{d=m+1}^{\infty} e^{-td}d \to 0, \quad (\text{as } m \to \infty).$$

Let $x = \sum_{d=0}^{\infty} \sum_{|s|=d} a(s) \otimes \lambda_s$ be an element in $B(\ell_2) \otimes C_r^*(\mathbb{F}_2)$ with finitely many $a(s)$ nonzero. Then we have

$$\left\|(\mathrm{id}_{B(\ell_2)} \otimes \Phi_t - \mathrm{id}_{B(\ell_2)} \otimes \Phi_{t,m})(x)\right\| = \left\|\sum_{d=m+1}^{\infty} \sum_{|s|=d} a(s) \otimes \Phi_t(\lambda_s)\right\|$$

$$\leq \sum_{d=m+1}^{\infty} \left\|\sum_{|s|=d} a(s) \otimes e^{-td}\lambda_s\right\| = \sum_{d=m+1}^{\infty} e^{-td}\left\|\sum_{|s|=d} a(s) \otimes \lambda_s\right\|$$

$$= \sum_{d=m+1}^{\infty} e^{-td}\left\|(\mathrm{id}_{B(\ell_2)} \otimes P_d)(x)\right\| \leq \sum_{d=m+1}^{\infty} e^{-td}\|P_d\|_{cb}\|x\|.$$

Since $\|P_d\|_{cb} \leq 2d$, we conclude that

$$\|\Phi_t - \Phi_{t,m}\|_{cb} \leq \sum_{d=m+1}^{\infty} e^{-td}\|P_d\|_{cb} \leq 2\sum_{d=m+1}^{\infty} e^{-td}d.$$

Since $\sum_{d=1}^{\infty} e^{-td}d$ is a convergent positive infinite series, its remainder part converges to 0. Thus, we can conclude that

$$\|\Phi_t - \Phi_{t,m}\|_{cb} \leq 2\sum_{d=m+1}^{\infty} e^{-td}d \to 0.$$

By the above calculation, for each $t > 0$, we can find $m \in \mathbb{N}$ such that $\|\Phi_t - \Phi_{t,m}\|_{cb}$ is sufficiently small. Thus, we can choose a sequence of (monotone decreasing) $t_k \to 0$ and a sequence of (monotone increasing) $m_k \to \infty$ such that

$$\lim_{k\to\infty}\left\|\Phi_{t_k} - \Phi_{t_k,m_k}\right\|_{cb} \leq \lim_{k\to\infty}\sum_{d=m_k+1}^{\infty} e^{-t_k d}d$$

$$= \lim_{k\to\infty} \frac{e^{-t_k(m_k+2)}}{(1-e^{-t_k})^2} + \frac{(m_k+1)e^{-t_k(m_k+1)}}{1-e^{-t_k}} = 0.$$

For example, we can choose $t_k = 1/\sqrt{k}$ and $m_k = k$, and we get

$$\lim_{k\to\infty}\sum_{d=k+1}^{\infty} e^{-d/\sqrt{k}}d = 0.$$

In this case, the finite-rank maps $\{\Phi_{1/\sqrt{k},k}\}_{k=1}^{\infty}$ (with $\sup_k \|\Phi_{1/\sqrt{k},k}\|_{cb} < \infty$) converge to the identity map on $C_r^*(\mathbb{F}_2)$ in the norm topology. Let $\{\delta_s\} \in B_r(\mathbb{F}_2) = C_r^*(\mathbb{F}_2)^*$ be the biorthogonal functionals of $\{\lambda_s\}$. For $k = 1$, we set

$$\Psi_1(x) = \Phi_{1,1}(x) = \sum_{|s|\leq 1} e^{-|s|}\delta_s(x)\lambda_s = \sum_{|s|\leq 1}(\lambda_s \otimes e^{-|s|}\delta_s)(x). \tag{2.5}$$

There are 5 terms in Eq. (2.5). If we list these 5 terms by the index $1 \leq j \leq 5$ and use $y_{1,j}$ (resp, $g_{1,j}$) for the corresponding λ_s (resp, $e^{-|s|}\delta_s$) in each term, we can write

$$\Psi_1 = \sum_{|s| \leq 1} (\lambda_s \otimes e^{-|s|}\delta_s) = \sum_{j=1}^{5} (y_{1,j} \otimes g_{1,j}). \tag{2.6}$$

For $k \geq 2$, we set

$$\Psi_k(x) = (\Phi_{1/\sqrt{k},k} - \Phi_{1/\sqrt{k-1},k-1})(x)$$

$$= \sum_{|s| \leq k-1} (e^{-|s|/\sqrt{k}} - e^{-|s|/\sqrt{k-1}})\delta_s(x)\lambda_s + \sum_{|s|=k} e^{-|s|/\sqrt{k}}\delta_s(x)\lambda_s$$

$$= \sum_{|s| \leq k-1} (\lambda_s \otimes (e^{-|s|/\sqrt{k}} - e^{-|s|/\sqrt{k-1}})\delta_s)(x) + \sum_{|s|=k} (\lambda_s \otimes e^{-|s|/\sqrt{k}}\delta_s)(x). \tag{2.7}$$

There are $2 \cdot 3^k - 1 = 1 + 4 + 4 \cdot 3 + \cdots + 4 \cdot 3^{k-1}$ terms in Eq. (2.7). So if we list these terms by the index $1 \leq j \leq 2 \cdot 3^k - 1$ and we use $y_{k,j}$ (resp, $g_{k,j}$) for the corresponding λ_s (resp, $(e^{-|s|/\sqrt{k}} - e^{-|s|/\sqrt{k-1}})\delta_s$ or $e^{-|s|/\sqrt{k}}\delta_s$) in each term, then we can write

$$\Psi_k = \sum_{|s| \leq k-1} \left(\lambda_s \otimes (e^{-|s|/\sqrt{k}} - e^{-|s|/\sqrt{k-1}})\delta_s\right) + \sum_{|s|=k} (\lambda_s \otimes e^{-|s|/\sqrt{k}}\delta_s)$$

$$= \sum_{j=1}^{2 \cdot 3^k - 1} (y_{k,j} \otimes g_{k,j}). \tag{2.8}$$

This is a sequence of completely bounded maps on $C_r^*(\mathbb{F}_2)$ with

$$\|\Psi_k\|_{cb} = \left\| \Phi_{1/\sqrt{k},k} - \Phi_{1/\sqrt{k-1},k-1} \right\|_{cb} \leq 2 \sup_k \|\Phi_{\frac{1}{\sqrt{k}},k}\|_{cb} \tag{2.9}$$

and for each $x \in C_r^*(\mathbb{F}_2)$, we have

$$x = \lim_{k \to \infty} \Phi_{1/\sqrt{k},k}(x) = \Phi_{1,1}(x) + \sum_{k=2}^{\infty} \left(\Phi_{1/\sqrt{k},k}(x) - \Phi_{1/\sqrt{k-1},k-1}(x)\right)$$

$$= \sum_{k=1}^{\infty} \Psi_k(x). \tag{2.10}$$

Now to get a frame, we need to further modify the terms in Eq. (2.6) and (2.8) by defining $x_{k,i} = y_{k,j}$ and $f_{k,i} = \frac{g_{k,j}}{(2 \cdot 3^k - 1)^2}$ when $i = p(2 \cdot 3^k - 1) + j$ with $0 \leq p \leq (2 \cdot 3^k - 1)^2 - 1$ and $1 \leq j \leq 2 \cdot 3^k - 1$. In this case, we can write

$$\Psi_k = \sum_{j=1}^{2\cdot 3^k-1} (y_{k,j} \otimes g_{k,j}) = \sum_{i=1}^{(2\cdot 3^k-1)^3} (x_{k,i} \otimes f_{k,i}) \tag{2.11}$$

for all $k \in \mathbb{N}$, and thus for each $x \in C_r^*(\mathbb{F}_2)$, we have

$$x = \sum_{k=1}^{\infty} \Psi_k(x) = \sum_{k=1}^{\infty} \Big(\sum_{j=1}^{2\cdot 3^k-1} (y_{k,j} \otimes g_{k,j})(x) \Big) = \sum_{k=1}^{\infty} \Big(\sum_{i=1}^{(2\cdot 3^k-1)^3} (x_{k,i} \otimes f_{k,i})(x) \Big).$$

Now, we need to set up an appropriate order to relate each pair (k, i) with a positive integer n. For $k = 1$, we have 5^3 terms related to $(1, i)$. So we simply let $n = i$. For $k \geq 2$, we let $n = \sum_{r=1}^{k-1}(2 \cdot 3^r - 1)^3 + i$ with $1 \leq i \leq (2 \cdot 3^k - 1)^3$.

Step 2. If we let $x_n = x_{k,i}$ and $f_n = f_{k,i}$, then $\{x_n, f_n\}_{n=1}^{\infty}$ is a frame for $C_r^*(\mathbb{F}_2)$.

It is sufficient to show that for every $x \in C_r^*(\mathbb{F}_2)$, the infinite series $\sum_{n=1}^{\infty}(x_n \otimes f_n)(x)$ converges to x. Since $x = \sum_{k=1}^{\infty} \Psi_k(x)$ is a convergent series in $C_r^*(\mathbb{F}_2)$, for arbitrary $\varepsilon > 0$ there exists $k_0 \geq 2$ such that for any $k \geq k_0$

$$\Big\| \sum_{r=k}^{\infty} \Psi_r(x) \Big\| + \|\Psi_k(x)\| + \frac{\|x\|}{2 \cdot 3^k - 1} < \varepsilon.$$

For any $m > m_0 = \sum_{r=1}^{k_0-1}(2 \cdot 3^r - 1)^3$, we can write $m = \sum_{r=1}^{k-1}(2 \cdot 3^r - 1)^3 + j$ for some $k \geq k_0$ and $1 \leq j \leq (2 \cdot 3^k - 1)^3$. In this case, there exist $0 \leq p \leq (2 \cdot 3^k - 1)^2 - 1$ and $1 \leq q \leq 2 \cdot 3^k - 1$ such that

$$\Big\| x - \sum_{n=1}^{m}(x_n \otimes f_n)(x) \Big\|$$

$$= \Big\| x - \sum_{r=1}^{k-1} \sum_{i=1}^{(2\cdot 3^r-1)^3} (x_{r,i} \otimes f_{r,i})(x) - \sum_{t=1}^{j}(x_{k,t} \otimes f_{k,t})(x) \Big\|$$

$$= \Big\| \sum_{r=1}^{\infty} \Psi_r(x) - \sum_{r=1}^{k-1} \Psi_r(x) - \frac{p}{(2 \cdot 3^k - 1)^2} \Psi_k(x) - \frac{1}{(2 \cdot 3^k - 1)^2} \sum_{i=1}^{q}(y_{k,i} \otimes g_{k,i})(x) \Big\|$$

$$\leq \Big\| \sum_{r=k}^{\infty} \Psi_r(x) \Big\| + \|\Psi_k(x)\| + \frac{\|x\|}{2 \cdot 3^k - 1} < \varepsilon.$$

This shows that $x = \sum_{n=1}^{\infty}(x_n \otimes f_n)(x)$ for every $x \in C_r^*(\mathbb{F}_2)$.

Step 3.   The sequence $\{x_n, f_n\}$ is a cb-frame for $C_r^*(\mathbb{F}_2)$,

It is equivalent to show that the initial sums $S_m = \sum_{n=1}^m (x_n \otimes f_n)$ are completely bounded maps on $C_r^*(\mathbb{F}_2)$ with $\sup \|S_m\|_{cb} < \infty$. Let $m \in \mathbb{N}$ be a positive integer. As we have shown that we can write $m = \sum_{r=1}^{k-1}(2 \cdot 3^r - 1)^3 + j$ for some $k \in \mathbb{N}$ and $1 \le j \le (2 \cdot 3^k - 1)^3$, and there exist $0 \le p \le (2 \cdot 3^k - 1)^2 - 1$ and $1 \le q \le 2 \cdot 3^k - 1$ such that

$$
\|S_m\|_{cb} = \Big\| \sum_{n=1}^m x_n \otimes f_n \Big\|_{cb} = \Big\| \sum_{r=1}^{k-1} \sum_{i=1}^{(2 \cdot 3^r - 1)^3} (x_{r,i} \otimes f_{r,i}) + \sum_{t=1}^j (x_{k,t} \otimes f_{k,t}) \Big\|_{cb}
$$

$$
= \Big\| \sum_{r=1}^{k-1} \Psi_r + \frac{p}{(2 \cdot 3^k - 1)^2} \Psi_k + \frac{1}{(2 \cdot 3^k - 1)^2} \sum_{i=1}^q (y_{k,i} \otimes g_{k,i}) \Big\|_{cb}
$$

$$
\le \Big\| \Phi_{\frac{1}{\sqrt{k-1}}, k-1} \Big\|_{cb} + \frac{p}{(2 \cdot 3^k - 1)^2} \|\Psi_k\|_{cb} + \frac{2 \cdot 3^k - 1}{(2 \cdot 3^k - 1)^2}
$$

$$
\le 3 \cdot \sup_k \Big\| \Phi_{\frac{1}{\sqrt{k}}, k} \Big\|_{cb} + 1.
$$

Finally, we let us recall from Eq. (2.6), (2.8) and (2.11) that we can write

$$
\Psi_1 = \sum_{n=1}^{5^3} (x_n \otimes f_n) = \sum_{i=1}^{5^3} (x_{1,i} \otimes f_{1,i}) = \sum_{l=1}^{5^2} \Big(\sum_{j=1}^5 \Big(y_{1,j}^l \otimes \frac{g_{1,j}^l}{5^2} \Big) \Big)
$$

with $y_{1,j}^l = y_{1,j} = \lambda_s$ and $g_{1,j}^l = g_{1,j} = \frac{e^{-|s|}}{5^2} \delta_s$ for corresponding $|s| \le 1$. For $k \ge 2$, we can write

$$
\Psi_k = \sum_{n=\sum_{r=1}^{k-1}(2\cdot 3^r - 1)^3 + 1}^{\sum_{r=1}^k (2\cdot 3^r - 1)^3} (x_n \otimes f_n) = \sum_{i=1}^{(2\cdot 3^k - 1)^3} (x_{k,i} \otimes f_{k,i})
$$

$$
= \sum_{l=1}^{(2\cdot 3^k - 1)^2} \Big(\sum_{j=1}^{2\cdot 3^k - 1} \Big(y_{k,j}^l \otimes \frac{g_{k,j}^l}{(2 \cdot 3^k - 1)^2} \Big) \Big)
$$

with $y_{k,j}^l = y_{k,j} = \lambda_s$ and

$$
g_{k,j}^l = g_{k,j} = \frac{(e^{-|s|/\sqrt{k}} - e^{-|s|/\sqrt{k-1}})}{(2 \cdot 3^k - 1)^2} \delta_s
$$

or $\frac{e^{-|s|/\sqrt{k}}}{(2 \cdot 3^k - 1)^2} \delta_s$ for corresponding $|s| \le k$. So for each $s \in \mathbb{F}_2$, there are many positive integers $n \in \mathbb{N}$ such that $x_n = \lambda_s$. We let $\varphi : n \in \mathbb{N} \to s \in \mathbb{F}_2$ be the map such that

$x_n = \lambda_s$, and let a_n be the coefficient for the corresponding δ_s. Then, it is easy to see that statements (1) and (2) in the theorem hold true.   □

Remark 2.4 The cb-frame $\{x_n, f_n\}$ for $C_r^*(\mathbb{F}_2)$ in Theorem 2.13 is not unconditional. Indeed, if it is unconditional (see definition given in [52]), then for each $x \in C_r^*(\mathbb{F}_2)$ the infinite series

$$x = \sum_{n=1}^{\infty}(x_n \otimes f_n)(x) = \sum_{n=1}^{\infty} a_n \delta_{\varphi(n)}(x)\lambda_{\varphi(n)}$$

converges unconditionally. In this case, we can rearrange its order such that

$$x = \sum_{n=1}^{\infty} a_n \delta_{\varphi(n)}(x)\lambda_{\varphi(n)} = \sum_{k=0}^{\infty}\left(\sum_{|s|=k}\left(\sum_{\varphi(n)=s} a_n\right)\delta_s(x)\lambda_s\right). \tag{2.12}$$

Note that for any $s \neq e$ in $\mathbb{F}_2$ the summation $\sum_{\varphi(n)=s} a_n$ in the last term of Eq. (2.12) is a positive infinite series with $\sum_{\varphi(n)=s} a_n = 1$. The last equality makes sense since we can apply a ε-argument to replace such an infinite sum with a finite sum if necessary. Thus, we can conclude from Eq. (2.12) that

$$x = \sum_{k=0}^{\infty}\left(\sum_{|s|=k}\left(\sum_{\varphi(n)=s} a_n\right)\delta_s(x)\lambda_s\right) = \sum_{k=0}^{\infty}\left(\sum_{|s|=k}\delta_s(x)\lambda_s\right) = \sum_{k=0}^{\infty} P_k(x).$$

This implies that for any $x \in C(\mathbb{T}) \cong C_r^*(\mathbb{Z}) \hookrightarrow C_r^*(\mathbb{F}_2)$

$$x = \sum_{k=0}^{\infty} P_k(x) = \lim_{K\to\infty}\sum_{k=0}^{K} P_k(x) = \lim_{K\to\infty}\sum_{j=-K}^{K} a_j z^j$$

converges uniformly in $C(\mathbb{T})$. This is impossible. At this moment, it is not known whether there is any unconditional frame for $C_r^*(\mathbb{F}_2)$.

2.3   Lipschitz Frames in Banach Spaces

In the past years, there has been an increased interest in investigating the "Nonlinear theory of Banach spaces" including stability for ε-isometries, linearization for isometries on spheres, and Lipschitz maps. This section presents some recent work on Lipschitz frames – the nonlinear version of Schauder frames. In particular, we use the linearization techniques in Lipschitz algebras [211] to establish the equivalence between the existence of Schauder frames and Lipschitz frames.

Let f be a map from a Banach space X. The Lipschitz norm of f is defined by

$$\|f\|_{Lip} := \sup\left\{\frac{\|f(x) - f(y)\|}{\|x - y\|} : x, y \in X, x \neq y\right\}.$$

If $\|f\|_{Lip}$ is finite, then f is called a Lipschitz operator. We use $Lip(X, Y)$ to denote all set of all the Lipschitz operators from X to Y, and write $Lip(X) := Lip(X, X)$ and $lip(X) := Lip(X, \mathbb{R})$ We also $Lip_0(X)$ to denote the set of all $f \in Lip(X)$ with $f(0) = 0$. It is well-known that $Lip(X)$ is a dual space, the predual is called free space or Arens-Eells space of X. Lipschitz Free spaces connect non-linear theory with the linear theory, and have gained a lot of interest in the last decades, cf. [94, 99, 211].

Similar to the Schauder frames in Definition 2.6 and unconditional Schauder frame, or just framing in Definition 2.4, Liu et al. [157] introduced the following concept of Lipschitz frames.

Definition 2.17 Let X he a separable Banach space and $\{x_n, f_n\}_{n=1}^{\infty} \subseteq X \times lip(X)$. Then $\{x_n, f_n\}_{n=1}^{\infty}$ is a called a *Lipschitz frame* of X if

$$\sup\left\{\left\|\sum_{k=1}^{n} f_k(\cdot)x_k\right\|_{Lip}\right\}_{n=1}^{\infty} < \infty$$

and $x = \sum_{n=1}^{\infty} f_n(x)x_n$ for all $x \in X$. Furthermore, a Lipschitz frame $\{x_n, f_n\}_{n=1}^{\infty}$ is an unconditional Lipschitz frame if $x = \sum_{n=1}^{\infty} f_n(x)x_n$ converges unconditionally for all $x \in X$ and

$$\sup_{E \subseteq \mathbb{N}}\left\|\sum_{k \in E} f_k(\cdot)x_k\right\|_{Lip} < \infty.$$

Remark 2.5

1. In Definition 2.17, we can assume $f_n(0) = 0$ for all $n \in \mathbb{N}$ by a simple translation. 2. The assumption

$$\sup_{n}\left\|\sum_{k=1}^{n} f_k(\cdot)x_k\right\|_{Lip} < \infty$$

is necessary, and this is different from the Schauder frame case. In the Schauder frame case, by the Uniformly Bounded Principle, we have $\sup_n \|\sum_{k=1}^{n} f_k(\cdot)x_k\| < \infty$ automatically. But we do not have this for Lipschitz operators. The following is a simple example.

Example 2.3.1 Let $g_n : \mathbb{R} \to \mathbb{R}$, $\mathrm{supp}(g_n) = [n, n+1]$,

$$
g_n(x) = \begin{cases} n(x-n), & n \le x \le n + \frac{1}{2} \\ n(n+1-x), & n + \frac{1}{2} < x \le n+1, \end{cases}
$$

and $\sum_{k=1}^{n} f_k(\cdot) := g_n + I_{\mathbb{R}}$ for each $n \in \mathbb{N}$. Then $x = \sum_{n=1}^{\infty} f_n(x)$ for all $x \in \mathbb{R}$ but

$$
\sup_{n \in \mathbb{N}} \left\| \sum_{k=1}^{n} f_k(\cdot) \right\|_{Lip} = \infty.
$$

To establish the connection between the existence of Lipschitz frames and the existence of Schauder frames, we need the following lemma from [211].

Lemma 2.1 *Let V and W be Banach spaces and $f \in Lip(V, W)$. Suppose V is separable. Then f has a linear surrogate, that is, there exists an $F \in B(V, W)$ such that*

1. $\|F\| \le \|f\|_{Lip}$ and
2. for any Banach space U and any $T \in B(W; U)$, if $T \circ f$ is linear, then $T \circ f = T \circ F$.

Theorem 2.14 *Let X be a separable Banach space. The following are equivalent:*

1. X has an (unconditional) Lipschitz frame.
2. X has an (unconditional) Schauder frame.

Proof It is sufficient to prove that (1) implies that X is a complemented subspace of a Banach space with the (unconditional) Schauder basis. We only need to prove the unconditional case since the proof of the classical case is essentially the same. Let $\{x_n, f_n\}_{n=1}^{\infty}$ be an unconditional Lipschitz frame of X. Without loss of generality, we can assume $\|x_n\| = 1$ for all $n \in \mathbb{N}$. Let $\{z_k\}_{k=1}^{\infty}$ be a copy of $\{x_k\}_{k=1}^{\infty}$. Let c_{00} be the collection of all sequences of real numbers with all but finite nonzero elements. Let Z be the completion of the vector space $\mathrm{span}\{\sum_{k=1}^{\infty} a_k z_k : \{a_k\}_{k \in \mathbb{N}} \in c_{00}\}$ under the norm

$$
\left\| \sum_{k=1}^{\infty} a_k z_k \right\|_Z := \sup_{E \subseteq \mathbb{N}} \left\| \sum_{k \in E} a_k x_k \right\|.
$$

It is easy to see that Z is a Banach space and $\{z_k\}_{k=1}^{\infty}$ is an unconditional basis of Z. Let

$$
S\left(\sum_{i \in \mathbb{N}} a_i z_i \right) := \sum_{i \in \mathbb{N}} a_i x_i, \quad \forall \sum_{i \in \mathbb{N}} a_i z_i \in Z
$$

and

$$T(x) := \sum_{i \in \mathbb{N}} f_i(x) z_i, \quad \forall x \in X.$$

It is trivial that S is a bounded linear operator. Let $x = \sum_{n=1}^{\infty} f_n(x) x_n \in X$ then

$$Tx = T\left(\sum_{n=1}^{\infty} f_n(x) x_n\right) = \sum_{n=1}^{\infty} f_n(x) z_n.$$

Then for any $x, y \in X$

$$\frac{\|Tx - Ty\|}{\|x - y\|} = \frac{\left\| \sum_{n=1}^{\infty} f_n(x) z_n - \sum_{n=1}^{\infty} f_n(y) z_n \right\|}{\|x - y\|}$$

$$= \sup_{E \subseteq \mathbb{N}} \frac{\left\| \sum_{n \in E} f_n(x) x_n - \sum_{n \in E} f_n(y) x_n \right\|}{\|x - y\|}.$$

Since $\{x_n, f_n\}_{n=1}^{\infty}$ is an unconditional Lipschitz frame, T is a Lipschitz operator. And for every x in X,

$$S \circ T(x) = S\left(\sum_{n=1}^{\infty} f_n(x) z_n\right) = \sum_{n=1}^{\infty} f_n(x) x_n = x.$$

So $S \circ T = I_X$. By Lemma 2.1, let F be a linear surrogate of T such that $\|F\| \leq \|T\|_{Lip}$ and $S \circ T = S \circ F = I_X$. Clearly, F is an injective operator and S is a surjective operator. Then for every x in X,

$$\|Fx\| \leq \|F\| \|x\|$$

and $F(X)$ is a subspace of Z. Next, we claim that $F(X)$ is a complemented subspace of Z. Since $\|S \circ F(x)\| = \|x\| \leq \|S\| \|Fx\|$, then

$$\|Fx\| \geq \frac{1}{\|S\|} \|x\|.$$

Hence, $F(X)$ is a closed subspace of Z. Because S is a surjective operator then $F \circ S : Z \to F(Z)$ and

$$(F \circ S) \circ (F \circ S) = F \circ S$$

$F(X)$ is a complemented subspace of Z. Then X is isomorphic to a complemented subspace $F(X)$ of Z that has an unconditional Schauder basis.   □

Remark 2.6 As shown above, for a separable Banach space X, it has an (unconditional) Lipschitz frame if and only if it has the (unconditional) Schauder frame hence X can be complementably embedded into a Banach space with an (unconditional) Schauder basis, which is exactly the equivalent characterization for a Banach space with the bounded approximation property [174] and will be demonstrated in next chapter. Thus, from the viewpoint of dilations from frames to bases, there is a close connection between the frame and bounded approximation property (resp, Lipschitz frame and Lipschitz bounded approximation property), all of these allow us to study metric embeddings on both Banach and metric spaces.

Frames and Approximation Properties

3

The approximation property originally introduced by Banach in his book [11] in 1932 plays a fundamental role in the structure theory of Banach spaces. The first systematic study of the variants of the approximation property was initiated by Grothendieck [100] in 1955. At that time, the main properties investigated were the approximation property (AP), the bounded approximation property (BAP), the metric approximation property (MAP) and the basis property. Trivially, a Banach space X having a Schauder basis implies X having both AP and BAP. In 1973, Enflo [77] constructed the first famous example of a separable Banach space that fails to have the AP. Furthermore, in 1973, Figiel and Johnson [85] constructed examples to show the connections between MAP, BAP and AP. In 1987, Szarek [198] gave an example showing that not every Banach space with BAP has a Schauder basis. For more details on Banach space approximation properties, we refer the reader to Casszza's comprehensive survey [43]. Meanwhile, various approximation properties for Banach spaces are also extended and investigated to operator spaces. In summary, there has been a rich theory surrounding the approximation property of both classical and non-linear theory for Banach spaces as well as for operator spaces [177]. Considering the nature of this investigation, it is not surprising that frame theory also plays an important role in the study of Banach space approximation properties. Combining the results of Johnson, Rosenthal and Zippin [146] and Pełczyński [174] in 1971, Casszza, Han and Larson [52] introduced the Schauder frame and established the connections between Schauder frames and BAP. Later, Liu and Ruan in [156] introduced the concept of completely bounded frames for operator spaces to obtain natural operator space analogues of corresponding Banach space results. Most recently, Liu et al. extended the Schauder frames to the Lipschitz version [157]. This chapter is devoted to discussing some connections between frame theory and various types of Banach space approximation properties.

3.1   Approximation Properties

Banach in 1932 in his book [11] formulated the following basis problem: *Does every separable Banach space have a basis?* This problem motivated a great deal of research over the next 40 years, in particularly, the approximation properties. Apart from the initial systematic studies on the variants of the approximation property of Grothendieck [100] and various highly technical counterexamples constructed by [77, 85, 198], there are many other variants, including the commuting bounded approximation property, the π-property, the finite-dimensional decomposition property, the uniform approximation property, the compact approximation property etc. (see [43]). In this section, we discuss some of them including some representative proofs illustrating the basic facts and investigate the bounded approximation properties from the Banach space frame perspective.

We begin with the definition of the approximation property from [43].

Definition 3.1 A Banach space X is said to have *the approximation property* (AP for short) if for every compact set K in X and every $\varepsilon > 0$, there is a finite-rank operator $T : X \rightarrow X$ such that $\|Tx - x\| < \varepsilon$ for every $x \in K$.

In other words, a Banach space X has AP if and only if the identity operator on X can be uniformly approximated on every compact set by finite-rank operators. The next result shows that the approximability of operators either from X or into X on compact sets is equivalent to the approximability of the identity operator on X on compact sets.

Proposition 3.1 *Let X be a Banach space. Then the following are equivalent:*

1. *For every compact subset K of X and every $\varepsilon > 0$ there exists a finite-rank operator $S : X \rightarrow X$ such that $\|x - Sx\| \leq \varepsilon$ for every $x \in K$.*
2. *For every Banach space Y, every operator $T : X \rightarrow Y$, every compact subset K of X and every $\varepsilon > 0$, there exists a finite-rank operator $S : X \rightarrow Y$ such that $\|Tx - Sx\| \leq \varepsilon$ for every $x \in K$.*
3. *For every Banach space Y, every operator $T : Y \rightarrow X$, every compact subset K of Y and every $\varepsilon > 0$, there exists a finite-rank operator $S : Y \rightarrow X$ such that $\|Ty - Sy\| \leq \varepsilon$ for every $y \in K$.*

Proof (1) $\implies$ (2): Let $T \in B(X, Y)$, $T \neq 0$, let $K \subset X$ be compact and let $\varepsilon > 0$. By (1), there exists a finite-rank operator $R : X \rightarrow X$ such that $\|x - Rx\| \leq \varepsilon/\|T\|$ for every $x \in K$. Setting $S = TR$, then S is a finite-rank operator from X into Y and $\|Tx - Sx\| \leq \varepsilon$ for every $x \in K$.

(1) $\implies$ (3): Let $T \in B(Y, X)$, let $K \subset Y$ be compact and let $\varepsilon > 0$. Applying (1) to the compact subset $T(K)$ of X, there exists a finite-rank operator $R : X \rightarrow X$ such that $\|x - Rx\| \leq \varepsilon$ for every $x \in T(K)$. Then $S = RT$ is a finite-rank operator from Y into X and $\|Ty - Sy\| \leq \varepsilon$ for every $y \in K$.

Since (2) and (3) clearly imply (1), the proof is complete.   □

The following is a simple sufficient condition for the AP.

Proposition 3.2 *Let X be a Banach space. Suppose that there exists a uniformly bounded net $\{T_\alpha\}_\alpha$ of finite-rank operators on X such that $T_\alpha x \to x$ for every $x \in X$. Then X has the approximation property.*

Proof Let $\{T_\alpha\}$ be a net of finite-rank operators such that $\sup_\alpha \|T_\alpha\| = C < \infty$ and $T_\alpha x \to x$ for every $x \in X$. Let K be a compact subset of X and let $\varepsilon > 0$. Choose a δ-net, $\{x_1, \ldots, x_n\}$ for K, where $\delta = \min\{\varepsilon/3, \varepsilon/3C\}$. There exists α_0 such that if $\alpha \geq \alpha_0$, then $\|x_i - T_\alpha x_i\| \leq \varepsilon/3$ for each i. Let $x \in K$ and choose i such that $\|x - x_i\| < \delta$. Then

$$\|x - T_{\alpha_0} x\| \leq \|x - x_i\| + \|x_i - T_{\alpha_0} x_i\| + \|T_{\alpha_0} x_i - T_{\alpha_0} x\| < \varepsilon.$$

Thus X has the AP.   □

Example 3.1.1 Every Banach space with a Schauder basis has the approximation property.   □

Proof Let $\{e_n\}_{n=1}^\infty$ be a Schauder basis for X and let $\{e_n^*\}_{n=1}^\infty$ be the associated coordinate functionals. For each n, let P_n be the finite-rank operator on X given by $P_n x = \sum_{i=1}^n e_i^*(x) x_i$. Then the operators P_n are uniformly bounded, and $P_n x \to x$ for every $x \in X$. Therefore, X has the AP.   □

Remark 3.1 It follows from the above example that the spaces c_0 and ℓ_p, for $1 \leq p < \infty$, $C[0, 1]$ all have the AP. In particular, every Hilbert space has the AP. From [188], $C(K)$, the space of continuous functions on the compact space K, and $L^p(\mu)$, where μ is any measure, has the approximation property for every $p \in [1, \infty)$. Moreover, The spaces $C(K)^*$, $L^\infty(\mu)$ and $L^\infty(\mu)^*$ also have the AP. Thus, all the classical Banach spaces of sequences and functions encountered thus far, along with their duals, have the AP. However, not every separable Banach space has the AP due to a deep result of Enflo [77] showing that the space $B(\ell_2)$ does not have the AP, and there is a separable reflexive Banach space which fails to have the AP. This also implies that there is a separable Banach space that fails to have a Schauder basis.

Recall that an operator is said to be *approximable* if it is the limit of a sequence of finite-rank operators in the operator norm. We can now give a formulation of the approximation property in terms of the uniform approximation of compact operators.

Proposition 3.3 *Let X be a Banach space. Then X has the approximation property if and only if for every Banach space Y, every compact operator from Y into X is approximable.*

Proof Suppose that X has the AP and $T : Y \to X$ is a compact operator. Let $\varepsilon > 0$. Then there exists a finite-rank operator $R : X \to X$ such that $\|x - Rx\| < \varepsilon$ for every x in the relatively compact subset $T(B_Y)$ of X. Let $S = RT$. Then S has finite-rank and $\|T - S\| \leq \varepsilon$.

Conversely, suppose that every compact operator with values in X is approximable. Let K be a compact subset of X and let $\varepsilon > 0$. By Lemma 4.11 in [188], there exists a convex, balanced, compact set L containing K such that K is compact in the Banach space X_L, where X_L is the subspace of X spanned by L, that is the union of all the positive scalar multiples of L, and normed by the Minkowski functional of L:

$$\|x\|_L = \inf\{\lambda > 0 : x \in \lambda L\}$$

Now, since the injection $I : X_L \to X$ is compact, there exists a finite-rank operator $R : X_L \to X$ such that $\|I - R\| < \varepsilon/2$. Taking $R = \sum_{i=1}^{n} \psi_i \otimes x_i$, where $\psi \in (X_L)^*$ and $x_i \in X$. Since $I : X_L \to X$ is injective, following the property of adjoint operators, $I^*(X^*)$ is dense in $(X_L)^*$ for the topology of uniform convergence on the compact subsets of X_L. Therefore, given $\eta > 0$, there exists $\varphi_i \in X^*$ such that

$$\|\varphi_i(x) - \psi_i(x)\| < \eta, \forall x \in K, i.$$

Let $S : X \to X$ be the finite-rank operator $\sum_{i=1}^{n} \varphi_i \otimes x_i$. If $x \in K$, then

$$\|Rx - Sx\| \leq \sum_{i=1}^{n} |\varphi_i(x) - \psi_i(x)| \|x_i\| < \eta \left(\sum_{i=1}^{n} \|x_i\| \right) < \varepsilon,$$

taking $\eta < \varepsilon/(2 \sum_{i=1}^{n} \|x_i\|)$. Since $\|x - Rx\| < \varepsilon/2$ for every $x \in K$, we have

$$\|x - Sx\| < \varepsilon, \forall x \in K.$$

Therefore, X has the AP. $\square$

It is worth pointing out that possession of AP leads to the resolution of several outstanding issues concerning projective and injective tensor products [188], including the approximation, reflexivity and bases in the tensor product of Banach spaces. For more details, we refer the reader to the references [5, 188].

Note that Definition 3.1 of the AP puts no restriction on the norm of the finite-rank operator which approximates the identity on a compact set. We have seen that in the case of X having a basis, the basis projections provide the required approximation and are uniformly bounded (independent of the compact set K). Now, we introduce this stronger variant of the AP.

Definition 3.2 Let X be a Banach space, and $1 \leq \lambda < \infty$. We say that X has the λ-*bounded approximation property* (λ-BAP for short) if for every compact subset K of X and every $\varepsilon > 0$, there exists a finite-rank operator $S : X \to X$ such that $\|S\| \leq \lambda$ and $\|x - Sx\| \leq \varepsilon$ for every $x \in K$. And X has *bounded approximation property* (BAP for short) if X has λ-BAP for some λ. If this holds with $\lambda = 1$, then X is said to have the *metric approximation property* (MAP for short).

Remark 3.2 The compact sets in Definition 3.2 can be replaced by finite sets. To see this, let K be a compact set in a Banach space X and let $\varepsilon > 0$ be given. We can find a finite set $\{x_i\}_{i=1}^n$ so that $K \subset \bigcup_{i=1}^n B(x_i, \varepsilon/3\lambda)$. Now, if T is a finite-rank operator with $\|Tx_i - x_i\| \leq \varepsilon/3$, for $1 \leq i \leq n$, then $\|Tx - x\| \leq \varepsilon$, for every $x \in K$.

Remark 3.3 It is immediate that a space with a basis has the BAP and that a space with the BAP has the AP. In particular, if X has λ-BAP and if P is a bounded linear projection from X onto Y, then Y has $\lambda\|P\|$-BAP. Thus, the BAP passes to complemented subspaces.

Next, we will discuss some equivalent conditions for BAP. To do that, we need the following useful lemma from [43] stating that finite-rank operators approximating the identity close enough on a finite-dimensional subspace can be perturbed to equal the identity on that subspace.

Lemma 3.1 *Let X be a Banach space and E be an n-dimensional subspace of X and let $T : X \to E$ be onto. Let $k \leq n$ and let F be a k-dimensional subspace of X such that $\|T|_F - I_F\| < \varepsilon < 1$, where $(1 - \varepsilon)^{-1}\varepsilon k < 1$. Then there is a rank n operator S on X such that $S|_F = I_F$, $\|S - T\| < (1 - \varepsilon)^{-1}\varepsilon k\|T\|$, and $S^*(X^*) = T^*(X^*)$. Moreover, if T is a projection, then S can be chosen as a projection.*

As an immediate consequence of Lemma 3.1, we have the following:

Theorem 3.1 *Let X be a Banach space. Then the following are equivalent:*

1. *X has the bounded approximation property.*
2. *There is a uniformly bounded net $\{T_\alpha\}$ of finite-rank operators on X that convergences strongly to the identity on X.*
3. *There is a $\lambda \geq 1$ such that for every finite dimensional subspace $E \subset X$ there is a finite-rank operator T on X such that $\|T\| \leq \lambda$ and $Tx = x$, for all $x \in E$.*

In the case of separable Banach spaces, if X has the BAP, then this net can be replaced by a sequence $\{T_n\}$. Thus, we can use induction to show the following corollary.

Corollary 3.1 *Let X be a separable Banach space. Then X has the λ-bounded approximation property if and only if there is a sequence of finite-rank operators $\{T_n\}$ on X*

converging strongly to the identity so that $T_m T_n = T_n$, for all $n < m$, and $\limsup_n \|T_n\| \leq \lambda$. We will call such a sequence $\{T_n\}$ an approximating sequence.

By the previous remark, the BAP passes to complemented subspaces. Since there are spaces with bases whose duals fail to have the AP and also BAP, thus BAP does not pass to the dual spaces. However, it passes from the dual space to the space, specifically, by Grothendieck's characterization by tensor product [188], if X^* has the BAP, then so does X.

Having introduced the BAP and AP, next we recall some important results concerning AP, BAP, MAP and the basis property. The first result is due to Grothendieck [100] which shows the connection between AP and MAP or BAP.

Theorem 3.2 *If a Banach space X is a separable dual space with the approximation property, then X has the metric approximation property. Moreover, every reflexive Banach space with the approximation property also has the metric approximation property.*

In 1973, Enflo [77] proved the following remarkable result.

Theorem 3.3 *There is a separable reflexive Banach space that fails to have the approximation property.*

Therefore, there exists a separable Banach space which fails to have a Schauder basis. Inspired by Enflo's ideas, Davies [59] and Szankowski [196, 197] provided significant modifications which made the construction of Enflo much simpler. Additionally, Figiel and Johnson [85] in 1973 constructed an example of a Banach space that proves the following important theorem.

Theorem 3.4 *1. There is a Banach space that has the bounded approximation property but fails the metric approximation property.*
2. There is a separable Banach space (which even has a separable dual) with the approximation property but fails to have the bounded approximation property.

In 1987, Szarek [198] constructed a separable superreflexive Banach space to distinguish BAP and basis property.

Theorem 3.5 *There exists a separable superreflexive Banach space with bounded approximation property, which does not have a Schauder basis.*

The following figure illustrates the historical events of the approximation properties (Fig. 3.1).

We have seen that there is a rich theory surrounding the approximation property for Banach spaces and the counter-examples merely illustrate the delicate nature of the

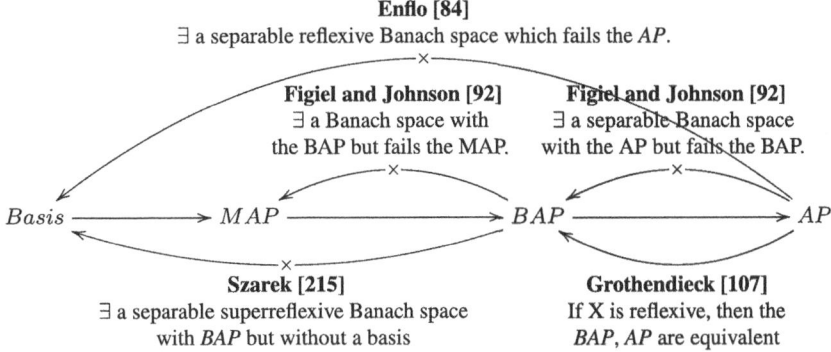

Fig. 3.1 Historical events on approximation properties

relationships on the chain of approximation properties. Next, we will particularly focus on BAP from the aspect of connections between space structure and frame theory.

Apart from the Definition 3.2 of the BAP, the following definition is useful to help us have a better understanding of the concept of BAP.

Definition 3.3 A sequence of nonzero finite-rank operators $\{A_i\}$ from a Banach space X into itself is called an *(unconditional) finite dimensional expansion of the identity* if for all $x \in X$,

$$x = \sum_i A_i x,$$

(and the series converges unconditionally). Moreover, if $A_i A_n = 0$, for all $i \neq n$, then $\{A_i\}$ is called an (unconditional) orthogonal expansion of the identity of X.

Note that if a sequence $\{A_i\}$ of finite-dimensional operators is an (unconditional) finite-dimensional expansion of the identity of X, then the span of $\cup_i A_i X$ is dense in X. In 1971, Pełczyński [174] and independently Johnson, Rosenthal and Zippin [146] established an important relationship between bases and the BAP.

Theorem 3.6 A Banach space X has the bounded approximation property if and only if X is isomorphic to a complemented subspace of a Banach space with a basis.

Combining the characterizations of atomic decomposition with the above results of Pełczyński [174] and Johnson, Rosenthal and Zippin [146], we proceed to examine the general relationship between atomic decompositions and several forms of the approximation property.

Theorem 3.7 ([52]) *For a Banach space* X, *the following are equivalent:*

1. X *has an atomic decomposition.*
2. X *has a finite-dimensional expansion of the identity.*
3. X *is complemented in a Banach space with a basis.*
4. X *has the bounded approximation property.*

Proof (1) $\implies$ (3): Let $(\{x_i, y_i\}, X_d)$ be an atomic decomposition for the Banach space X. First, we assume that all the x_i are nonzero. Let c_{00} denote the linear space of scalar sequences with only finitely many nonzero terms. Let $\{e_i, e_i^*\}_{i=1}^{\infty}$ be the unit vectors in c_{00} and define a norm on c_{00} by:

$$\left\| \sum_i a_i e_i \right\| = \sup_n \left\| \sum_{i=1}^n a_i x_i \right\|_X, \tag{3.1}$$

for every sequence of scalars $(a_i) \in c_{00}$. It follows immediately from Theorem 1.2 that $\{e_i\}$ is a basis for its Banach space completion, which we denote by Z. We define an operator $\theta : X \to Z$ by

$$\theta x = \sum_{i=1}^{\infty} \langle x, y_i \rangle e_i.$$

By assumption (3) of Definition 2.1, it follows that θ is an isomorphism taking X onto a subspace of Z. Also, we define $\Gamma : Z \to X$ by

$$\Gamma \left(\sum_{i=1}^{\infty} a_i e_i \right) = \sum_{i=1}^{\infty} a_i x_i,$$

then by Eq. (3.1), we have that Γ is a linear contraction from Z to X, and by (3) of Definition 2.1, we have that $\Gamma \theta = I_X$. It follows that $P = \theta \Gamma$ is a bounded linear projection of Z onto θX, which takes e_i to θx_i. Renorming X by $\sup_n \| \sum_{i=1}^n \langle x, y_i \rangle x_i \|$ for $x \in X$, by Uniform Bounded Principle, it is an equivalent norm, in this norm for X, θ is an (into) isometry and then identify X with θX and x_i with θx_i. Thus X is complemented in Z with the basis $\{e_i\}$. Next, if some of the x_i are zero, set $J = \{i : x_i \neq 0\}$. Then, we can do our above construction on $(x_i)_{i \in J}$. For $i \in J^c$, we may assume that $y_i = 0$. So we consider the larger space $Z \oplus \ell_2^M$ with $M = |J^c|$. Let $(f_i)_{i \in M}$ be any orthonormal basis for ℓ_2^M and we define a projection Q on $Z \oplus \ell_2^M$ by $Q(z \oplus y) = P(z) \oplus 0$. Now, we have the desired space $Z \oplus \ell_2^M$ with a basis so that x_i is the image under a projection of a basis for this space.

(3) $\implies$ (1) Let Z be the Banach space with basis $\{e_i\}$. We define the space of scalar-valued sequences by:

$$X_d = \left\{ (a_i) : \sum_{i=1}^{\infty} a_i e_i \text{ converges in } X \right\}$$

If we equip X_d with the norm:

$$\|(a_i)\|_{X_d} = \left\| \sum_{i=1}^{\infty} a_i e_i \right\|_X$$

then X_d becomes a Banach space which is isometric to X, which is called *associated Banach space of scalar-valued sequences* of Z. Let P be the continuous linear projection (an idempotent: $P^2 = P$) from Z to X. Set $x_i = Pe_i$, $y_i = P^*e_i^*$, then it can be seen that $(\{x_i, y_i\}, X_d)$ is an atomic decomposition for X.

(3) $\Longleftrightarrow$ (4) : It follows directly from Theorem 3.6 above.

(2) $\Longrightarrow$ (4) : If $\{A_i\}$ is a finite dimensional expansion of the identity on X, then

$$T_n = \sum_{i=1}^{n} A_i,$$

is a sequence of finite-rank operators on X with the property that $T_n x \to x$, for all $x \in X$. Hence, X has the bounded approximation property.

(4) $\Longrightarrow$ (2) : If $\{T_n\}$ is a sequence of finite-rank operators on X with $T_n x \to x$, for all $x \in X$, then it is immediate that $A_n = T_{n+1} - T_n$ is a finite dimensional expansion of the identity on X. $\square$

As mentioned before, Schauder frames for Banach spaces were developed in [52] to create a procedure to represent vectors using quantized redundant coefficients, which generalizes the frames for Hilbert spaces and Schauder bases for Banach spaces but also shows the relationship with the BAP. An easy review of Definition 2.1 and (1) $\Longrightarrow$ (3) in Theorem 3.7 shows that a sequence $\{x_i, f_i\}_{i=1}^{\infty}$ is a Schauder frame for a Banach space if and only if there is a Banach sequence space of scalar-valued sequences X_d such that $(\{x_i, f_i\}, X_d)$ is an atomic decomposition for X.

Combining Theorem 3.7 and 3.6, the following result, due to Casazza, Han and Larson [52], establishes the connection between frames and the BAP of Banach spaces.

Proposition 3.4 *For a separable Banach space X, the following are equivalent:*

1. *X has a Schauder frame;*
2. *X has the bounded approximation property;*
3. *X is isomorphic to a complemented subspace of a Banach space with a Schauder basis.*

Combining the above result with Szarek's example [198], we get that there is a Banach space with Schauder frames which has no Schauder basis. The next result is from unconditionalizing Proposition 3.4.

Corollary 3.2 *For a Banach space X, the following are equivalent:*

1. *X has a framing.*
2. *X is complemented in a Banach space with an unconditional basis.*

In [52, 122], Han and Larson investigated the frame theory from the dilation or geometric point of view. That is, Parseval frames (general frames or even dual frame pairs) are exactly the compressions of orthonormal bases (resp, Riesz bases), and (unconditional) Schauder frames for Banach spaces are a compression of (resp, unconditional) Schauder bases of Banach spaces. In [45], it is shown that every Schauder frame has an associated space, which is referred to as *the minimal associated space* in [153] or can be referred to Definition 2.7. Furthermore, the minimal associated basis will be unconditional if and only if the Schauder frame is unconditional. It is easy to see that if a Schauder frame has a shrinking (boundedly complete) associated basis, then the frame must be shrinking (resp, boundedly complete) as well. Given some desirable property for a basis to have, it is natural to consider the problem of whether a particular Schauder frame has an associated basis with that property. One result of such a kind is due to Beanland et al. [31] who proved the following interesting result.

Theorem 3.8 *Suppose that $\{x_i, y_i\}_{i=1}^{\infty}$ is a Schauder frame for a reflexive Banach space X. Then there exist a reflexive Banach space Z complementedly containing X and a Schauder basis $\{z_i, z_i^*\}_{i=1}^{\infty}$ for Z such that $x_i = Pz_i$ and $y_i = P^*z_i^*$, where P is a bounded projection from Z onto X.*

The above result leads to the following problem.

Problem 3.1 Suppose that $\{x_i, y_i\}_{i=1}^{\infty}$ is an unconditional Schauder frame for a reflexive Banach space X, is there a reflexive Banach space Z with an unconditional basis $\{z_i, z_i^*\}_{i=1}^{\infty}$, such that $x_i = Pz_i$ and $y_i = P^*z_i^*$ for a bounded projection P from Z onto X?

Finally, we would like to point out that various approximation properties also have some connections with the following concept.

Definition 3.4 A sequence $\{E_i\}_{i=1}^{\infty}$ of finite dimensional subspaces of a Banach space X is called a *finite dimensional decomposition (FDD)* for X if for each $x \in X$ there is a unique sequence of vectors $x_i \in E_i$ so that

$$x = \sum_i x_i. \tag{3.2}$$

In this case, we write:

$$X = \sum_{n=1}^{\infty} \oplus E_i. \tag{3.3}$$

If the convergence in Eq. (3.2) is unconditional, then we call Eq. (3.3) an *unconditional finite dimensional decomposition (UFDD)* of X.

We refer the reader to [42, 53] for a detailed treatment of FDD's for Banach spaces. Clearly, a basis is an FDD for X with dim $E_i = 1$, for all i, and an unconditional basis for X is a UFDD of this type. If $\sum \oplus E_i$ is a finite-dimensional decomposition for X, we have a natural sequence P_n of commuting finite-rank projections on X given by:

$$P_n \left(\sum_{k=1}^{\infty} x_k \right) = \sum_{k=1}^{n} x_k.$$

We call $K = \sup_n \| P_n \|$ the FDD constant of X, and the projections $Q_n = P_n - P_{n-1}$ the coordinate projections. Moreover, by using the projections P_n, we have that FDD implies BAP and AP.

3.2   Cb-frames and Completely Bounded Approximation Property

In [52], Casazza, Han and Larson showed that a separable Banach space has the BAP if and only if it has a frame. It is natural to consider the operator space analogues of corresponding results. For this purpose, we first introduce the completely bounded approximation property for operator spaces.

Definition 3.5 Let X be an operator space. If there exists a net of finite-rank maps $\Phi_\alpha : X \to X$ such that $\sup_\alpha \| \Phi_\alpha \|_{cb} < \infty$ and

$$\Phi_\alpha(x) \to x, \forall x \in X,$$

then we call that X has the *completely bounded approximation property* (CBAP for short).

Clearly, if a separable operator space X has a cb-basis $\{x_i, f_i\}_{i=1}^{\infty}$, we define a sequence of finite-rank maps $S_m : X \to X$ by

$$S_m(x) = \sum_{i=1}^{m} f_i(x) x_i,$$

then it is easy to see that $\sup_m \| S_m \|_{cb} < \infty$ and $S_m(x) \to x$ for every $x \in X$. Thus every separable operator space X with a cb-basis has CBAP. It is natural to ask whether the converse is true or not. However, this is not true in general. Indeed, let X be a separable Banach space X which has the BAP and does not admit any Schauder basis. Then $Min(X)$, the space X equipped with the Min-operator space structure, is an operator space with CBAP, but does not have any cb-basis. However, the problem is still open for separable C^*-algebras. In particular, it is not known whether $C_r^*(\Gamma)$ has a cb-basis if Γ is a weakly amenable discrete group. All these considerations led us to investigate the connections between the cb-frames and CBAP.

3.2.1 Cb-Frames for Operator Spaces

The first main result is considered to be an operator space analogue of corresponding Banach space results by Casazza, Han and Larson [52], Johnson, Rosenthal and Zippin [146], and Pełczński [174].

Theorem 3.9 *[156] Let X be a separable operator space. Then the following properties are equivalent:*

1. *X has the completely bounded approximation property.*
2. *X has a cb-frame.*
3. *X is completely isomorphic to a completely complemented subspace of an operator space with a cb-basis.*

Proof (1) $\Longleftrightarrow$ (2) The "only if" part is obvious. We only need to prove the "if" part. The proof is motivated by Pełczński's decomposition technique. Suppose that X has the CBAP. Then there is a sequence of finite-rank maps $\{\Phi_k\}$ on X such that $\sup_k \| \Phi_k \|_{cb} \leq K < \infty$ and $x = \lim_{k \to \infty} \Phi_k(x)$ for all $x \in X$. Let

$$\Psi_1 = \Phi_1 = \Phi_1, \quad \text{and} \quad \Psi_k = \Phi_k - \Phi_{k-1}, \ \forall k \geq 2.$$

Then $\{\Psi_k\}$ is a sequence of finite-rank maps on X such that $\sup_k \| \Psi_k \|_{cb} \leq 2K < \infty$ and $x = \sum_{k=1}^{\infty} \Psi_k(x)$ for all $x \in X$. Let $m(k)$ be the dimension of $\Psi_k(X)$. It is known from Auerbach theorem that there exists a biorthogonal basis $\{y_{k,j}, y_{k,j}^*\}_{1 \leq j \leq m(k)}$ for $\Psi_k(X)$ such that $\| y_{k,j} \| \leq 1$, $\| y_{k,j}^* \| \leq 1$ and $\langle y_{k,i}, y_{k,j}^* \rangle = \delta_{i,j}$. For each $1 \leq j \leq m(k)$, $g_{k,j} = y_{k,j}^* \circ \Psi_k$ is a bounded linear functional on X such that $\| g_{k,j} \| \leq 2K$. We get $\{y_{k,j}, g_{k,j}\}_{1 \leq j \leq m(k)}$ in $X \times X^*$ such that $\Psi_k = \sum_{j=1}^{m(k)} y_{k,j} \otimes g_{k,j}$ satisfying $\| y_{k,j} \| \leq 1$ and $\| g_{k,j} \| \leq 2K$. Set

$$x_{k,i} = y_{k,j}, \quad \text{and} \quad f_{k,i} = \frac{g_{k,j}}{m(k)^2},$$

when $i = p \cdot m(k) + j$ with $0 \leq p \leq m(k)^2 - 1$ and $1 \leq j \leq m(k)$. Then we can use a similar argument as that given in the proof of Theorem 2.13 to show that if we let $x_n = x_{k,i}$ and $f_n = f_{k,i}$ when $n = \sum_{r=0}^{k-1} m(r)^3 + i$ and $1 \leq i \leq m(k)^3$, then $\{x_n, f_n\}$ is a cb-frame for X.

(1) $\Longleftrightarrow$ (3) The "if" part is obvious since every completely complemented subspace of an operator space with the CBAP always has the CBAP. We only need to prove the "only if" part. Let $\{x_i, f_i\} \in X \times X^*$ be a cb-frame of X with $\|x_i\| = 1$ for all $i \in \mathbb{N}$. Let c_{00} be the linear space of all sequences of complex numbers with finitely many nonzeros, and $\{e_i\}_{i=1}^{\infty}$ be the canonical basis of c_{00}. For any $u \in \mathbb{M}_n(c_{00}) \cong c_{00}(\mathbb{M}_n)$, there is a unique linear expression $u = \sum u_i \otimes e_i$ with finitely many $u_i \neq 0$ in $\mathbb{M}_n$. We define a norm $\|\|\cdot\|\|_n$ on $\mathbb{M}_n(c_{00})$ as follows:

$$\|\|u\|\|_n = \sup_{m \geq 1} \left\| \sum_{i=1}^{m} u_i \otimes x_i \right\|_n = \max_{m \geq 1} \left\| \sum_{i=1}^{m} u_i \otimes x_i \right\|_n < \infty. \tag{3.4}$$

More precisely for any $u, v \in \mathbb{M}_n(c_{00})$, we have

$$\|\|u + v\|\|_n = \max_{m \geq 1} \left\| \sum_{i=1}^{m} \left(u_i + v_i \right) \otimes x_i \right\|_n$$

$$\leq \max_{m \geq 1} \left(\left\| \sum_{i=1}^{m} u_i \otimes x_i \right\|_n + \left\| \sum_{i=1}^{m} v_i \otimes x_i \right\|_n \right)$$

$$\leq \|\|u\|\|_n + \|\|v\|\|_n.$$

If $\|\|u\|\|_n = 0$, then $\|\sum_{i=1}^{m} u_i \otimes x_i\|_n = 0$ for all $m \in \mathbb{N}$. By induction, we can conclude $u = 0$. This shows that each $\|\|\cdot\|\|_n$ is a norm on $\mathbb{M}_n(c_{00})$. Moreover, $\|\|\cdot\|\|_n$ satisfies the following properties:

(N1) $\|u \otimes e_i\|_n = \|u \otimes x_i\|_n = \|u\|_n$ for all $u \in \mathbb{M}_n$ and $i \in \mathbb{N}$.
(N2) For any $u_i \in \mathbb{M}_n$ and $m \leq l$, we have

$$\|\| \sum_{i=1}^{m} u_i \otimes e_i \|\|_n \leq \|\| \sum_{i=1}^{l} u_i \otimes e_i \|\|_n.$$

(N3) For any $x \in X$, $(\sum_{i=1}^{n} f_i(x)e_i)_{n=1}^{\infty}$ is a Cauchy sequence in $(c_{00}, \|\|\cdot\|\|_1)$.

Since (N1) and (N2) are obvious, we only need to prove (N3). Since $\{x_i, f_i\}$ is a frame for X, for any $x \in X$, we have $x = \sum_{i=1}^{\infty} f_i(x)x_i$. Then for any $\varepsilon > 0$, there is n_0 such that, for any $m > n \geq n_0$, we have $\|\sum_{i=n+1}^{m} f_i(x)x_i\| < \varepsilon$. Now if we let $y_n = \sum_{i=1}^{n} f_i(x)e_i$

for each $n \in \mathbb{N}$, the we get

$$\|y_m - y_n\|_1 = \left\|\left\|\sum_{i=n+1}^{m} f_i(x)e_i\right\|\right\|_1 = \max_{n+1 \leq k \leq m} \|\sum_{i=n+1}^{k} f_i(x)x_i\| < \varepsilon.$$

This shows (N3).

Now we prove that $\|\cdot\|_n$ is actually an operator space matrix norm. For $u = \sum u_i \otimes e_i \in \mathbb{M}_n(c_{00})$, $w = \sum w_i \otimes e_i \in \mathbb{M}_m(c_{00})$, and $\alpha \in \mathbb{M}_n, \beta \in \mathbb{M}_n$, we have

$$\|u \oplus w\|_{n+m} = \left\|\left\|\sum (u_i \oplus w_i) \otimes e_i\right\|\right\|_{n+m}$$

$$= \max_{l \geq 1} \left\|\sum_{i=1}^{l} (u_i \oplus w_i) \otimes x_i\right\|_{n+m}$$

$$= \max_{l \geq 1} \left\|\left(\sum_{i=1}^{l} u_i \otimes x_i\right) \oplus \left(\sum_{i=1}^{l} w_i \otimes x_i\right)\right\|_{n+m} \tag{3.5}$$

$$= \max_{l \geq 1} \max\left\{\left\|\sum_{i=1}^{l} u_i \otimes x_i\right\|_n, \left\|\sum_{i=1}^{l} w_i \otimes x_i\right\|_m\right\}$$

$$= \max\left\{\max_{l \geq 1}\left\|\sum_{i=1}^{l} u_i \otimes x_i\right\|_n, \max_{l \geq 1}\left\|\sum_{i=1}^{l} w_i \otimes x_i\right\|_m\right\}$$

$$= \max\{\|u\|_n, \|w\|_m\}$$

and

$$\|\alpha u \beta\|_n = \left\|\left\|\alpha\left(\sum u_i \otimes e_i\right)\beta\right\|\right\|_n = \left\|\left\|\sum(\alpha u_i \beta) \otimes e_i\right\|\right\|_n$$

$$= \max_{l \geq 1}\left\|\sum_{i=1}^{l}(\alpha u_i \beta) \otimes x_i\right\|_n = \max_{l}\left\|\alpha\left(\sum_{i=1}^{l} u_i \otimes x_i\right)\beta\right\|_n \tag{3.6}$$

$$\leq \max_{l \geq 1} \|\alpha\|_n\left\|\sum_{i=1}^{l} u_i \otimes x_i\right\|_n \|\beta\|_n = \|\alpha\|_n\|u\|_n\|\beta\|_n.$$

Thus, by the abstract characterization theorem given in [186] or Theorem 2.12, this newly defined matricial norm $\{\|\cdot\|_n\}$ determines an operator space structure on c_{00}. We let $Y = c_{00}^{-\|\cdot\|_n}$ denote the completion. It is known from [84, Fact 6.3] that $\{e_i\}_{i=1}^{\infty}$ is a basis for Y. According to (N2), $\{e_i\}$ is a cb-basis for Y since for any $u = \sum u_i \otimes e_i \in \mathbb{M}_n(c_{00})$, the natural projections P_m satisfy

$$\|(P_m)_n(u)\|_n = \left\|\left\|\sum_{i=1}^{m} u_i \otimes e_i\right\|\right\|_n \leq \|\|u\|\|_n.$$

Now, let us define a linear map

$$Q : \sum \alpha_i e_i \in c_{00} \rightarrow \sum \alpha_i x_i \in X.$$

For any $u = \sum u_i \otimes e_i \in \mathbb{M}_n(c_{00})$, we get

$$(Q)_n(u) = (Q)_n\left(\sum u_i \otimes e_i\right) = \sum u_i \otimes x_i \tag{3.7}$$

and by Eq. (3.4), we have

$$\|(Q)_n(u)\|_n = \left\|\sum u_i \otimes x_i\right\|_n \leq \|\|u\|\|_n.$$

Together with (N1), we get $\|Q\|_{cb} = 1$, and Q can be uniquely extended to the whole space Y. On the other hand, we can define a linear map

$$T : X \rightarrow Y, \quad \sum_{i=1}^{\infty} f_i(x)x_i \mapsto \sum_{i=1}^{\infty} f_i(x)e_i. \tag{3.8}$$

Then, by (N3), T is well-defined. For any $x = [x_{j,k}] \in \mathbb{M}_n(X)$, we have

$$(T)_n(x) = [T(x_{j,k})] = \left[\sum_{i=1}^{\infty} f_i(x_{j,k})e_i\right] = \sum_{i=1}^{\infty}[f_i(x_{j,k})] \otimes e_i$$

and thus

$$\|\|(T)_n(x)\|\|_n = \left\|\left\|\sum_{i=1}^{\infty}[f_i(x_{j,k})] \otimes e_i\right\|\right\|_n = \lim_{m\to\infty}\left\|\left\|\sum_{i=1}^{m}[f_i(x_{j,k})] \otimes e_i\right\|\right\|_n$$

$$= \lim_{m\to\infty}\sup_{1\leq l\leq m}\left\|\sum_{i=1}^{l}[f_i(x_{j,k})] \otimes x_i\right\|_n \leq \sup_{l\geq 1}\|S_l\|_{cb}\|x\|_n.$$

This shows $\|T\|_{cb} \leq \sup_{l\geq 1}\|S_l\|_{cb} < \infty$. Moreover, for all $x \in X$,

$$QT(x) = Q(\sum f_i(x)e_i) = \sum f_i(x)x_i = x.$$

That is, $QT = I_X$. It follows that Q is a surjection from Y onto X and that T is injective from X into Y. Since

$$\|x\|_n = \|(QT)_n(x)\|_n = \|(Q)_n(T)_n(x)\|_n \le \||(T)_n(x)\||_n$$

for all $x \in \mathbb{M}_n(X)$, thus T is a complete isomorphism from X onto $T(X) \subset Y$. Moreover, TQ is a completely bounded projection from Y onto $TQ(Y) = T(X)$. This completes the proof.   □

Remark 3.4 Using the above theorem we can show the existence of cb-frames for some operator spaces.

(i) Let G be a countable discrete group. Then G is weakly amenable, or equivalently, the reduced group C*-algebra $C_r^*(G)$ has the CBAP, and thus $C_r^*(G)$ has a cb-frame.
(ii) Let $A(G)$ be the Fourier algebra of G with the canonical operator space structure as the operator predual of the left group von Neumann algebra $VN(G)$. If G is weakly amenable, then $A(G)$ has the CBAP and thus has a cb-frame.
(iii) Assume that G is weakly amenable. Then, by taking the complex interpolation, for each $1 < p < \infty$ the non-commutative L^p-space $L^p(VN(G)) = (A(G), VN(G))^{\frac{1}{p}}$ (with the canonical operator space structure introduced by Pisier) has the CBAP and thus has a cb-frame. The converse statement is not necessarily true for non-commutative L^p-spaces.
(iv) It is known from [145, Proposition 5.2] that if a countable residually finite discrete group G has the AP, then $L^p(VN(G))$ has a cb-basis. This contains a very interesting class of groups, such as many weakly amenable groups, such as $\mathbb{F}_n$, $SL(2, \mathbb{Z})$ and $Sp(1, n)$, as well as some non-weakly amenable groups including $\mathbb{Z}^2 \rtimes SL(2, \mathbb{Z})$.

Remark 3.5 Similar to the minimal associated sequence space or minimal associated basis for Schauder frames defined in [159, Definition 3.5] or Definition 2.7, the operator space Y and the cb-basis $\{e_i\}_{i=1}^{\infty}$ constructed in Theorem 3.9 is the minimal choice in the following sense: Suppose that we have another operator space Z with a cb-basis $\{z_i\}_{i=1}^{\infty}$ and completely bounded maps satisfying Eqs. (3.7) and (3.8). i.e.

$$Q_Z : Z \to X, \quad \sum a_i z_i \mapsto \sum a_i x_i$$

and

$$T_Z : X \to Z, \quad \sum f_i(x)x_i \mapsto \sum f_i(x)z_i.$$

Then for any $n \in \mathbb{N}$ and $u_i \in \mathbb{M}_n$, we have

$$\left\|\left\|\sum u_i \otimes e_i\right\|\right\|_n = \max_{m \geq 1}\left\|\left\|\sum_{i=1}^{m} u_i \otimes x_i\right\|\right\|_n = \max_{m \geq 1}\left\|\left\|\sum_{i=1}^{m} u_i \otimes Q_Z(z_i)\right\|\right\|_n$$

$$\leq \|Q_Z\|_{cb} \max_{m \geq 1}\left\|\left\|\sum_{i=1}^{m} u_i \otimes z_i\right\|\right\|_n \leq \left\|Q_Z\right\|_{cb} K_Z \left\|\left\|\sum u_i \otimes z_i\right\|\right\|_n,$$

where $K_Z = \sup_m \|P_m^Z\|_{cb}$ is the cb-constant of the cb-basis $\{z_i\}$. This shows that there is a constant $K = \|Q_Z\|_{cb} K_Z > 0$ such that

$$\left\|\left\|\sum u_i \otimes e_i\right\|\right\|_n \leq K \left\|\left\|\sum u_i \otimes z_i\right\|\right\|_n.$$

This shows that the cb-basis $\{e_i\}$ in Y is completely dominated by such a cb-basis $\{z_i\}$ in Z. Therefore, the cb-basis $\{e_i\}$ in Y constructed in Theorem 3.9 is a minimal choice associated to the cb-frame $\{x_i, f_i\}$ in X.

As in Banach space theory, we also consider the unconditional case. For an operator space, it is necessary to demand the partial-sum operator for an unconditional basis to be completely bounded, thus, we say that an unconditional basis $\{e_i\}_{i=1}^{\infty}$ of an operator space Y is *completely unconditional* if

$$\sup_{E \subset \mathbb{N}, \#E < \infty} \|P_E\|_{cb} < \infty,$$

where E is a finite subset of $\mathbb{N}$ and P_E is the natural projection defined by

$$P_E\left(\sum_{i=1}^{\infty} \alpha_i e_i\right) = \sum_{i \in E} \alpha_i e_i.$$

We say that $\{e_i\}$ is *completely 1-unconditional* if $\sup_{E \subset \mathbb{N}, \#E < \infty} \|P_E\|_{cb} = 1$, or equivalently, $\|P_E\|_{cb} = 1$ for any finite subset $E \subset \mathbb{N}$ (see [170]). And an unconditional frame $\{x_i, f_i\}_{i=1}^{\infty}$ of an operator space X (see Remark 2.4) is completely unconditional if

$$\sup_{E \subset \mathbb{N}, \#E < \infty} \|S_E\|_{cb} < \infty,$$

where S_E is the natural partial sum operator defined by $S_E(x) = \sum_{i \in E} f_i(x) x_i$.

The following result unconditionalizes Theorem 3.9, and the proof is omitted due to its similarity, which is also the operator space version of Theorem 3.6 in [52].

Proposition 3.5 *An operator space X has a completely unconditional cb-frame if and only if X is completely isomorphic to a completely complemented subspace of an operator space with a completely 1-unconditional cb-basis.*

3.2.2   Subspaces of Operator Spaces with Cb-basis

Notice that Liu and Ruan showed that a separable operator space X has a cb-frame if and only if it has the CBAP if and only if it is completely isomorphic to a completely complemented subspace of an operator space with a cb-basis. The Banach-Mazur theorem shows that every separable Banach space can be isometrically embedded into $C[0, 1]$. Therefore, every separable Banach space can be isometrically embedded into a Banach space with a Schauder basis. Then they asked in [156] the following natural question: "What is the local characterization of a separable operator space which can be a subspace of an operator space with a cb-basis?" The following result settles this question, which also implies that the corresponding result is not true for operator spaces (see Theorem 3.10).

Proposition 3.6 *Let X be a separable operator space. Then the following are equivalent:*

(i) *X is completely isomorphic to a subspace of an operator space with a cb-basis;*
(ii) *X is completely isomorphic to a subspace of an operator space with a cb-frame*
(iii) *X is completely isomorphic to a subspace of a (not necessarily separable) operator space with the CBAP.*

Proof It is obvious that (i) $\implies$ (ii) and (i) $\implies$ (iii).

(ii) $\implies$ (i): Suppose that X is completely isomorphic to a subspace of an operator space Y with a cb-frame. By Theorem 3.9, Y is completely isomorphic to a completely complemented subspace of an operator space Z with a cb-basis. Then, we can conclude from the following completely isomorphic embeddings

$$X \hookrightarrow Y \hookrightarrow Z$$

that X is completely isomorphic to a subspace of Z with a cb-basis.

(iii) $\implies$ (i): If the operator space with the CBAP is separable, then it has a cb-frame by Theorem 3.9 and thus we can get the result by (i) $\iff$ (ii). If the operator space with the CBAP is non-separable, then by [156, Lemma 4.4], which states that if X is an operator space with the λ-CBAP, then for every separable subspace Y of X, there is a separable subspace $\widetilde{Y}$ of X which contains Y and has the λ-CBAP, then we still obtain the same results. $\square$

Now, we turn to the counterexample in the operator space case. In [170], Oikhberg and Ricard constructed a separable Hilbertian operator space X_0, which is isometrically isomorphic to $\ell_2(\mathbb{N})$, but every infinite-dimensional closed subspace of X_0 fails to have the *operator space approximation property (OAP)*.

Theorem 3.10 *The Oikhberg-Ricard space X_0 cannot be completely isomorphic to any subspace of an operator space with a cb-basis (or with a cb-frame).*

Proof Suppose that X_0 is completely isomorphic to an (infinite-dimensional) subspace Y of an operator space Z with a cb-basis (or with a cb-frame, then by Proposition 3.6, also with a cb-basis). Similar to the Banach space theory, then Y has a cb-basis sequence $\{y_i\}_{i=1}^{\infty}$, that is $\{y_i\}$ is a cb-basis for $\overline{\text{span}}\{y_i\}$, see [156, Proposition 4.2] for detailed structure. It follows from that X_0 has a cb-basis sequence $\{e_i\}$. Then $\overline{\text{span}}\{e_i\}$ is an infinite-dimensional closed subspace of X_0 having the cb-basis hence CBAP. This contradicts Oikhberg-Ricard's result since CBAP implies OAP (see [79]).   □

Remark 3.6 Recall from [78] that an operator space X is 1-exact if every finite-dimensional subspace F of X is 1-exact, meaning for every $\varepsilon > 0$ there exist $n \in \mathbb{N}$ and a completely contractive injective map $\varphi : F \to M_n$ such that $\|\varphi^{-1}\|_{cb} \le 1 + \varepsilon$. Note that every 1-exact separable operator space is completely isomorphic to a subspace of an operator space with a cb-basis. To see this, if X is a 1-exact separable operator space, then it is completely isometric to a subspace of a 1-nuclear separable operator space Y. Then, by Proposition 3.6, X is completely isomorphic to a subspace of an operator space with a cb-basis.

3.3   Lipschitz Frames and Lipschitz Bounded Approximation Property

In this section, we consider the general (Lipschitz) bounded approximation property for operators defined in [157], which will be compatible with the canonical (Lipschitz) bounded approximation property for Banach spaces. And we combine the technique of dilations for Banach space frames and linearization of Lipschitz operators to include the theorem of Godefroy and Kalton [94] that the Lipschitz bounded approximation property and the bounded approximation property are equivalent for every Banach space.

3.3.1   Lipschitz Bounded Approximation Property for Operators

The definition of BAP for Banach space has been introduced before. Now we give the definition of the BAP or Lipschitz BAP for operators which were first introduced in [157].

Definition 3.6 Let X, Y be Banach spaces. Let $\lambda > 0$ and $T \in B(X, Y)$.

(i) T is said to have the λ-*bounded approximation property* (λ-BAP) if there exists a net $\{S_\alpha\}_{\alpha \in \Gamma} \subseteq B(X; Y)$ with

$$\sup_{\alpha \in \Gamma} \|S_\alpha\| \le \lambda$$

such that both $\dim(S_\alpha X) < \infty$ for all $\alpha \in \Gamma$ and $\lim_\alpha \|S_\alpha x - Tx\| = 0$ for all $x \in X$. We say that T has the *bounded approximation property* (BAP) if there exists a $\lambda > 0$ such that T has the λ-BAP.

(ii) T is said to have the λ-*Lipschitz bounded approximation property* (λ-*Lip BAP)* if there exists a net $\{S_\alpha\}_{\alpha \in \Gamma} \subseteq Lip(X, Y)$ with

$$\sup_{\alpha \in \Gamma} \|S_\alpha\|_{Lip} \leq \lambda$$

such that $\dim(\mathrm{span}\{S_\alpha X\}) < \infty$ for all $\alpha \in \Gamma$ and $\lim_n \|S_\alpha x - Tx\| = 0$ for all $x \in X$. And T has the *Lipschitz bounded approximation property (Lip BAP)* if there exists a $\lambda > 0$ such that T has the λ-Lip BAP.

Remark 3.7 It follows from the definition that a Banach space X has the BAP if and only if the identity operator I_X on X has the BAP. When X is separable, the net used in the above definition can be replaced by a sequence.

Definition 3.7 ([157]) Let X be a separable Banach space, Y be a Banach space and $T \in B(X, Y)$. Then T is said to be *Lipschitz decomposable* if there exists $\{y_n, f_n\}_{n=1}^\infty \subseteq Y \times lip(X)$ with

$$\sup_{n \in \mathbb{N}} \left\| \sum_{k=1}^n f_k(\cdot) y_k \right\|_{Lip} < \infty$$

so that $Tx = \sum_{n=1}^\infty f_n(x) y_n$ for all $x \in X$. We say that T is *decomposable* if the sequence $\{f_n\}$ can be chosen in X^*.

Next, we prove that when X is separable, for every $T \in B(X, Y)$ having the Lip BAP is equivalent to being (Lipschitz) decomposable.

Theorem 3.11 ([157]) *Let X be a separable Banach space, Y be a Banach space and $T \in B(X, Y)$. Then T is (Lipschitz) decomposable if and only if T has the (Lip) bounded approximation property.*

Proof We only prove the Lipschitz case since the proof of the linear case is essentially the same. It is easy to see from the definition that if T is Lipschitz decomposable, then T has the Lip BAP. So, we only prove the "if" part.

Suppose that T has the Lip BAP. By definition, there exist $\lambda > 0$ and $\{S_n\}_{n=1}^\infty \subseteq Lip(X; Y)$ with $\sup\{\|S_n\|_{Lip}\}_{n=1}^\infty \leq \lambda$ such that $\dim(\mathrm{span}\{S_n X\}) < \infty$ for all $n \in \Gamma$ and $\lim_n \|S_n x - Tx\| = 0$ for all $x \in X$.

Let $S_0 := 0$, $A_n := S_n - S_{n-1}$, $E_n := \mathrm{span}\{A_n X\}$, $m_n := \dim(E_n)$ and $m_0 := 0$ for every n in $\mathbb{N}$. Then $m_n < \infty$ and

$$\sup_{n \in \mathbb{N}} \left\| \sum_{k=1}^{n} A_k \right\|_{Lip} =: K \leq \lambda.$$

It follows from the Auerbach lemma (see [174]) that there are one-dimensional operators $B_j^{(n)} : E_n \to E_n$ with $\|B_j^{(n)}\| = 1$ for $j = 1, \ldots, m_n$ such that $\sum_{j=1}^{m_n} B_j^{(n)}(y) = y$ for every y in E_n. Let $C_i^{(n)} := m_n^{-1} B_j^{(n)}$, if $i = rm_n + j$, where $r = 0, \ldots, m_n - 1$ and $j = 1, \ldots, m_n$ For each $n \in \mathbb{N}$, we have

$$\sum_{i=1}^{m_n^2} C_i^{(n)}(y) = y \quad \text{for } y \in E_n$$

and for each well-defined $q = rm_n + j \leq m_n^2$ and $y \in E_n$, we have

$$\frac{\left\| \sum_{i=1}^{q} C_i^{(n)}(y) \right\|}{\|y\|} \leq \frac{\|ry\|}{m_n \|y\|} + \frac{\left\| \sum_{i=1}^{j} B_i^{(n)}(y) \right\|}{m_n \|y\|} \leq \frac{r}{m_n} + \frac{j}{m_n} \leq 2.$$

Thus we have

$$\max_{1 \leq q \leq m_n^2} \left\| \sum_{i=1}^{q} C_i^{(n)} \right\| \leq 2.$$

Let $D_s := C_i^{(n)} A_n$ for $s = m_0^2 + m_1^2 + \cdots + m_{n-1}^2 + i$, where $i = 1, 2, \ldots, m_n^2$ and $n \in \mathbb{N}$. For each well-defined $s = m_0^2 + m_1^2 + \cdots + m_{n-1}^2 + i$ and $x_1, x_2 \in X$, we have

$$\frac{\left\| \sum_{k=1}^{s} D_k(x_1) - \sum_{k=1}^{s} D_k(x_2) \right\|}{\|x_1 - x_2\|}$$

$$= \frac{\left\| \sum_{k=1}^{n-1} A_k(x_1) - \sum_{k=1}^{n-1} A_k(x_2) + \sum_{k=1}^{i} C_k^{(n)}(A_n(x_1) - A_n(x_2)) \right\|}{\|x_1 - x_2\|}$$

$$\leq \frac{\left\| \sum_{k=1}^{n-1} A_k(x_1) - \sum_{k=1}^{n-1} A_k(x_2) \right\|}{\|x_1 - x_2\|} + 2 \frac{\left\| A_n(x_1) - A_n(x_2) \right\|}{\|x_1 - x_2\|}$$

$$\leq K + 2 \frac{\left\| \sum_{k=1}^{n} A_k(x_1) - \sum_{k=1}^{n} A_k(x_2) - \sum_{k=1}^{n-1} A_k(x_1) + \sum_{k=1}^{n-1} A_k(x_2) \right\|}{\|x_1 - x_2\|}$$

$$\leq K + 2K + 2K$$

$$= 5K.$$

Then we know that

$$\sup_{n\in\mathbb{N}}\left\{\left\|\sum_{k=1}^{n}D_k\right\|_{Lip}\right\}\le 5K<\infty.$$

Moreover,

$$Tx=\sum_{n=1}^{\infty}D_n(x).$$

Next, we construct the Lipschitz decomposition of T. For every finite-dimensional space E_n, let its basis be $\{y_1^n, y_2^n, \ldots, y_{m_n}^n\}$ and its bi-orthogonal basis be

$$\{(y_1^n, y_1^{n*}), (y_2^n, y_2^{n*}), \ldots, (y_{m_n}^n, y_{m_n}^{n*})\}.$$

Then

$$Tx=\sum_{n=1}^{\infty}D_n(x)=\sum_{n=1}^{\infty}\sum_{j=1}^{m_{p_n}}y_j^{p_n*}(D_n(x))y_j^{p_n},$$

where $D_n(x)\in E_{p_n}$ in the above equality is uniquely determined. Let C_n be the basis constant for $\{(y_1^n, y_1^{n*}), (y_2^n, y_2^{n*}), \ldots, (y_{m_n}^n, y_{m_n}^{n*})\}$. Let $M_n\in\mathbb{Z}$ be such that $M_n\ge C_n$. Define

$$y_i:=y_j^{p_n}, \quad \text{if } i=\sum_{k=1}^{n-1}M_{p_k}m_{p_k}+rm_{p_n}+j,$$

where $r=0,\ldots,M_{p_n}-1$ and $j=1,\ldots,m_{p_n}$ and

$$f_i:=\frac{y_j^{p_n*}\circ D_n}{M_{p_n}} \quad \text{if } i=\sum_{k=1}^{n-1}M_{p_k}m_{p_k}+rm_{p_n}+j,$$

where $r=0,\ldots,M_{p_n}-1$ and $j=1,\ldots,m_{p_n}$.

Now we verify that $\{y_k, f_k\}_{k=1}^{\infty}$ is a Lipschitz decomposition of T. Trivially,

$$\{y_k, f_k\}_{k\in\mathbb{N}}\subseteq Y\times lip(X)$$

and

$$Tx=\sum_{n=1}^{\infty}\sum_{j=1}^{m_{p_n}}y_j^{p_n*}(D_n(x))y_j^{p_n}=\sum_{n=1}^{\infty}M_{p_n}\sum_{j=1}^{m_{p_n}}\frac{y_j^{p_n*}(D_n(x))y_j^{p_n}}{M_{p_n}}=\sum_{n=1}^{\infty}f_n(x)y_n.$$

And for every well-defined $s = \sum_{k=1}^{n-1} M_{p_k} m_{p_k} + r m_{p_n} + j$ and every $x_1, x_2 \in X$,

$$\frac{\left\| \sum_{k=1}^{s} f_k(x_1) y_k - \sum_{k=1}^{s} f_k(x_2) y_k \right\|}{\|x_1 - x_2\|}$$

$$= \frac{\left\| \sum_{k=1}^{n-1}(D_k(x_1) - D_k(x_2)) + \frac{r}{M_{p_n}}(D_n(x_1) - D_n(x_2)) + \sum_{k=1}^{j} \frac{[y_k^{p_n*}(D_n(x_1)) - y_k^{p_n*}(D_n(x_2))] y_k^{p_n}}{M_{p_n}} \right\|}{\|x_1 - x_2\|}$$

$$\leq 3 \sup_{n \in \mathbb{N}} \left\{ \left\| \sum_{k=1}^{n} D_k \right\|_{Lip} \right\} + \frac{\left\| \sum_{k=1}^{j} [y_k^{p_n*}(D_n(x_1)) - y_k^{p_n*}(D_n(x_2))] y_k^{p_n} \right\|}{M_{p_n} \|x_1 - x_2\|}$$

$$\leq 15K + \frac{C_{p_n} \|D_n(x_1) - D_n(x_2)\|}{M_{p_n} \|x_1 - x_2\|}$$

$$\leq 15K + \frac{\left\| \sum_{k=1}^{n} D_k(x_1) - \sum_{k=1}^{n} D_k(x_2) - \sum_{k=1}^{n-1} D_k(x_1) + \sum_{k=1}^{n-1} D_k(x_2) \right\|}{\|x_1 - x_2\|}$$

$$\leq 25k.$$

So

$$\sup_{n \in \mathbb{N}} \left\{ \left\| \sum_{k=1}^{n} f_k(\cdot) y_k \right\|_{Lip} \right\} < \infty$$

and $\{y_k, f_k\}_{k=1}^{\infty}$ is a Lipschitz decomposition of T. $\qquad \square$

Remark 3.8 By the proof of Theorem 3.11 if T has the λ-Lip BAP, then T has a Lipschitz decomposition $\{y_k, f_k\}_{k=1}^{\infty}$ with

$$\sup_{n \in \mathbb{N}} \left\| \sum_{k=1}^{n} f_k(\cdot) y_k \right\|_{Lip} \leq 25\lambda.$$

3.3.2 Linearization for Lipschitz Bounded Approximation Property

The next theorem shows that the Lip BAP coincides with the BAP for operators with separable domains.

Theorem 3.12 ([157]) *Let X be a separable Banach space, Y be a Banach space and $T \in B(X, Y)$. Then T has the bounded approximation property if and only if T has the Lipschitz bounded approximation property.*

To prove the above theorem, we need the following lemma from [157].

Lemma 3.2 *Let X be a separable Banach space, Y be a Banach space, $T \in B(X, Y)$, and $\{y_n, f_n\}_{n=1}^{\infty}$ be a Lipschitz decomposition of T. Then there exist a Banach space Z with a Schauder basis $\{z_n\}_{n=1}^{\infty}$,*

$$S \in B(Z, Y), \quad where \quad S\left(\sum_{n=1}^{\infty} a_i z_i\right) = \sum_{n=1}^{\infty} a_i y_i$$

and

$$R \in Lip(X, Z), \quad where \quad R(x) = \sum_{i \in \mathbb{N}} f_i(x) z_i$$

such that $S \circ R = T$.

Proof We first construct the Banach space Z. Without loss of generality, we can assume $\|y_n\| - 1$ for all $n \in \mathbb{N}$. Let $\{z_k\}_{k=1}^{\infty}$ be a copy of $\{y_k\}_{k=1}^{\infty}$. Let c_{00} be the collection of all the sequences of real numbers with all but finitely many nonzero elements. Let Z be the completion of the vector space span$\{z_k\}_{k \in \mathbb{N}}$ under the norm

$$\left\|\sum_{k=1}^{\infty} a_k z_k\right\|_Z := \sup_{n \leq m} \left\|\sum_{k=n}^{m} a_k y_k\right\|,$$

where $(a_k) \in c_{00}$. It is easy to see that Z is a Banach space and $\{z_k\}_{k=1}^{\infty}$ is monotone Schauder basis of Z. Hence, S is a well-defined, bounded linear operator. We only need to check that $R \in Lip(X; Z)$. For any $x_1, x_2 \in X$,

$$\frac{\|Rx_1 - Rx_2\|}{\|x_1 - x_2\|} = \frac{\|\sum_{n=1}^{\infty} f_n(x_1)z_n - \sum_{n=1}^{\infty} f_n(x_2)z_n\|}{\|x_1 - x_2\|}$$

$$= \sup_{t \leq s} \frac{\|\sum_{n=t}^{s} f_n(x_1)y_n - \sum_{n=t}^{s} f_n(x_2)y_n\|}{\|x_1 - x_2\|}$$

$$\leq 2 \sup_{n \in \mathbb{N}} \left\|\sum_{k=1}^{n} f_k(\cdot)y_k\right\|_{Lip}.$$

Since $\sup\{\|\sum_{k=1}^{n} f_k(\cdot)y_k\|_{Lip} : n \in \mathbb{N}\} < \infty$, R is a Lipschitz operator. Finally, for all $x \in X$,

$$S \circ R(x) = S\left(\sum_{n=1}^{\infty} f_n(x)z_n\right) = \sum_{n=1}^{\infty} f_n(x)y_n = Tx,$$

so $S \circ R = T$.   □

Proof of Theorem 3.12 The "only if" part is trivial, and so we only need to prove the "if" part. By Theorem 3.11, it is sufficient to prove that T has a linear decomposition. By Lemma 3.2, there exists a Banach space Z, $R \in Lip(X, Z)$ and $S \in B(Z, Y)$ such that $S \circ R = T$. Then by Lemma 2.1, R has a linear surrogate F such that $\|F\| \le \|R\|_{Lip}$ and $S \circ F = S \circ R = T$. Since Z has a Schauder basis $\{z_i\}_{i=1}^{\infty}$ and $F \in B(X; Z)$, let $\{z_i, z_i^*\}_{i=1}^{\infty}$ be the biorthogonal basis of $\{z_i\}_{i=1}^{\infty}$. Then

$$F(x) = \sum_{i \in \mathbb{N}} z_i^*(F(x))z_i.$$

Moreover,

$$S \circ F(x) = \sum_{i \in \mathbb{N}} z_i^*(F(x))y_i = Tx.$$

Thus $\{y_i, z_i^* \circ F\}_{i=1}^{\infty}$ is a linear decomposition of T. We can conclude that T is bounded approximable if and only if T is Lipschitz bounded approximable. □

For operators with non-separable domains, we need a very useful tool, represented as the following lemma, whose proof can be found in [43, Proposition 9.5] or the "Lindenstrauss subspace" defined in [152].

Lemma 3.3 *Let X be a Banach space and let E be a finite-dimensional subspace of X. Let k be an integer and let $\varepsilon > 0$. Then there is a finite-dimensional subspace F of X containing E such that for every subspace Y of X containing E with $\dim Y/E = k$ there is a linear operator $T : Y \to F$ with $\|T\| \le 1 + \varepsilon$ and $Tx = x$ for every $x \in E$.*

The next theorem generalized the [43, Theorem 9.7] that a non-separable Banach space X has the λ-BAP if and only if every separable subspace of X is contained in a separable subspace with the λ-BAP.

Theorem 3.13 ([157]) *Let X and Y be two Banach spaces, $T \in B(X, Y)$ and $\lambda \ge 0$. Then T has the λ-bounded approximation property if and only if for any separable subspace U of X, there exists a separable subspace V with $U \subseteq V \subseteq X$ such that $T|_V : V \to Y$ has the λ-bounded approximation property.*

Proof Suppose that T has λ-BAP and is approximated by the net $\{S_n\}_{n \in \Gamma} \subseteq B(X, Y)$. Then for any separable subspace U of X, $T|_U$ can be approximated by the net $\{S_n|_U\}_{n \in \Gamma} \subseteq B(U, Y)$. On the other hand, let E be any finite-dimensional subspace of X. By Lemma 3.3, there exists a separable subspace Z of X so that for any given finite-dimensional subspace F with $E \subseteq F$, there exists an operator $S_F : F \to Z$ with

$$\|S_F\| \le 1 + \frac{1}{\dim F},$$

and $S_F|_E = I_E$. By hypotheses, there exists a separable subspace with $Z \subseteq W \subseteq X$ such that $T|_W$ has the λ-BAP. Choose a sequence of finite-rank operators $L_{E,n} : W \to Y$ so that $\|L_{E,n}\| \leq \lambda$ and

$$L_{E,n} \xrightarrow{sot} T|_W, \quad n \to \infty.$$

Since $\{S_F\}$ is a net, we can take a subnet of S_F which pointwisely converges to a linear operator $S : X \to Z$ with $\|S\| \leq 1$. Moreover, $S|_E = I_E$. Let $T_{E,n} = L_{E,n} \circ S$. Then for all $x \in E$, we have

$$T_{E,n}x = L_{E,n}x \to T|_W x = Tx$$

as $n \to \infty$. Define an order on $\{(E, n)\}$ as follows:

$$(E, n) \leq (F, m), \quad \text{if } E \subseteq F \text{ and } n \leq m.$$

Then $\{T_{(E,n)}\}$ is a net which satisfies $\|T_{(E,n)}\| \leq \lambda$, $\dim(T_{(E,n)}X) < \infty$ and

$$\lim_{(E,n)} \|T_{E,n}x - Tx\| = 0$$

for all $x \in X$. So T has the λ-BAP.   □

Remark 3.9 By an argument essentially the same as the proof of Theorem 3.13, one can also show that for any separable subspace U of X, there exists a separable subspace V with $U \subseteq V \subseteq X$ such that $T|_V : V \to Y$ has the λ-Lip BAP if and only if T has the λ-Lip BAP.

Corollary 3.3 ([157]) *Let X and Y be arbitrary Banach spaces and $T \in B(X, Y)$. Then T has the bounded approximation property if and only if T has the Lipschitz bounded approximation property.*

Proof The "only if" part is trivial, and we only prove the "if" part. Suppose that T has the λ-Lip BAP. By Remark 3.9, for any separable subspace U of X, there exists a separable subspace V with $U \subseteq V \subseteq X$ such that $T|_V : V \to Y$ has the λ-Lip BAP. By Theorem 3.12, $T|_V$ has the BAP. Moreover, by Remarks 3.8 and the proof of Lemma 3.2, $T|_V$ has the 50λ-BAP. By Theorem 3.13, T has the 50λ-BAP.   □

Definition 3.8 Let X be a Banach space and $\lambda \in \mathbb{R}$. Then X has the λ-*Lip BAP* if there exists a net $\{S_\alpha\}_{\alpha \in \Gamma} \subseteq Lip(X)$ such that $\dim(\text{span}\{S_\alpha X\}) < \infty$ for all $\alpha \in \Gamma$, $\sup_{\alpha \in \Gamma} \|S_\alpha\|_{Lip} \leq \lambda$ and $\lim_\alpha \|S_\alpha x - x\| = 0$ for all $x \in X$. X has the *Lip BAP* if there exists a $\lambda \in \mathbb{R}$ such that X has the λ-Lip BAP.

Remark 3.10 Let X be a separable Banach space. Then X has the Lip BAP if and only if the identity operator on X has the Lip BAP. So, an immediate consequence of Corollary 3.3 yields the main result in [94].

Corollary 3.4 *A Banach space X has the bounded approximation property if and only if X has the Lipschitz bounded approximation property.*

Banach Space Dilations

4

The idea of dilation played an important role in the development of functional analysis, especially in the study of operator theory and operator algebras. The celebrated dilation theorems include (1) Sz-Nagy's theorem: unitary dilation for a contraction, (2) Naimark's dilation theorem: orthogonal projection-valued dilation for positive regular operator-valued measure and (3) Stinespring's dilation theorem: *-homomorphism dilation for completely positive map on C^*-algebras (c.f. [14, 30, 61, 129, 171]). All these theorems are established in the context of Hilbert space operators and the dilation spaces are all Hilbert spaces, which are referred to as *Hilbertian dilations*, and they have important applications in many areas of research such as operator algebras, system control and quantum information theory [98, 165].

The dilation theory of operator-valued measures has a natural connection with frame theory, as discovered by Casazza, Han and Larson in their work [52, 122], which established the dilation theory of frames (framings) for both Hilbert and Banach spaces. The dilation of various kinds of frames (framings) can be interpreted as a special type of dilation for a special type of (not necessarily positive) operator-valued measures over a purely atomic measure space. The bounded linear maps induced by a Hilbert space framing are not necessarily completely bounded, and hence the dilation space is not necessarily a Hilbert space. These considerations led to extensive research activities with the goal of establishing a more general Banach space dilation theory for frames, operator-valued measures, and more general bounded linear maps, mostly acting on Banach spaces. It is essential to point out that since the dilation spaces are not Hilbert spaces even when the original underlying spaces are Hilbert spaces, Banach space techniques are essential and heavily involved in establishing such a general theory. In this chapter, we will report some recent results on this general dilation theory [120, 124–127].

© The Author(s), under exclusive license to Springer Nature Switzerland AG 2026
D. Han et al., *Banach Space Aspects of Frame Theory with Applications*, Synthesis
Lectures on Mathematics & Statistics, https://doi.org/10.1007/978-3-032-10174-7_4

4.1   Classical Dilation Theorems

To begin with, we first briefly review two Hilbertian dilations: Naimark's dilation and Stinespring's dilation.

4.1.1   Naimark's Dilation Theorem

Let Σ be the σ-field of Borel subsets of a compact Hausdorff space Ω, and $\mathcal{H}$ be a Hilbert space. A $B(\mathcal{H})$-valued measure on Ω is a map $E : \Sigma \to B(\mathcal{H})$ that is weakly countably additive, (or equivalently, strongly countably additive), i.e., if $\{B_i\}_{i=1}^\infty$ is a countable collection of disjoint Borel set with union B, then

$$\langle E(B)x, y \rangle = \sum_{i=1}^\infty \langle E(B_i)x, y \rangle, \quad \forall\, x, y \in \mathcal{H}.$$

A $B(\mathcal{H})$-valued measure E is called

1. *bounded* if $\sup\{\|E(B)\| : B \in \Sigma\} < \infty$.
2. *regular* if the complex measure given by $\mu_{x,y}(B) := \langle E(B)x, y \rangle$ is regular.
3. *spectral* if $E(B_1 \cap B_2) = E(B_1) \cdot E(B_2)$ for any $B_1, B_2 \in \Sigma$.
4. *positive* if $E(B) \geq 0$ for all $B \in \Sigma$.
5. *self-adjoint* if $E(B)^* = E(B)$ for all $B \in \Sigma$.
6. *normalized or probability* if $E(\Omega) = I$, where I is the identity map on H.

We call E a positive operator-valued measure (POVM) if it is positive and $E(\Omega) = I_{\mathcal{H}}$ and E a *projection-valued measure* if E is spectral and self-adjoint.

Example 4.1.1 Let $\{x_n\}_{n=1}^\infty$ be a frame for a separable Hilbert space $\mathcal{H}$. Let Σ be the σ-algebra of all subsets of $\mathbb{N}$. Define the map

$$E : \Sigma \to B(\mathcal{H}), \quad E(B) = \sum_{n \in B} x_n \otimes x_n, \quad \forall B \in \Sigma,$$

where $x \otimes y$ is the map on $\mathcal{H}$ defined by $(x \otimes y)(u) = \langle u, y \rangle x$. Then E is a regular, positive $B(\mathcal{H})$-valued measure.

The celebrated Naimark's dilation theorem deals with (orthogonal) projection-valued dilations for positive operator-valued measures on Hilbert spaces.

Theorem 4.1 ([171, Theorem 4.6] Naimark) *Let E be a regular, positive, $B(\mathcal{H})$-valued measure on a compact Hausdorff space Ω. Then there exist a Hilbert space $\mathcal{K}$, a bounded* *linear operator $V : \mathcal{H} \to \mathcal{K}$, and a regular, self-adjoint, spectral, $B(\mathcal{K})$-valued measure F on Ω, such that*

$$E(B) = V^* F(B) V,$$

for all $B \in \Sigma$.

4.1.2 Stinespring's Dilation Theorem

We have already introduced the completely bounded (CB) maps on operator spaces, now, we introduce the completely positive maps on C^*-algebras. Let $\mathcal{A}$ be a unital C^*-algebra. A linear map $\phi : \mathcal{A} \to B(\mathcal{H})$ is said to be *positive* if $\phi(a^*a) \geq 0$ for every $a \in \mathcal{A}$, and it is called *completely positive* if for every n-tuple $a_1, \ldots, a_n$ of elements in $\mathcal{A}$, the matrix $(\phi(a_i^* a_j))$ is positive in the usual sense that for every n-tuple of vectors $\xi_1, \ldots, \xi_n \in \mathcal{H}$, we have

$$\sum_{i,j=1}^{n} \langle \phi(a_i a_j^*) \xi_j, \xi_i \rangle \geq 0,$$

or equivalently, $(\phi(a_i^* a_j))$ is a positive operator on the Hilbert space $\mathcal{H} \otimes \mathbb{C}^n$.

The well-known Stinespring's dilation theorem characterizes completely positive maps on C^*-algebras in terms of their $*$-homomorphism dilations.

Theorem 4.2 ([171, Theorem 4.1] Stinespring's Dilation Theorem) *Let $\mathcal{A}$ be a unital C^* algebra, and let $\phi : \mathcal{A} \to B(\mathcal{H})$ be a completely positive map. Then there exists a Hilbert space $\mathcal{K}$, a unital $*$-homomorphism $\pi : \mathcal{A} \to B(\mathcal{K})$, and a bounded operator $V : \mathcal{H} \to \mathcal{K}$ with $\|\phi(1)\| = \|V\|^2$ such that*

$$\phi(a) = V^* \pi(a) V,$$

for all $a \in \mathcal{A}$.

We call a triple $\{\pi, V, \mathcal{K}\}$ a Stinespring representation for ϕ and refer to [171] for details. Let $\mathcal{K}_1$ be the closed linear span of $\pi(\mathcal{A}) V \mathcal{H}$. It is easily verified that $\mathcal{K}_1$ reduces $\pi(\mathcal{A})$ so that the restriction of π to $\mathcal{K}_1$ defines a $*$-homomorphism, $\pi_1 : \mathcal{A} \to B(\mathcal{K}_1)$. Note that if $V \mathcal{H} \subseteq \mathcal{K}_1$, we still have that $\phi(a) = V^* \pi_1(a) V$, i.e., that $(\pi_1, V, \mathcal{K}_1)$ is also a Stinespring representation. If $\mathcal{K}_1$ is the closed linear span of $\pi_1(A) V \mathcal{H}$, the representation has one additional property, such a triple is called a *minimal Stinespring representation*. The following result summarizes the importance of this minimality condition.

Proposition 4.1 ([171]) *Let $\mathcal{A}$ be a C^*-algebra, let $\phi : \mathcal{A} \to B(\mathcal{H})$ be completely positive, and let*

$$(\pi_i, V_i, \mathcal{K}_i), \quad i = 1, 2$$

be two minimal Stinespring representations for ϕ. Then there exists a unitary $U : \mathcal{K}_1 \to \mathcal{K}_2$ satisfying $U V_1 = V_2$ and $U \pi_1 U^* = \pi_2$.

Proof If U exists, then necessarily,

$$U\left(\sum_i \pi_1(a_i)V_1 h_i\right) = \sum_i \pi_2(a_i)V_2 h_i.$$

We first show that U is a well-defined isometry from $\mathcal{K}_1$ to $\mathcal{K}_2$. Note that

$$\left\|\sum_i \pi_1(a_i)V_1 h_i\right\|^2 = \sum_{i,j}\langle\phi(a_i^* a_j)h_j, h_i\rangle = \left\|\sum_i \pi_2(a_i)V_2 h_i\right\|^2.$$

So U is isometric and consequently well-defined. By the minimality condition, U will have a dense range and hence be onto, thus, U is a unitary from $U : \mathcal{K}_1 \to \mathcal{K}_2$, which completes the proof. $\qquad\square$

The following fact is also well known and useful for the commutative C^*-algebra case [171].

Proposition 4.2 *Let $\mathcal{A}$ be a commutative C^*-algebra and $\phi : \mathcal{A} \to B(\mathcal{H})$ be a bounded linear map. If ϕ is positive, then ϕ is completely positive.*

In view of the connections between operator-valued measures and bounded linear maps [171], Stinespring's dilation theorem provides an alternative proof for Naimark's dilation theorem.

Given a regular bounded $B(\mathcal{H})$-valued measure E on Ω. We obtain a bounded linear map

$$\varphi_E : C(\Omega) \to B(\mathcal{H}), \quad \text{by} \quad \langle\varphi_E(f)x, y\rangle = \int_\Omega f \, d\mu_{x,y}, \tag{4.1}$$

where $\mu_{x,y}$ is the complex measure induced by E with $\mu_{x,y}(B) = \langle E(B)x, y\rangle$. Conversely, for a bounded linear map $\phi : C(\Omega) \to B(\mathcal{H})$, then it would be necessary to have that $\phi = \varphi_E$ for some regular bounded $B(\mathcal{H})$-valued measures E. Let $\varphi_E : C(\Omega) \to B(\mathcal{H})$ be the bounded linear map induced by a positive regular measure E via Eq. (4.1), by Proposition 4.2, φ_E is completely positive. Then, by Stinespring's dilation theorem, there exists a Hilbert space $\mathcal{K}$, a *-homomorphism $\pi : C(\Omega) \to B(\mathcal{K})$ and a bounded linear map $V : \mathcal{H} \to \mathcal{K}$ such that $\varphi_E(f) = V^*\pi(f)V$ for all $f \in C(\Omega)$. Let F be the $B(\mathcal{K})$-valued measure corresponding to π. Then, F is the desired regular, projection-valued measure dilation for E.

Completely positive maps are completely bounded in the above definition. In the other direction, we have the following Wittstock's decomposition theorem [171, 176].

Proposition 4.3 *Let $\mathcal{A}$ be a unital C^*-algebra, and let $\phi : \mathcal{A} \to B(\mathcal{H})$ be a completely bounded map. Then, ϕ is a linear combination of four completely positive maps.*

The following is a generalization of Stinespring's dilation theorem.

Theorem 4.3 ([171]) *Let $\mathcal{A}$ be a unital C^*-algebra, and let $\phi : \mathcal{A} \to B(\mathcal{H})$ be a completely bounded map. Then there exists a Hilbert space $\mathcal{K}$, a *-homomorphism $\pi : \mathcal{A} \to B(\mathcal{K})$, and bounded operators $V_i : \mathcal{H} \to \mathcal{K}, i = 1, 2$, with $\|\phi\|_{cb} = \|V_1\| \cdot \|V_2\|$ such that*

$$\phi(a) = V_1^*\pi(a)V_2$$

for all $a \in \mathcal{A}$. Moreover, if $\|\phi\|_{cb} = 1$, then V_1 and V_2 may be taken to be isometries.

4.2   Dilations for Operator-Valued Measures

Similar to the operator-valued measure on Hilbert spaces, we can define the general operator-valued measure on a Banach space. For a Banach space X, the weak-operator topology (*wot* in short) or strong-operator topology (*sot* in short) of $B(X)$ and $B(X)$-valued measures can be similarly defined as that of $B(\mathcal{H})$.

Definition 4.1 Let X and Y be Banach spaces, and let (Ω, Σ) be a measurable space. A $B(X, Y)$-*valued measure on* Ω is a map $E : \Sigma \to B(X, Y)$ that is countably additive in the weak operator topology (*wot*, by Orlicz-Pettis theorem [66], equivalently, in the strong operator topology (*sot*)), that is, if $\{B_i\}$ is a disjoint countable collection of members of Σ with union B, then

$$y^*(E(B)x) = \sum_i y^*(E(B_i)x)$$

for all $x \in X$ and $y^* \in Y^*$. And a $B(X)$-valued measure E is called:

(1) *bounded* if $\sup\{\|E(B)\| : B \in \Sigma\} < \infty$.
(2) a *projection-valued measure* if $E(B)$ is a projection on X for all $B \in \Sigma$.

(3) a *spectral operator-valued measure* if for all

$$E(A \cap B) = E(A) \cdot E(B), \forall A, B \in \Sigma.$$

(4) an *idempotent-valued measure* if for all

$$E(A) \cdot E(A) = E(A), \forall A \in \Sigma.$$

To simplify, we use quadruple $(\Omega, \Sigma, E, B(X, Y))$ to denote this operator-valued measure. Note that spectral operator-valued measures are clearly idempotent, and the converse is also true (c.f. [171]), so we will also use the term *idempotent-valued measure* to mean a spectral-valued measure. Compared with the projection-valued measure for Hilbert space $\mathcal{H}$, for a general $B(X)$-valued measure, we say that E is projection-valued if it is idempotent-valued and bounded.

4.2.1   Minimal and Maximal Dilations

4.2.1.1 Minimal Dilation

For general operator-valued measures, we establish the following dilation theorem: Every operator-valued measure can be dilated to a projection-valued measure.

Theorem 4.4 *Let $E : \Sigma \to B(X, Y)$ be an operator-valued measure. Then there exists a Banach space Z, bounded linear operators $S : Z \to Y$ and $T : X \to Z$, and a projection-valued probability measure $F : \Sigma \to B(Z)$ such that*

$$E(B) = SF(B)T, \ \forall B \in \Sigma.$$

Namely, for all $B \in \Sigma$, the diagram commutes:

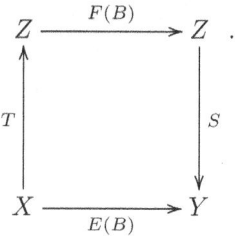

The space Z is called the dilation space and S, T the corresponding analysis operator and synthesis operator, respectively. For convenience, we call (F, Z, S, T) in the above theorem a projection-valued measure dilation system.

Note from [125] that the strategy of constructing the dilation is to introduce an appropriate norm on the elementary dilation space. Let

$$M_E = \text{span}\{E_{B,x} : x \in X, B \in \Sigma\},$$

where $E_{B,x} : \Sigma \to Y$ are defined to be a vector-valued measure on (Ω, Σ) by

$$E_{B,x}(A) = E(B \cap A)x, \ \forall A \in \Sigma.$$

And we often call M_E the *elementary dilation space*.

Before proving this theorem, we need the following necessary definition.

Definition 4.2 Let M_E be the elementary dilation space induced by operator-valued measure $(\Omega, \Sigma, E, B(X, Y))$. Let $\| \cdot \|$ be a norm on M_E and its completion is denoted by $\widetilde{M}_{E, \|\cdot\|}$. The norm on $\widetilde{M}_{E, \|\cdot\|}$, with $\| \cdot \| := \| \cdot \|_{\mathcal{D}}$ given by a norming function $\mathcal{D}$,[1] is called a *dilation norm* of E if the following conditions are satisfied:

(1) The map $S_{\mathcal{D}} : \widetilde{M}_{E, \mathcal{D}} \to Y$ defined on M_E by

$$S_{\mathcal{D}}\Big(\sum_{i=1}^{N} C_i E_{B_i, x_i} \Big) = \sum_{i=1}^{N} C_i E(B_i) x_i$$

is bounded.
(2) The map $T_{\mathcal{D}} : X \to \widetilde{M}_{E, \mathcal{D}}$ defined by

$$T_{\mathcal{D}}(x) = E_{\Omega, x}$$

is bounded.
(3) The map $F_{\mathcal{D}} : \Sigma \to B(\widetilde{M}_{E, \mathcal{D}})$ defined by

$$F_{\mathcal{D}}(B)\Big(\sum_{i=1}^{N} C_i E_{B_i, x_i} \Big) = \sum_{i=1}^{N} C_i E_{B \cap B_i, x_i}$$

is an operator-valued measure, where $\{C_i\}_{i=1}^{N} \subset \mathbb{C}, \{x_i\}_{i=1}^{N} \subset X, \{B_i\}_{i=1}^{N} \subset \Sigma$.

---

[1] In most cases, we use the traditional notation $\| \cdot \|$ for a norm. However, here, we also use the functional notation, typically $\mathcal{D}$ for a norm, because of the length of the verification of norm. In this case, we use $\| \cdot \|_{\mathcal{D}}$ to denote the norm induced by the norming functional $\mathcal{D}(\cdot)$.

Proof of Theorem 4.4 Define $\| \cdot \|_\alpha : M_E \to \mathbb{R}^+ \cup \{0\}$ by

$$\left\| \sum_{i=1}^{N} C_i E_{B_i, x_i} \right\|_m = \sup_{B \in \Sigma} \left\| \sum_{i=1}^{N} C_i E(B \cap B_i) x_i \right\|_Y,$$

for any $\sum_{i=1}^{N} C_i E_{B_i, x_i} \in M_E$. For any $\Phi \in M_E$, $\|\Phi\|_m \geq 0$. If If $\|\Phi\|_m = 0$, taking $\Phi = \sum_{i=1}^{N} C_i E_{B_i, x_i}$, then we have

$$\sup_{B \in \Sigma} \left\| \sum_{i=1}^{N} C_i E(B \cap B_i) x_i \right\|_Y = 0,$$

then $\sum_{i=1}^{N} C_i E(B \cap B_i) x_i = 0$ for any $B \in \Sigma$, then $\Phi = 0$ as vector-measure. And it is routine to verify the homogeneity and the triangle inequality. Thus, $\| \cdot \|_m$ is a norm on M_E.

Next, we show that $\| \cdot \|_m$ is a dilation norm on M_E. Let $\widetilde{M}_{E,m}$ be its completion. Choose $\sum_{i=1}^{N} C_i E_{B_i, x_i} \in M_E$, where $N > 0$, $\{C_i\}_{i=1}^{N} \subset \mathbb{C}$, $\{x_i\}_{i=1}^{N} \subset X$ and $\{B_i\}_{i=1}^{N} \subset \Sigma$. Define the map $S_m : \widetilde{M}_{E,\alpha} \to Y$ by

$$S_m \left(\sum_{i=1}^{N} C_i E_{B_i, x_i} \right) = \sum_{i=1}^{N} C_i E(B_i) x_i.$$

It is easy to check that S_m is linear and well-defined and

$$\left\| S_m \left(\sum_{i=1}^{N} C_i E_{B_i, x_i} \right) \right\|_Y = \left\| \sum_{i=1}^{N} C_i E(B_i) x_i \right\|_Y$$

$$\leq \sup_{B \in \Sigma} \left\| \sum_{i=1}^{N} C_i E(B \cap B_i) x_i \right\|_Y = \left\| \sum_{i=1}^{N} C_i E_{B_i, x_i} \right\|_m,$$

which implies that S_m is bounded and $\|S_m\| \leq 1$.

Define the map $T_m : X \to \widetilde{M}_{E,m}$ by $T_m(x) = E_{\Omega, x}$. Obviously, it is a well-defined linear map and

$$\|T_m x\|_m = \|E_{\Omega, x}\|_m = \sup_{B \in \Sigma} \|E(B)x\|_Y \leq \|E\| \cdot \|x\|$$

and hence $\|T_m\| \leq \|E\|$.

Now we define the map $F_m : \Sigma \to B(\widetilde{M}_{E,\alpha})$ by

$$F_m(B) \left(\sum_{i=1}^{N} C_i E_{B_i, x_i} \right) = \sum_{i=1}^{N} C_i E_{B \cap B_i, x_i}.$$

Then, it is easy to see that F_m is a spectral operator-valued measure. Taking extension by density [5], we only need to prove F_m is strongly countably additive and uniform bounded on $M_{E,\alpha}$.

Note that

$$\left\| F_m(B) \left(\sum_{i=1}^{N} C_i E_{B_i,x_i} \right) \right\|_m$$

$$= \left\| \sum_{i=1}^{N} C_i E_{B \cap B_i,x_i} \right\|_m = \sup_{B' \in \Sigma} \left\| \sum_{i=1}^{N} C_i E(B_i \cap B \cap B')(x_i) \right\|_Y$$

$$\leq \sup_{B'' \in \Sigma} \left\| \sum_{i=1}^{N} C_i E(B_i \cap B')x_i \right\|_Y = \left\| \sum_{i=1}^{N} C_i E_{B_i,x_i} \right\|_Y,$$

thus, we get that $\| F_m(B) \| = 1$ for any $B \in \Sigma$.

For the strong countable additivity, let $\{A_j\}_{j=1}^{\infty}$ be a countable disjoint collection of members of Σ with union A. Then we have

$$\left\| \sum_{j=1}^{M} F_m(A_j) \left(\sum_{i=1}^{N} C_i E_{B_i,x_i} \right) - F_\alpha(A) \left(\sum_{i=1}^{N} C_i E_{B_i,x_i} \right) \right\|_m$$

$$= \left\| \sum_{j=1}^{M} \left(\sum_{i=1}^{N} C_i E_{A_j \cap A_i,x_i} \right) - \left(\sum_{i=1}^{N} C_i E_{A \cap B_i,x_i} \right) \right\|_m$$

$$= \left\| \sum_{i=1}^{N} C_i \left(\sum_{j=1}^{M} E_{A_j \cap B_i,x_i} - E_{A \cap B_i,x_i} \right) \right\|_m$$

$$= \sup_{B' \in \Sigma} \left\| \sum_{i=1}^{N} C_i \left(\sum_{j=1}^{M} E(A_j \cap B_i \cap B')x_i - E(A \cap B_i \cap B')x_i \right) \right\|_Y$$

$$= \sup_{B' \in \Sigma} \left\| \sum_{i=1}^{N} C_i E \left(\bigcup_{j=M+1}^{\infty} \left(A_j \cap B_i \cap B' \right) \right)x_i \right\|_Y$$

$$\leq \sum_{i=1}^{N} |C_i| \sup_{B' \in \Sigma} \left\| B \left(\bigcup_{j=M+1}^{\infty} \left(A_j \cap B_i \cap B' \right)x_i \right) \right\|_Y$$

$$= \sum_{i=1}^{N} |C_i| \sup_{B' \in \Sigma} \left\| E \left(\bigcup_{j=M+1}^{\infty} (A_j \cap B') \right)x_i \right\|_Y$$

$$= \sum_{i=1}^{N} |C_i| \sup_{B' \in \Sigma} \left\| \sum_{j=M+1}^{\infty} E(A_j \cap B') x_i \right\|_Y.$$

If $\sup_{B' \in \Sigma} \| \sum_{j=M+1}^{\infty} E(A_j \cap B') x_i \|_Y$ does not tend to 0 as $M \to \infty$, then we can find $\delta > 0$, a sequence of $n_1 \le m_1 < n_2 \le m_2 < n_3 \le m_3 < \dots$, and $\{B_l'\}_{l=1}^{\infty} \subset \Sigma$ such that

$$\left\| \sum_{j=n_l}^{m_l} E\left(A_j \cap B_l'\right) x_i \right\| \ge \delta, \quad \forall l \in \mathbb{N}.$$

Since for $l \in \mathbb{N}$ and $n_l \le j \le m_l$, $A_j \cap B_l'$ are disjoint from each other, we have

$$E\left(\bigcup_{l=1}^{\infty} \bigcup_{j=n_l}^{m_l} A_j \cap B_l' \right) x_i = \sum_{l=1}^{\infty} \sum_{j=n_l}^{m_l} E(B_j \cap B_l') x_i,$$

which implies $\| \sum_{j=n_i}^{m_i} E(B_j \cap B_i') x_i \| \to 0$, which is a contradiction. Hence,

$$F_m(A) \left(\sum_{i=1}^{N} C_i E_{B_i, x_i} \right) = \sum_{j=1}^{\infty} F_m(A_j) \left(\sum_{i=1}^{N} C_i E_{B_i, x_i} \right)$$

as expected.

To conclude, $\{F_m, \widetilde{M}_{E,m}, S_m, T_m\}$ is a projection-valued measure dilation system for $(\Omega, \Sigma, E, B(X, Y))$. $\qquad \square$

Given an operator-valued measure, there could be many projection-valued measure dilations, which means that the dilation spaces are not unique. The following result establishes the connections between any hypothetical existing dilation space and the elementary dilation space: any given projection-valued dilation space Y can induce a dilation norm on the elementary dilation space M_E so that E can be dilated to a projection-valued measure and $\| \cdot \|_m$ in the proof of Theorem 4.4 is a "minimal" norm on the elementary dilation space.

Theorem 4.5 *Let $E : \Sigma \to B(X, Y)$ be an operator-valued measure and (F, Z, S, T) be an injective projection-valued measure dilation system, which means*

$$\sum F(B_i) T(x_i) = 0 \quad \text{whenever} \quad \sum E_{B_i, x_i} = 0.$$

Then we have the following:

(1) There exist an elementary Banach space dilation system $(F_{\mathcal{D}}, \tilde{M}_{E,\mathcal{D}}, S_{\mathcal{D}}, T_{\mathcal{D}})$ of E and a linear isometric embedding

$$U : \tilde{M}_{E,\mathcal{D}} \to Z$$

such that

$$S_{\mathcal{D}} = SU, \ F(\Omega)T = UT_{\mathcal{D}}, U F_{\mathcal{D}}(B) = F(B)U, \quad \forall B \in \Sigma.$$

(2) The norm $\| \cdot \|_m$ is indeed a dilation norm. And if $\mathcal{D}$ is a dilation norm of E, then there exists a constant $C_{\mathcal{D}}$ such that for any $\sum_{i=1}^{N} C_i E_{B_i, x_i} \in M_{E,\mathcal{D}}$,

$$\sup_{B \in \Sigma} \Big\| \sum_{i=1}^{N} C_i E(B \cap B_i) x_i \Big\|_Y \leq C_{\mathcal{D}} \Big\| \sum_{i=1}^{N} C_i E_{B_i, x_i} \Big\|_{\mathcal{D}},$$

where $N > 0, \{C_i\}_{i=1}^{N} \subset \mathbb{C}, \{x_i\}_{i=1}^{N} \subset X$ and $\{B_i\}_{i=1}^{N} \subset \Sigma$. Consequently

$$\|f\|_m \leq C_{\mathcal{D}} \|f\|_{\mathcal{D}}, \quad \forall f \in M_E.$$

Proof

(1) Define $\mathcal{D} : M_E \to R^+ \cup \{0\}$ by

$$\mathcal{D}\Big(\sum_{i=1}^{N} C_i E_{B_i, x_i} \Big) = \Big\| \sum_{i=1}^{N} C_i F(B_i) T(x_i) \Big\|_Z, \quad \forall \sum_{i=1}^{N} C_i E_{B_i, x_i} \in M_E.$$

It is routine to verify that $\mathcal{D}$ is a norm on M_E and denote it by $\| \cdot \|_{\mathcal{D}}$, and let $\tilde{M}_{E,\mathcal{D}}$ be the completion of M_E under this norm. Next, we show that $\mathcal{D}$ is a dilation norm of E. By Definition 4.2 of dilation norm, it is easy to see that the map

$$S_{\mathcal{D}} : \tilde{M}_E \to Y, \quad S_{\mathcal{D}}\Big(\sum_{i=1}^{N} C_i E_{B_i, x_i} \Big) = \sum_{i=1}^{N} C_i E(B_i) x_i$$

and

$$T_{\mathcal{D}} : X \to \tilde{M}_{E,\mathcal{D}}, \quad T_{\mathcal{D}}(x) = E_{\Omega, x}$$

are both well-defined, and linear. Moreover,

$$\Big\| S_{\mathcal{D}}\Big(\sum_{i=1}^{N} C_i E_{B_i, x_i} \Big) \Big\|_Y = \Big\| S\Big(\sum_{i=1}^{N} C_i F(B_i) T x_i \Big) \Big\|_Y \leq \|S\| \cdot \Big\| \sum_{i=1}^{N} C_i E_{B_i, x_i} \Big\|_{\mathcal{D}}$$

and

$$\|T_{\mathcal{D}}(x)\|_{\mathcal{D}} = \|E_{\Omega,x}\|_{\mathcal{D}} = \|F(\Omega)T(x)\|_Z \le \|F(\Omega)\| \cdot \|T\| \cdot \|x\|_X.$$

That is, $S_{\mathcal{D}}$, $T_{\mathcal{D}}$ both are bounded.

Note that the map $F_{\mathcal{D}} : \Sigma \rightarrow B(\widetilde{M}_E)$ defined by

$$F_{\mathcal{D}}(B)\Big(\sum_{i=1}^{N} C_i E_{B_i,x_i}\Big) = \sum_{i=1}^{N} C_i E_{B \cap B_i, x_i}$$

is a (finite-additive) spectral operator-valued measure. For any $w = \sum_{i=1}^{N} C_i E_{B_i,x_i} \in M_E$ and $A \in \Sigma$,

$$\|F_{\mathcal{D}}(A)(w)\|_{\mathcal{D}} = \Big\| \sum_{i=1}^{N} C_i E_{A \cap B_i, x_i} \Big\|_{\mathcal{D}} = \Big\| \sum_{i=1}^{N} C_i F(A \cap B_i)T(x_i) \Big\|_Z$$

$$= \Big\| \sum_{i=1}^{N} C_i F(A)F(B_i)T(x_i) \Big\|_Z \le \|F(A)\| \Big\| \sum_{i=1}^{N} C_i F(B_i)T(x_i) \Big\|_Z$$

$$\le \sup_{A \in \Sigma} \|F(A)\| \|w\|_{\mathcal{D}}.$$

Thus, $F_{\mathcal{D}}$ is uniformly bounded on M_E, taking extension by the density [5], we only needed to show $F_{\mathcal{D}}$ is strongly countably additive on $M_{E,\mathcal{D}}$. If $\{A_j\}_{j=1}^{\infty}$ is a disjoint countable collection of members of Σ with union A, since F is an operator-valued measure, we have

$$F(A \cap B_i)T(x_i) = \sum_{j=1}^{\infty} F(A_j \cap B_i)T(x_i).$$

Thus, we have

$$F_{\mathcal{D}}(A)(E_{B_i,x_i}) = \sum_{j=1}^{\infty} F_{\mathcal{D}}(A_j)(E_{B_i,x_i}).$$

Then for any $\sum_{i=1}^{N} C_i E_{B_i,x_i} \in M_E$,

$$F_{\mathcal{D}}(A)\Big(\sum_{i=1}^{N} C_i E_{B_i,x_i}\Big) = \sum_{j=1}^{\infty} F_{\mathcal{D}}(A_j)\Big(\sum_{i=1}^{N} C_i E_{B_i,x_i}\Big).$$

Define a map $U : \tilde{M}_{E,\mathcal{D}} \to Z$ by

$$U\left(\sum_{i=1}^{N} C_i E_{B_i,x_i}\right) = \sum_{i=1}^{N} C_i F(B_i)T(x_i).$$

It is easy to see that U is a well-defined and linear isometric embedding map and for any $\phi \in M_E$ represented by $\sum_{i=1}^{N} C_i E_{B_i,x_i}$, then we have

$$S_{\mathcal{D}}(\phi) = \sum_{i=1}^{N} C_i E(B_i)x_i = S\left(\sum_{i=1}^{N} C_i F(B_i)T(x_i)\right) = SU(\phi)$$

and

$$UT_{\mathcal{D}}(x) = U(E_{\Omega,x}) = F(\Omega)T(x)$$

and

$$UF_{\mathcal{D}}(B)(\phi) = U\left(\sum_{i=1}^{N} C_i E_{B\cap B_i,x_i}\right) = \sum_{i=1}^{N} C_i F\left(B \cap B_i\right)T(x_i)$$

$$= F(B)\left(\sum_{i=1}^{N} C_i F(B_i)T(x_i)\right) = F(B)U(\phi).$$

Thus we get

$$S_{\mathcal{D}} = SU, \; F(\Omega)T = UT_{\mathcal{D}}, \; UF_{\mathcal{D}}(B) = F(B)U, \forall B \in \Sigma,$$

as claimed.

(2) Let $C_{\mathcal{D}} = \sup_{B \in \Sigma} \|S_{\mathcal{D}}F_{\mathcal{D}}(B)\|$. Then obviously $C_{\mathcal{D}} < +\infty$. We also have

$$\sup_{B \in \Sigma}\left\|\sum_{i=1}^{N} C_i E(B \cap B_i)x_i\right\|_Y = \sup_{B \in \Sigma}\left\|\sum_{i=1}^{N} C_i S_{\mathcal{D}}F_{\mathcal{D}}(B \cap B_i)T_{\mathcal{D}}(x_i)\right\|_Y$$

$$= \sup_{B \in \Sigma}\left\|\sum_{i=1}^{N} C_i S_{\mathcal{D}}F_{\mathcal{D}}(B)F_{\mathcal{D}}(B_i)T_{\mathcal{D}}(x_i)\right\|_Y$$

$$\leq \sup_{B \in \Sigma} \|S_{\mathcal{D}}F_{\mathcal{D}}(B)\|\left\|\sum_{i=1}^{N} C_i F_{\mathcal{D}}(B_i)T_{\mathcal{D}}(x_i)\right\|_{\mathcal{D}}$$

$$= C_{\mathcal{D}}\left\|\sum_{i=1}^{N} C_i E_{B_i,x_i}\right\|_{\mathcal{D}}.$$

Remark 4.1 This result justifies using the name of minimal dilation norm $\| \cdot \|_m$. Unlike Theorem 4.1 and 4.2 due to Naimark or Stinespring, which have been extensively investigated in the context of positive, or completely bounded maps on C^*-algebras, an operator-valued probability measure on Banach spaces, in general, does not admit a Hilbert space dilation. However, the dilation theory can be strengthened in the case that it does admit a Hilbert space dilation [125], which turns out to have some interesting applications, for example, the complete characterization of framings whose induced operator-valued measures are completely bounded.

As shown in Chap. 2, framings are the natural generalizations of the discrete frame theory (or more specifically, dual-frame pairs). Even if the underlying space is a Hilbert space, the dilation space for framing-induced operator-valued measures can fail to be a Hilbert space. So in the context of Hilbert spaces, the dilation theory for discrete framings from [52] induces a dilation theory for discrete operator-valued measures that may fail to be completely bounded in the sense of (c.f. [171]).

Theorem 4.6 *Let $\{x_i, y_i\}_{i \in \mathbb{N}}$ be a nonzero framing for a Hilbert space $\mathcal{H}$ and E be the operator-valued measure induced by $\{x_i, y_i\}_{i \in \mathbb{N}}$ defined by*

$$E(B) = \sum_{i \in B} x_i \otimes y_i, \quad \forall \, B \in 2^{\mathbb{N}},$$

where $x_i \otimes y_i(z) := \langle z, y_i \rangle x_i$. Then we have the following:

(1) E has a Hilbert dilation space $\mathcal{K}$ if and only if there exist $\alpha_i, \beta_i \in \mathbb{C}, i \in \mathbb{N}$ with $\alpha_i \bar{\beta}_i = 1$ such that $\{\alpha_i x_i\}_{i \in \mathbb{N}}$ and $\{\beta_i y_i\}_{i \in \mathbb{N}}$ both are the frames for the Hilbert space $\mathcal{H}$.

(2) E is a completely bounded map if and only if $\{x_i, y_i\}_{i \in \mathbb{N}}$ can be re-scaled to dual frames.

(3) If $\inf \|x_i\| \cdot \|y_i\| > 0$, then we can find $\alpha_i, \beta_i \in \mathbb{C}, i \in \mathbb{N}$ with $\alpha_i \bar{\beta}_i = 1$ such that $\{\alpha_i x_i\}_{i \in \mathbb{N}}$ and $\{\beta_i y_i\}_{i \in \mathbb{N}}$ both are frames for the Hilbert space $\mathcal{H}$. Hence the operator-valued measure induced by $\{x_i, y_i\}_{i \in \mathbb{N}}$ has a Hilbertian dilation.

4.2.1.2 Maximal Dilations

As shown in the previous subsection, the procedure of introducing the elementary dilation space and equipping the minimal dilation norm is usually called the minimal dilation technique, which has been extensively used [120, 124, 126, 127]. In addition to the minimal dilation norm, some other dilation norms were introduced in [125]. In this subsubsection, we introduce some structural dilation norms for the elementary dilation space.

Let $(\Omega, \Sigma, E, B(X, Y))$ be an operator-valued measure system. To define an appropriate dilation norm on the elementary dilation space M_E. If we have a norm, it is natural to require that the basic element $E_{B,x}$ satisfies

$$\|E_{B,x}\| \leq \sup_{B' \in \Sigma} \|E(B \cap B')x\|.$$

Now let f be any element of M_E with a representation $\sum_{i=1}^{N} C_i E_{B_i,x_i}$. Then it follows from the triangle inequality that

$$\|f\| \leq \sum_{i=1}^{N} \sup_{B \in \Sigma} \|C_i E(B_i \cap B)x_i\|.$$

Since this holds for every representation of f, we have

$$\|f\| \leq \inf \left\{ \sum_{i=1}^{N} \sup_{B \in \Sigma} \|C_i E(B_i \cap B)x_i\| \right\},$$

where the infimum is taken over all representations of f.

Define $\| \cdot \|_\omega : M_E \to \mathbb{R}^+ \cup \{0\}$ by

$$\|f\|_\omega = \inf \left\{ \sum_{i=1}^{N} \sup_{B \in \Sigma} \|C_i E(B_i \cap B)x_i\| : f = \sum_{i=1}^{N} C_i E_{B_i,x_i} \in M_E \right\}.$$

Han et al. [125] showed the homogeneity and the triangle inequality for $\| \cdot \|_\omega$. So we get:

Proposition 4.4 $\| \cdot \|_\omega$ *is a semi-norm on M_E and*

$$\|E_{B,x}\|_\omega = \sup_{E' \in \Sigma} \|E(B \cap B')x\|.$$

for every $B \in \Sigma$ and $x \in X$.

Thus, we can define an equivalence relation R_ω on M_E by $f \sim g$ if $\|f - g\|_\omega = 0$ to make $\| \cdot \|_\omega$ is a norm. Denote by $M_{E,\omega}$ the space of the R_ω-equivalence classes of M_E endowed with the norm $\| \cdot \|_\omega$, and by $\widetilde{M}_{E,\omega}$ for its completion.

Theorem 4.7 ([125, Theorem 2.43]) $\| \cdot \|_\omega$ *is a dilation norm of E.*

Proof Let $N > 0$, $\{C_i\}_{i=1}^{N} \subset \mathbb{C}$, $\{x_i\}_{i=1}^{N} \subset X$ and $\{B_i\}_{i=1}^{N} \subset \Sigma$. First, we show that the map $S_\omega : \widetilde{M}_{E,\omega} \to Y$ defined on $M_{E,\omega}$ by

$$S_\omega \left(\sum_{i=1}^{N} C_i E_{B_i,x_i} \right) = \sum_{i=1}^{N} E(B_i)x_i$$

is well-defined and $\|S_\omega\| \leq 1$. If $f = \sum_{i=1}^{N} C_i E_{B_i,x_i}$ is a representation of f, then

$$\|S_\omega(f)\| = \left\| \sum_{i=1}^{N} C_i E(B_i)x_i \right\| \leq \sum_{i=1}^{N} \sup_{B \in \Sigma} \|C_i E(B_i \cap B)x_i\|.$$

then it follows that $\|S_\omega(f)\| \leq \|f\|_\omega$. Therefore S_ω is well-defined and bounded with $\|S_\omega\| \leq 1$. It follows from the definition of ω that

$$\|T_\omega x\|_\omega = \|E_{\Omega,x}\|_\omega = \sup_{B \in \Sigma} \|E(B)x\| \leq \|E\| \cdot \|x\|.$$

Thus, T is bounded with $\|T_\omega\| \leq \|E\|$.

Finally, we define the map $F_\omega : \Sigma \to B(\widetilde{M}_{E,\omega})$ by

$$F_\omega(B)\left(\sum_{i=1}^{N} C_i E_{B_i,x_i} \right) = \sum_{i=1}^{N} C_i E_{B \cap B_i,x_i}.$$

It is easy to see that is F_ω is a finite additive idempotent-valued measure. Similar to the proof of Theorem 4.5, F_ω is strongly countably additive and uniform bounded on $M_{E,\omega}$, thus, F_ω is a projection-valued measure dilation for E.   □

Note that the idea of the construction of $\|\cdot\|_\omega$ comes from the idea of the projective tensor product of Banach spaces [188], so we call $\|\cdot\|_\omega$ the *projective dilation norm*.

Given an operator-valued measure system $(\Omega, \Sigma, E, B(X, Y))$, we define the following elementary dual dilation space by

$$W_E = \text{span}\{F_{B,y^*} : y^* \in Y^*, B \in \Sigma\}.$$

where $F_{B,x} : \Sigma \to X^*$ are defined to be a vector-valued measure on (Ω, Σ) by

$$F_{B,y^*}(A) = y^* E(B \cap A), \ \forall A \in \Sigma.$$

Similar to the minimal dilation norm on the elementary dilation norm, we define a norm on W_E by

$$\left\| \sum_i C_i F_{B_i,y_i^*} \right\|_D = \sup_{B \in \Sigma} \left\| \sum_i C_i y_i^* E(B_i \cap B) \right\|.$$

By Han et al. [125, Proposition 2.39], the element of W_E can be viewed as a bounded linear functional on M_E and is weak* dense in the dual space of the minimal dilation space. Then W_E can induce a functional on M_E by

$$\left\| \sum_{i=1}^{N} C_i E_{B_i, x_i} \right\|_M = \sup_{w \in M_E, \|w\|_D \leq 1} \left| w\left(\sum_{i=1}^{N} C_i E(B \cap B_i) x_i \right) \right|$$

for any $\sum_{i=1}^{N} C_i E_{B_i, x_i} \in M_E$. It can be verified that $\| \cdot \|_M$ is a well-defined norm on M_E. Furthermore, we have:

Theorem 4.8 *Let* $E : \Sigma \to B(X, Y)$ *be an operator-valued measure.* $\| \cdot \|_M$ *is a dilation norm on* M_E. *And if* α *is a dilation norm of* E, *then there exists a constant* C_α *such that for any* $\sum_{i=1}^{N} C_i E_{B_i, x_i} \in M_E$,

$$\left\| \sum_{i=1}^{N} C_i E_{B_i, x_i} \right\|_\alpha \leq C_\alpha \sup_{w \in M_E, \|w\|_D \leq 1} \left| w\left(\sum_{i=1}^{N} C_i E_{B_i, x_i} \right) \right|,$$

where $N > 0$, $\{C_i\}_{i=1}^{N} \subset \mathbb{C}$, $\{x_i\}_{i=1}^{N} \subset X$ *and* $\{B_i\}_{i=1}^{N} \subset \Sigma$. *Consequently,*

$$\|f\|_\alpha \leq C_\alpha \|f\|_M, \quad \forall f \in M_E. \tag{4.2}$$

The proof of this dilation norm is similar to the minimal dilation norm, we omit it and leave it to the interested readers. And we call it the maximal dilation norm in the sense of Eq. (4.2).

Even for Hilbert space operator-valued measures, the dilation spaces arising from the elementary dilation equipped with the minimal or maximal dilation norm are usually Banach spaces, i.e., non-Hilbertian dilations. But Hilbertian dilation exists when imposing additional properties (e.g., completely boundedness) on the operator-valued measures. Usually, when we consider the dilations of a special class of operator-valued measures on some Banach spaces with some desirable properties, we also hope that the dilation spaces preserve the same properties. In general, the minimal and maximal dilation spaces are not well-behaved, so we have the following problem:

Problem 4.1 Apart from the minimal and maximal dilation norms, are there any dilation norms with good structures? Furthermore, can we quantify their relationship with the minimum and maximum norms?

4.2.2 Imprimitive Systems

The concept of a system of imprimitivity was introduced by Mackey [160, 161] for his theory of induced representations of locally compact groups. It is used in algebra and analysis in the theory of group representations. In Mackey's theory, every infinite-dimensional system of imprimitivity is affiliated with a projection-valued measure on a Hilbert space. The Banach space dilation techniques for general operator-valued measures can be used to build a dilation theory of operator-valued systems of imprimitivity based on semigroups acting on both Hilbert and Banach spaces [126].

We first generalize the concept of infinite-dimensional systems of imprimitivity in Hilbert spaces to Banach space settings. The ingredients are a semigroup S, a σ-field Σ of subsets of a set Ω, and a measurable semigroup action

$$S \times \Sigma \to \Sigma, \quad (s, E) \mapsto sE,$$

that satisfies the conditions that for all $E \in \Sigma$ and $s, t \in S$:

(1) $s(tE) = (st)E$,
(2) $eE = E, s\emptyset = \emptyset, s\Omega = \Omega$,
(3) $s(\bigcup_{n=1}^{\infty} E_n) = \bigcup_{n=1}^{\infty} sE_n$,
(4) $s(\bigcap_{n=1}^{\infty} E_n) = \bigcap_{n=1}^{\infty} sE_n$.

We refer to this as a Σ-measurable S-space, denoted by (S, Σ), for example, $(\mathbb{N}, \mathbb{R}^+)$. The term semigroup will signify a semigroup with a unit. We shall usually write the operation multiplicatively and denote the unit by e. In particular, a subsemigroup S of a Σ-measurable group G is a Σ-measurable S-space with the relative measure. A multiplier on a semigroup S is a function ω, from the Cartesian product $S \times S$ to the unit circle $\mathbb{T}$ in the complex plane $\mathbb{C}$, such that for all elements s, t and u of S:

(1) $\omega(e, s) = \omega(s, e) = 1$;
(2) $\omega(s, t)\omega(st, u) = \omega(s, tu)\omega(t, u)$.

If S is a group, then it follows from (1) and (2) that for all $s \in S$, we have $\omega(s, s^{-1}) = \omega(s^{-1}, s)$. If a multiplier ω satisfies $\omega(s, s^{-1}) = \omega(s^{-1}, s) = 1$ for all $s \in S$, we say that ω is symmetric.

Let S be a semigroup and X be a Banach space. A projective isometric representation of S on X is a map, $W : s \mapsto W_s$, from S to $B(X)$ having the following properties for all elements $s, t \in S$:

(1) W_s is an isometry and $W_e = 1$;
(2) $W_s W_t = \omega(s, t)W_{st}$, where $\omega(s, t)$ are scalars of unit modulus.

The function $\omega : (s, t) \mapsto \omega(s, t)$ is a multiplier on S, and it is called the multiplier associated with W. If all the W_s are surjective isometries, we say that W is a projective isometric isomorphism representation. In this case, if the associated multiplier ω is symmetric, then for all $s \in SW_{s^{-1}} = W_s^{-1}$. A projective isometric representation or projective isometric isomorphism representation with ω an associated multiplier will sometimes be referred to as an isometric ω representation or, respectively, an isometric isomorphism ω-representation. We remark that if S is a group, then a projective isometric representation is automatically a projective isometric isomorphism representation.

Let $\mathcal{G}$ be a locally compact Hausdorff topological group acting on a measurable space (Ω, Σ). An (orthogonal) projection-valued system of imprimitivity (resp, positive operator-valued system of imprimitivity) based on $(\mathcal{G}, (\Omega, \Sigma))$ consists of a separable Hilbert space $\mathcal{H}$ and a pair consisting of a strongly-continuous unitary representation $U : g \mapsto U_g$ of $\mathcal{G}$ on $\mathcal{H}$, and an (orthogonal) projection-valued measure (resp, a positive operator-valued measure) π on the measurable subsets of Ω with values in the projections (resp, positive operators) on $\mathcal{H}$, which satisfy

$$U_g \pi(E) U_{g^{-1}} = \pi(g \cdot E) \quad \text{for all } g \in \mathcal{G} \text{ and } E \in \Sigma.$$

Similarly, we introduce the following definition of projective isometric operator-valued system of imprimitivity acting on Banach spaces.

Definition 4.3 Let S be a semigroup acting on a σ-field Σ of subsets of a set Ω. A projective isometric operator-valued system of imprimitivity based on (S, Σ) consists of a Banach space X and a pair consisting of:

(1) A projective isometric representation $W : s \mapsto W_s$ of S on X;
(2) An operator-valued measure φ on Σ with values in the operators on X, satisfying that for all $s \in S$ and $E \in \Sigma$

$$W_s \varphi(E) = \varphi(sE) W_s.$$

For a projective isometric operator-valued system (W, ρ) of imprimitivity based on (S, Σ), if ρ is also a spectral measure, then we call it a projective isometric spectral system of imprimitivity.

A projective isometric operator-valued system of imprimitivity or projective isometric spectral system of imprimitivity with ω an associated multiplier will be referred to as an operator-valued ω-system of imprimitivity or, respectively, a spectral ω-system of imprimitivity.

Definition 4.4 Let (W, φ) be an operator-valued ω-system of imprimitivity of (S, Σ) on a Banach space X and (V, ρ) a spectral ω-system of imprimitivity of (S, Σ) on a Banach space Z. Then (V, ρ) is said to be a dilation of (W, φ) if there are bounded operators $Q : Z \to X$ and $T : X \to Z$ such that for all $s \in S$ and $E \in \Sigma$:

(1) $\varphi(E) = Q\rho(E)T$;
(2) $QV_s = W_s Q$;
(3) $V_s T = T W_s$.

In this case, (V, ρ, Q, T) is called a dilation system of (W, φ). The representation of S on X can be viewed as a subrepresentation of the representation of S on Z.

Theorem 4.9 ([126]) *Let Σ be a σ-field of subsets of a set Ω and S a Σ measurable semigroup. Then, every projective isometric operator-valued system of imprimitivity of (S, Σ) can be dilated to a probability projective isometric spectral system of imprimitivity.*

Since there exists an example of a Hilbert space operator-valued measure which has a Banach space dilation but not a Hilbert space dilation, the dilation space in the above theorem is usually a Banach space. However, similar to Naimark's dilation theorem, if the condition of positivity is imposed on the isometric operator-valued system, then a Hilbert space dilation is possible.

Theorem 4.10 ([126]) *Let $\mathcal{G}$ be a locally compact Hausdorff topological group acting on a measurable space (Ω, Σ). Let (U, ϖ) be an isometric positive operator-valued system of imprimitivity of $(\mathcal{G}, \Sigma)$ on a Hilbert space $\mathcal{H}$. Then there is an isometric (orthogonal) projection-valued system of imprimitivity $(\widetilde{U}, \pi)$ of $(\mathcal{G}, \Sigma)$ on a Hilbert space $\mathcal{K}$ and a bounded linear operator $V : \mathcal{H} \rightarrow \mathcal{K}$ such that*

$$\widetilde{U}_g V = V U_g \quad \text{and} \quad \varpi(E) = V^* \pi(E) V$$

for all $g \in \mathcal{G}$ and $E \in \Sigma$.

4.2.3 OVMs with Bounded p-Variations

The variation of scalar measures or vector measures is well-known [188]. It is natural to define similar variations for operator-valued measures. A special class of operator-valued measures is the one with finite total variation. It is known that any operator-valued measure acting on a Hilbert space with finite total variation is completely bounded in the sense that its associated map defined by Paulsen in [171] is completely bounded. And Don Hadwin and Vern Paulsen constructed examples that are completely bounded but do not have finite total variations [110, 171]. This motivates us to examine a dilation preserving theory for operator-valued measures that have various types of finite total variations in both Hilbert and Banach space setting.

Let Σ be a σ-algebra of subsets of a set Ω and let $\mathcal{M}(\Sigma)$ denote the Banach space of scalar valued measures on Σ with the variation norm, $\|\mu\| = |\mu|(\Omega)$. We denote by $\mathcal{M}(\Sigma, X)$ the Banach space of vector measures on Σ with values in X. The semivariation norm is defined by $\|\mu\|_\infty = \|\mu\|(\Omega)$. Since $\varphi\mu$ is a scalar measure for every $\varphi \in X^*$ and so we may define a linear mapping $M_\mu : X^* \rightarrow \mathcal{M}(\Sigma)$ by $M_\mu \varphi = \varphi\mu$ with $\|\mu\|_\infty = \|\mu\|(\Omega) = \|M_\mu\|$.

We denote by $\mathcal{M}(\Sigma, X, Y)$ the vector space of operator-valued measures on Σ with values in $B(X, Y)$. The semivariation norm is defined on this space by

$$\|\mu\|_\infty = |\mu|_\infty(\Omega) = \|V_\mu\|.$$

Let μ be a vector measure on Σ with values in a Banach space X. We consider the definition of the p-variation for vector measures as follows

$$|\mu|_p(E) = \sup\left\{\left(\sum_{i=1}^n \|\mu(A_i)\|^p\right)^{1/p} : \{A_1, \ldots, A_n\} \text{ a partition of } E\right\}.$$

The vector measure μ is said to have bounded p-variation (or bounded variation for $p = 1$) if $|\mu|_p(E)$ is finite for every $E \in \Sigma$, or equivalently, if $|\mu|_p(\Omega)$ is finite. Moreover, it is not difficult to show that the set of vector measures that have bounded p-variation is a Banach space with respect to the p-variation norm:

$$\|\mu\|_p = |\mu|_p(\Omega).$$

We now consider the definition of the p-variation for operator-valued measures. Let

$$|\varphi|_p(E) = \sup_{\|x\|\leq 1} |\varphi_x|_p(E).$$

The operator-valued measure φ is said to have bounded p-variation (or bounded variation for $p = 1$) if $|\varphi|_p(E)$ is finite for every $E \in \Sigma$, or equivalently, if $|\varphi|_p(\Omega)$ is finite. By the Uniform Boundedness Principle, the operator-valued measure φ has bounded p-variation if and only if for all $x \in X$, the vector measure φ_x has bounded p-variation.

Example 4.2.1 Let (Ω, Σ, μ) be a measure space with $1 \leq p < \infty$, and $\rho : \Sigma \to B(L_p(\mu))$ be given by multiplication $\rho(E) = \chi_E$ for all $E \in \Sigma$. Then ρ is a projection-valued measure with bounded p-variation and $\|\rho\|_p = 1$.

Example 4.2.2 Recall that a sequence $\{x_n\}_{i=1}^\infty$ in a Hilbert space $\mathcal{H}$ is Bessel if there exists a constant $C > 0$ (a Bessel bound) such that $\sum_n |\langle x, x_n\rangle|^2 \leq C\|x\|^2$ holds for every $x \in \mathcal{H}$. So in this case the operator defined Θ_X by $\Theta_X(x) = (\langle x, x_n\rangle)_n$ is a bounded linear operator from $\mathcal{H}$ to $\ell^2(\mathbb{N})$. A pair of sequences $\{x_n, y_n\}_{i=1}^\infty$ is a dual frame pair for a Hilbert space $\mathcal{H}$ if both of them are Bessel sequences and $I = \sum_n x_n \otimes y_n$, where the convergence is in norm and unconditional. A dual frame pair $\{x_n, y_n\}$ naturally induces an operator-valued measure $\varphi : \varphi(A) = \sum_{n\in A} x_n \otimes y_n$ for any subset A of $\mathbb{N}$. From [124], it can be verified that every dual frame pair induces an operator-valued measure in Hilbert spaces has bounded 2-variation.

Example 4.2.3 Let $\mathcal{H}$ be a Hilbert space. Then, for any $p \geq 1$, there is a $B(\mathcal{H})$-valued measure with bounded q-variation for any $q > p$ but not bounded p-variation. Actually, let $\{e_n\}$ be the unit vector basis of c_0, and define the operator-valued measure on the subsets of $\mathbb{N}$ as follows

$$\varphi(E) = \sum_{n \in E} \frac{1}{n^{1/p}} e_n \otimes e_n, \quad \forall E \subset \mathbb{N}.$$

Proposition 4.5 *Let $\rho : \Sigma \to B(Z)$ be an operator-valued measure with bounded p-variation. If $S : Z \to X$ and $T : X \to Z$ are bounded linear maps, then the compressed operator-valued measure $\varphi = S\rho(\cdot)T : \Sigma \to B(X)$ has bounded p-variation with*

$$|\varphi|_p(E) \leq \|S\| \cdot |\rho|_p(E) \cdot \|T\|$$

for any $E \in \Sigma$.

Next theorem answers the question whether every operator-valued measure with bounded p-variation can be dilated to a projection-valued measure that also has bounded p-variation.

Theorem 4.11 ([124]) *Every operator-valued measure with bounded p-variation has a dilation to a projection-valued measure with bounded p-variation.*

We point out that the above dilation p-norm is completely new and different from the ones constructed in [43, 125]. This can be explained in the next theorem by showing that the minimal framing model of a redundant framing constructed in [125] always contains an isomorphic copy of c_0, and hence the dilation does not have the bounded p-variation for any $p \geq 1$.

Theorem 4.12 ([124]) *Let $\{x_i, f_i\}_{i=1}^{\infty}$ be a framing of a Banach space X satisfying property that $\sum_{i=1}^{\infty} a_i x_i$ converges if and only if $\sum_{i=1}^{\infty} a_i x_i$ converges unconditionally. Let $(E, \{e_i\}_{i=1}^{\infty})$ be the above minimal framing model for $(X, \{x_i, f_i\}_{i=1}^{\infty})$. If $\{x_i, f_i\}$ is not a near-unconditional basis, then $\{e_i\}$ contains a block (unconditional-basic) sequence equivalent to the unit vector basis of c_0. Thus E contains an isomorphic copy of c_0 as a subspace, and the projection-valued measure (minimal dilation) on E induced by $\{e_i\}$ does not have bounded p-variation for any $p \geq 1$.*

Assume that φ is a $B(\mathcal{H})$-valued measure on a measurable space (Ω, Σ, μ) with the property that φ is μ-continuous, where $\mathcal{H}$ is a Hilbert space and μ is a positive measure. We identify each $B \in \Sigma$ with the orthogonal projection $P_B \in L^{\infty}(\Omega, \mu)$ defined by $P_B f = \chi_B f$ for any $f \in L^2(\Omega, \mu)$. Then E extends (uniquely) to an ultraweakly continuous map $\Phi : L^{\infty}(\Omega, \mu) \to B(\mathcal{H})$. We say that E is completely bounded if Φ

is completely bounded (cf. [125]). The following result from [124] shows the connection between complete boundedness and bounded 2-variation.

Corollary 4.1 *Let φ be a $B(\mathcal{H})$-valued measure on a measurable space (Ω, Σ, μ). If it is completely bounded, then it has bounded 2-variation.*

Proof Since the induced map Φ is completely bounded, by Stinespring's dilation theorem there exist a Hilbert space $\mathcal{K}$, bounded linear operators $S : \mathcal{K} \to \mathcal{H}$ and $T : \mathcal{H} \to \mathcal{K}$ and a $*$-homomorphism $\pi : L^\infty(\Omega, \mu) \to B(\mathcal{K})$ such that $\Phi(\cdot) = S\pi(\cdot)T$. Let ρ be the induced operator-valued measure by π. Then it is an orthogonal projection valued measure which has bounded 2-variation. Thus, by Proposition 4.5, we have that φ is 2-variation bounded. $\qquad\square$

Paulsen et al. [110, 171] used the following definition for total variations: Let φ be a Hilbert space operator-valued measure. The total variation is defined to be

$$\tau(\varphi) = \sup\left\{ \left\| \sum |\varphi(A_i)| \right\| : A_i \text{ disjoint partition of } \Omega \right\}$$

where $|T| = (T^*T)^{1/2}$. It is known that φ is completely bounded if $\tau(\varphi) < \infty$ (and hence if $|\varphi|_1(\Omega) < \infty$). Completely boundedness does not imply the finiteness of $\tau(\varphi)$, i.e. there exists an example such that $|\varphi|_1(\Omega) = \infty$ but it completely bounded. Since we have $|\varphi|_2(\Omega) \leq |\varphi|_1(\Omega)$, we ask the following question:

Problem 4.2 Is every Hilbert space operator-valued measure with bounded 2-variation completely bounded?

Next, we use purely atomic operator-valued measures to discuss the connections between the p-summable property and the bounded p-variation property.

A pair of sequences $\{x_n, f_n\}_{n=1}^\infty \subset X \times X^*$ is said to have bounded p-variation if the induced operator-valued measure has bounded p-variation, that is,

$$\sup_{\|x\| \leq 1} \left\{ \left(\sum_j \left\| \sum_{n \in N_j} \langle x, f_n \rangle x_n \right\|^p \right)^{1/p} : N_j \text{'s a partition of } \mathbb{N} \right\} < \infty.$$

It is called p-summable if

$$\sum_{n=1}^\infty \|\langle x, f_n \rangle x_n\|^p = \sum_n (|\langle x, f_n \rangle| \|x_n\|)^p < \infty$$

for all $x \in X$. By the Uniform Boundedness Principle, we have

$$\sup_{\|x\| \leq 1} \sum_{n=1}^{\infty} (|\langle x, f_n \rangle| \|x_n\|)^p < \infty.$$

It is obvious that bounded p-variation $\implies$ p-summability. However, the reverse direction does not hold. For example, for each nonzero $x \in X$ and $f \in X^*$, let the pair of sequences be $x_n = \frac{1}{n}x$ and $f_n = f$. Then it is easy to check that $\{x_n, f_n\}_{n=1}^{\infty}$ is 2-summable but does not have bounded 2-variation. Furthermore, the next example shows that it is possible to construct a sequence with unbounded p-variation, but still p-summable for every $p < 2$.

Example 4.2.4 Let $\{e_n\}_{n=1}^{\infty}$ be an orthonormal basis of a Hilbert space $\mathcal{H}$. Define

$$f_{j,n} = \frac{1}{2^n}e_n, \quad 1 \leq j \leq 4^n, \quad n \in \mathbb{N}.$$

Then $\{f_{j,n}\}_{j,n}$ is a Parseval frame of $\mathcal{H}$.

We first show that it has bounded 2-variation. Set $K = \{(j, n) : 1 \leq j \leq 4^n, n \in \mathbb{N}\}$, then for any partition K_m of K, let $K_m^n = \{j : (j, n) \in K_m\}$ and $\sharp K_m^n$ as the cardinality of the set K_m^n. Then,

$$\sum_m \left(\left\| \sum_{(j,n) \in K_m} \langle x, f_{j,n} \rangle f_{j,n} \right\| \right)^2 = \sum_m \left\| \sum_n \frac{\sharp K_m^n}{4^n} \langle x, e_n \rangle e_n \right\|^2$$

$$= \sum_m \sum_n \frac{(\sharp K_m^n)^2}{16^n} |\langle x, e_n \rangle|^2$$

$$= \sum_n \frac{\sum_m (\sharp K_m^n)^2}{16^n} |\langle x, e_n \rangle|^2$$

$$\leq \sum_n \frac{(\sum_m \sharp K_m^n)^2}{16^n} |\langle x, e_n \rangle|^2$$

$$\leq \sum_n \frac{(4^n)^2}{16^n} |\langle x, e_n \rangle|^2$$

$$\leq \|x\|^2.$$

While for any $x \in \mathcal{H}$ with $\|x\| \leq 1$, we have

$$\sum_{j,n} |\langle x, f_{j,n}\rangle|^p \|f_{j,n}\|^p = \sum_{n=1}^{\infty} \sum_{1 \leq j \leq 4^n} \frac{1}{4^{pn}} |\langle x, e_n\rangle|^p = \sum_{n=1}^{\infty} \frac{1}{4^{(p-1)n}} |\langle x, e_n\rangle|^p$$

$$\leq \max_{n \in \mathbb{N}} |\langle x, e_n\rangle|^p \sum_{n=1}^{\infty} \left(\frac{1}{4^{(p-1)}}\right)^n \leq \frac{1}{4^{(p-1)} - 1}.$$

That is, $\{f_{j,n}\}_{j,n}$ is p-summable for every $p < 2$.

However, taking the following partition, we have

$$\sum_{n=1}^{\infty} \left\| \sum_{1 \leq j \leq 4^n} \langle x, f_{j,n}\rangle f_{j,n} \right\|^p = \sum_{n} \left\| \sum_{1 \leq j \leq 4^n} \langle x, \frac{1}{2^n} e_n\rangle \frac{1}{2^n} e_n \right\|^p = \sum_{n=1}^{\infty} |\langle x, e_n\rangle|^p.$$

As $\ell^p (1 \leq p < 2)$ is a proper subset of ℓ^2, it follows that $\{f_{j,n}\}_{j,n}$ has unbounded p-variation.

4.2.4 Quantum Measures

Quantum measure theory has its origin in the mathematical formalism of quantum mechanics and is often viewed as a "noncommutative" or "quantum" analogue of classical measure theory. In classical measure theory, the basic concept is a measure on a σ-algebra of sets, while in quantum measure theory, the measure acts on the projection lattice of a von Neumann algebra, or equivalently, on a lattice of closed subspaces of a Hilbert space. Intuitively, the transition from a σ-algebra to a projection lattice includes replacing the disjointness of subsets with the orthogonality of subspaces. But from an abstract viewpoint, unlike the σ-algebra of sets, a projection lattice is not necessarily a Boolean algebra. A typical example is the lattice of projections in $B(\mathcal{H})$, which is never distributive (unless $\mathcal{H}$ is one-dimensional). Consequently, it usually requires a substantially different set of techniques in order to establish a non-commutative dilation theory for operator-valued quantum measures.

We remark that every bounded linear map from a von Neumann algebra $\mathcal{M}$ to a Banach space X induces a vector-valued quantum measure on the projection lattice $\mathcal{P}(\mathcal{M})$ of $\mathcal{M}$. However, the converse is not necessarily true since there exist counterexamples of scalar-valued measures that fail to extend to linear functionals when $\mathcal{M}$ is the algebra of 2×2 matrices. A von Neumann algebra $\mathcal{M}$ is said to be σ-finite if every family of nonzero pairwise orthogonal projections in $\mathcal{M}$ is at most countable [178]. To simplify calculations, we will be mainly focused on the σ-finite von Neumann algebras without a Type I_2 direct summand.

The celebrated Gleason theorem [97] asserts that every bounded completely additive measure on the projection lattice $\mathcal{P}(\mathcal{H})$, where $\mathcal{H}$ is a Hilbert space with $\dim \mathcal{H} \geq 3$, extends uniquely to a normal functional on the algebra $B(\mathcal{H})$. Bunce and Wright's generalization to vector measures [40] states that if a von Neumann algebra $\mathcal{M}$ has no summand of a Type I_2, then every bounded finitely additive measure from $\mathcal{P}(\mathcal{M})$ to a Banach space X extends uniquely to a bounded linear operator from $\mathcal{M}$ to X. In particular, every finitely additive $B(X)$-valued measure extends to a bounded linear map. By the dilation theorem [125,126] for bounded linear maps, this linear map has a bounded homomorphism dilation. Therefore, every finitely additive operator-valued quantum measure can be dilated to a finitely additive projection-valued measure. However, ultraweak continuity (resp. countable additivity) often occurs when dealing with linear maps (resp. measures) on von Neumann algebras, it is natural to ask whether there exists a dilation that preserves such a continuity (resp. additivity). In this subsection, we show that this can be achieved when imposing additional assumptions either on X or on the quantum measure.

A (finitely additive) quantum measure μ on a projection lattice $\mathcal{P}(\mathcal{M})$ of a von Neumann algebra $\mathcal{M}$ is a map $\mu : \mathcal{P}(\mathcal{M}) \to \mathbb{C}$ such that

$$\mu(P_1 \vee P_2) = \mu(P_1) + \mu(P_2)$$

whenever $P_1 \perp P_2$, where $P_1 \vee P_2$ denotes the supremum of two projections P_1 and P_2 and $P_1 \perp P_2$ denotes that projections P_1 and P_2 are orthogonal. Similarly, we give the following definition of operator-valued quantum measure.

Definition 4.5 Let $\mathcal{P}(\mathcal{M})$ be the lattice of projections of a von Neumann algebra $\mathcal{M}$ and X be a Banach space. If

$$U : \mathcal{P}(\mathcal{M}) \to B(X)$$

is a map such that

$$U(P_1 \vee P_2) = U(P_1) + U(P_2), \text{ whenever } P_1 \perp P_2,$$

then U is said to be a (finitely additive) $B(X)$-valued measure on $\mathcal{P}(\mathcal{M})$. We say that U is

(1) countably additive (or σ additive) if

$$U\left(\bigvee_{n=1}^{\infty} P_n\right) = \sum_{n=1}^{\infty} U(P_n)$$

for any sequence (P_n) of pairwise orthogonal elements in $\mathcal{P}(\mathcal{M})$ and the series is unconditionally convergent in the weak operator topology (wot) of $B(X)$; completely additive if $U(\bigvee_{\alpha \in I} P_\alpha) = \sum_{\alpha \in I} U(P_\alpha)$ for any system $(P_\alpha)_{\alpha \in I}$ of pairwise orthogonal elements in $\mathcal{P}(\mathcal{M})$ and $U(\bigvee_{\alpha \in I} P_\alpha)$ converges unconditional in the wot of $B(X)$.

(2) bounded if $\sup\{\|U(P)\| : P \in \mathcal{P}(\mathcal{M})\} < \infty$ and the supremum is called the norm of U and denoted as $\|U\|$.

(3) projection-valued if $U(P)$ is a projection (i.e. idempotent) on X for every $P \in \mathcal{P}(\mathcal{M})$.

(4) self-adjoint if X is a Hilbert space $\mathcal{H}$ and $U(P)^* = U(P)$ for all $P \in \mathcal{P}(\mathcal{M})$ and positive if $U(P)$ is positive operators in $\mathcal{M}$ for all $P \in \mathcal{P}(\mathcal{M})$.

Completely additive or countably additive bounded linear maps from $\mathcal{M}$ to $B(X)$ are defined similarly. When dealing with the operator-valued quantum measure U on a σ-finite von Neumann algebra, countable additivity of U is equivalent to being completely additive [178]. Moreover, if U is a countably additive operator-valued quantum measure, then by the Orlicz-Pettis theorem [5], then it is equivalent that U is strong countably additive on X, that is, $\sum_{n=1}^{\infty} U(P_n)x = U(P)x$ for all $x \in X$.

Definition 4.6 Let U be a $B(X)$-valued quantum measure on the projection lattice of a von Neumann algebra $\mathcal{M}$. If there exist a Banach space Y, bounded linear maps S, T and an operator-valued quantum measure $V : \mathcal{P}(\mathcal{M}) \to B(Y)$ such that

$$U(P) = SV(P)T,$$

for every $P \in \mathcal{P}(\mathcal{M})$, then we say that V is a dilation of U and Y is a dilation space. We call V a projection-valued dilation if $V(P)$ is an idempotent on Y for every $P \in \mathcal{P}(\mathcal{M})$. The quadruple (S, V, T, Y) is called a projection-valued quantum measure dilation system of U.

Similar to the commutative domain case introduced in [125], our approach to establish a dilation theory for quantum measures is to introduce an "elementary dilation space." To construct such a dilation space, we will need the Generalized Gleason Theorem [40] stated in the following lemma.

Lemma 4.1 *Let $\mathcal{M}$ be a von Neumann algebra without a Type I_2 direct summand and let X be a Banach space. Then each bounded, finitely additive X-valued measure μ on $\mathcal{P}(\mathcal{M})$ extends uniquely to a bounded linear operator T from $\mathcal{M}$ to X with $\|\mu\| \leq \|T\| \leq 4\|\mu\|$.*

With Lemma 4.1 in hand, we now outline the algebraic structure of our dilation space. Let X be a Banach space and $\mathcal{M}$ be a von Neumann algebra without a Type I_2 direct summand, $\mathcal{P}(\mathcal{M})$ be the projection lattice of $\mathcal{M}$ and $L(\mathcal{M}, X)$ be the linear space of all vector-valued linear maps from $\mathcal{M}$ to X. Let U be a $B(X)$-valued quantum measure on $\mathcal{P}(\mathcal{M})$ and by Lemma 4.1, we use $\bar{U}$ to denote its extension from $\mathcal{M}$ to $B(X)$. Let $B_{\mathcal{M}}$ be the closed unit ball of $\mathcal{M}$, for any $Q \in B_{\mathcal{M}}$ and $x \in X$, we define the linear map on $\mathcal{M}$ by

$$\bar{U}_{Q,x} : \mathcal{M} \to X, \quad \bar{U}_{Q,x}(R) = \bar{U}(RQ)x, \quad \forall R \in \mathcal{M}.$$

Let

$$W_U = \mathrm{span}\{\bar{U}_{Q,x} : Q \in B_{\mathcal{M}}, x \in X\} \subset \mathcal{L}(\mathcal{M}, X).$$

We will refer to W_U as the elementary dilation space induced by U. Now, we define the following fundamental linear maps by

$$S : W_U \to X, S(\Phi) = \Phi(I), \quad \forall \, \Phi \in W_U \tag{4.3}$$

and

$$T : X \to W_U, \quad T(x) = \bar{U}_{I,x}, \quad \forall \, x \in X. \tag{4.4}$$

For any $\sum_{i=1}^{N} C_i \bar{U}_{Q_i, x_i} \in W_U$, we define

$$V : \mathcal{P}(\mathcal{M}) \to L(W_U), \quad V(P)\Big(\sum_{i=1}^{N} C_i \bar{U}_{Q_i, x_i} \Big) = \sum_{i=1}^{N} C_i \bar{U}_{PQ_i, x_i} \tag{4.5}$$

for any $P \in \mathcal{P}(\mathcal{M})$, where $\mathcal{L}(W_U)$ denotes the space of all linear maps on W_U. Then, it can be verified that $V : \mathcal{P}(\mathcal{M}) \to L(W_U)$ is finitely additive and idempotent valued. Obviously, we have

$$U(P) = SV(P)T, \quad \forall P \in \mathcal{P}(\mathcal{M}).$$

Observe that V is only an algebraic dilation of U. So, we need to impose an appropriate norm on the elementary dilation space. Thus, we introduce the following definition:

Definition 4.7 Let W_U be the elementary dilation space induced by U. Assume that $\| \cdot \|_V$ is a norm on W_U, and denotes its completion by $\widetilde{W}_U$. Then the norm $\| \cdot \|_V$ on W_U is called a dilation norm of U if the above linear maps S, T and $V(P)$ for any $P \in \mathcal{P}(\mathcal{M})$ are all bounded.

Given an operator-valued quantum measure, there could be many projection-valued dilations. The following result establishes the connections between any hypothetical existing dilation space and the elementary dilation space: Any given projection-valued dilation space Y can induce a dilation norm on the elementary dilation space W_U so that U can be dilated to a projection-valued quantum measure from $\mathcal{P}(M)$ to $B(W_U)$, which extends from the algebraic dilation and generalizes the minimal dilation for operator-valued measures over commutative domains.

Theorem 4.13 ([120]) *Let X be a Banach space and M be a von Neumann algebra without a Type I_2 direct summand. Let U be a $B(X)$-valued quantum measure on $\mathcal{P}(M)$. Suppose that (A, W, B, Y) is a projection-valued quantum measure dilation system. Then there exists an induced dilation norm $\|\cdot\|_{\mathcal{D}}$ on W_U (denote its completion by $\widetilde{W}_{U,\mathcal{D}}$) such that the linear maps $S, V(P), T$ defined in Eq. (4.3),(4.4), (4.5) respectively, can be extended to bounded linear maps and still denoted by $S, V(P), T$. Then $(S, \widetilde{W}_{U,\mathcal{D}}, V, T)$ is a projection-valued quantum measure dilation system of U. Furthermore, there exists a linear contraction*

$$C : \widetilde{\mathcal{M}}_{U,\mathcal{D}} \to Y/\ker A$$

such that $S = AC, CT = qW(1)B$, where q is the quotient map form Y to $Y/\ker A$ by $q(x) = [x]$.

Similar to the dilation of operator-valued measure on a measure space [125], define

$$\left\| \sum_{i=1}^{N} C_i \bar{U}_{Q_i,x_i} \right\| = \sup_{R \in B_M} \left\| \sum_{i=1}^{N} C_i \bar{U}\left(R Q_i \right) x_i \right\|_X$$

on W_U. It is routine to check that this norm is a dilation norm and let $\widetilde{W}_U$ be its completion. To obtain projection-valued measure dilations that preserve countable additivity, we impose additional assumptions either on X or on the quantum measure. In particular, unlike the traditional treatment of disjoint partitions of sets to define the variation or more generally p-variation of operator-valued measure, in [124], we use the orthogonal representation of finite tree in [120] to define the p-variation of quantum measure which is consistent with the definition for the commutative domain case. The following is one of the main results obtained in our effort of building a noncommutative dilation theory.

Theorem 4.14 *Let $\mathcal{M}$ be a σ-finite von Neumann algebra without a Type I_2 direct summand and $U : \mathcal{P}(\mathcal{M}) \to B(X)$ be a countably additive quantum measure, where X is a Banach space. Suppose that one of the following conditions is satisfied:*

(i) $X = \ell^p (1 < p < 2)$ or has the Schur property,
(ii) U has bounded p-variation for some $1 \le p < \infty$.

Then there exists a Banach space Z, bounded linear operators $S : Z \to X$ and $T : X \to Z$, and a countably additive projection-valued quantum measure $V : \mathcal{P}(\mathcal{M}) \to B(Z)$ such that

$$U(P) = SV(P)T$$

for every $P \in \mathcal{P}(\mathcal{M})$.

4.3   Homomorphism Dilation for Bounded Linear Operators

In this section, we present our homomorphism dilation theory for Banach space operator-valued bounded (or other types of continuous) linear maps acting on Banach algebras which can be considered as a natural extension of Stinespring's dilation theorem.

4.3.1   Banach Algebra Case

Similar to the minimal dilation for operator-valued measure, we first prove a dilation result for bounded linear maps between Banach algebras.

Theorem 4.15 ([125]) *Let $\mathcal{A}$ be a Banach algebra, and let $\phi : \mathcal{A} \to B(X)$ be a bounded linear operator, where X is a Banach space. Then there exists a Banach space Z, a bounded linear unital homomorphism $\pi : \mathcal{A} \to B(Z)$, and bounded linear operators $T : X \to Z$ and $S : Z \to X$ such that*

$$\phi(a) = S\pi(a)T$$

for all $a \in \mathcal{A}$.

Proof Consider the algebraic tensor product space $\mathcal{A} \otimes \mathcal{H}$, and define a $B(\mathcal{A}, \mathcal{H})$ operator tensor norm with respect to ϕ as follows: for any $a \in \mathcal{A}$ and $x \in \mathcal{H}$, identity $a \otimes x$ with the map $\mathcal{A} \to \mathcal{H}$ defined by

$$(a \otimes x)(b) = \phi(ba)x.$$

Then $a \otimes x$ is a bounded linear operator from $\mathcal{A}$ to $\mathcal{H}$ with $\|a \otimes x\|_{B(\mathcal{A}, \mathcal{H})} \leq \|\phi\| \|a\| \|x\|$. So this defines a quasi-norm on $\mathcal{A} \otimes \mathcal{H}$. Let

$$\mathcal{N} = \left\{ \sum_i c_i a_i \otimes x_i : \left\| \sum_i c_i a_i \otimes x_i \right\|_{B(\mathcal{A}, \mathcal{H})} = 0 \right\}$$

and $\mathcal{A} \widehat{\otimes}_\phi \mathcal{H}/\mathcal{N}$ be the completion of the norm space $\mathcal{A} \otimes \mathcal{H}/\mathcal{N}$.

For any $a \in \mathcal{A}$, let us define $\pi(a) : \mathcal{A} \otimes \mathcal{H}/\mathcal{N} \to \mathcal{A} \otimes \mathcal{H}/\mathcal{N}$ by

$$\pi(a) \left(\sum_i a_i \otimes x_i \right) = \sum_i (a a_i) \otimes x_i.$$

Assume that $f = \sum_{i=1}^n a_i \otimes x_i = \sum_{j=1}^m b_j \otimes y_i$. Then we have

$$\pi(a)\left(\sum_i a_i \otimes x_i\right)(b) = \left(\sum_i (aa_i) \otimes x_i\right)(b) = \sum_i \phi(baa_i)x_i$$

$$= \left(\sum_i a_i \otimes x_i\right)(ba) = f(ba)$$

and

$$\pi(a)\left(\sum_i b_j \otimes_j\right)(b) = \left(\sum_i (ab_j) \otimes_y y_j\right)(b) = \sum_i \phi(bab_j)y_j$$

$$= \left(\sum_i b_j \otimes y_j\right)(ba) = f(ba).$$

Therefore, $\pi(a)$ is well-defined.

The boundedness $\pi(a)($ with $\|\pi(a)\| \le \|a\|)$ follows from

$$\|\pi(a)(f)\| = \left\|\pi(a)\left(\sum_i a_i \otimes x_i\right)\right\| = \left\|\sum_i (aa_i) \otimes x_i\right\|$$

$$= \sup_{b \in B_A}\left\|\left(\sum_i (aa_i) \otimes x_i\right)(b)\right\| = \sup_{b \in B_A}\left\|\sum_i \phi(baa_i)x_i\right\|$$

$$= \|a\| \sup_{b \in B_A}\left\|\sum_i \phi\left(b\frac{a}{\|a\|}a_i\right)x_i\right\| = \|a\| \sup_{b \in B_A}\left\|\left(\sum_i a_i \otimes x_i\right)(b)\right\|$$

$$= \|a\|\left\|\sum_i a_i \otimes x_i\right\| = \|a\| \cdot \|f\|.$$

Extend $\pi(a)$ to a bounded linear operator on $A \widehat{\otimes}_\phi \mathcal{H}/\mathcal{N}$ which we still denote it by $\pi(a)$. Thus, $\pi : A \to A \widehat{\otimes}_\phi \mathcal{H}/\mathcal{N}$ is a bounded linear operator. Moreover,

$$\pi(ab)f = \pi(ab)\left(\sum_i a_i \otimes x_i\right) = \sum_i (aba_i) \otimes x_i$$

$$= \pi(a)\left(\sum_i (ba_i) \otimes x_i\right) = \pi(a)\pi(b)\left(\sum_i a_i \otimes x_i\right) = \pi(a)\pi(b)f.$$

Hence

$$\pi(ab) = \pi(a)\pi(b)$$

and therefore π is a homomorphism.

Define $T : \mathcal{H} \to A \widetilde{\otimes}_\phi \mathcal{H}/\mathcal{N}$ by $T(x) = I \otimes x$. Since

$$\|T(x)\| = \sup_{a \in B_A} \|(I \otimes x)(a)\| = \sup_{a \in B_A} |\phi(a)x\| \leq \sup_{a \in B_A} \|a\| \|\phi\| \|x\| = \|\phi\| \|x\|,$$

we have that T is a bounded linear operator with $\|T\| \leq \|\phi\|$.

Define $S : A \otimes \mathcal{H}/\mathcal{N} \to \mathcal{H}$ by $S(a \otimes x) = E(a)x$ and linearly extend S to $A \otimes \mathcal{H}/\mathcal{N}$. It is easy to check that S is well-defined. Since

$$\|S(a \otimes x)\| = \|\phi(a)x\| \leq \sup_{b \in B_A} \|\phi(ba)x\| = \sup_{b \in B_A} \|(a \otimes x)(b)\| = \|a \otimes x\|,$$

we have that S is a bounded linear operator with $\|S\| \leq 1$. Extend S to a bounded linear operator from $A \widetilde{\otimes}_\phi \mathcal{H}/\mathcal{N}$ to $\mathcal{H}$, which we still denote it by S.

Finally, for any $x \in \mathcal{H}$ we have

$$S\pi(a)T(x) = S\rho(a)(I \otimes x) = S(a \otimes x) = \phi(a)x.$$

Thus $\phi(a) = S\pi(a)T$.   □

When A is a C^*-algebra and $X = \mathcal{H}$ is a Hilbert space, it is well-known that a bounded linear map $\phi : A \to B(\mathcal{H})$ can be dilated to a $*$-homomorphism on an enlarge Hilbert space $\mathcal{K}$ if and only if Φ is completely bounded (cf. [171, 176]). However, we don't know what would happen if we drop the $*$-homomorphism. This leads to the following problem.

Problem 4.3 Is there a bounded linear map on a C^*-algebra that has a bounded homomorphism dilation on a Hilbert space but does not admit a $*$-homomorphism dilation? Or, more generally, is there a classification theory for bounded linear maps that is based on the dilation space Z?

Example 4.3.1 Let $A = B(\mathcal{H})$ and ϕ be the transpose map with respect to an orthonormal basis of $\mathcal{H}$. Then we know by Theorem 4.15 that ϕ can be dilated to a bounded homomorphism $\Phi \cdot B(\mathcal{H}) \to B(Z)$ for some Banach space Z. However, Z can never be a Hilbert space. Since if Z can be taken as a Hilbert space, then Φ will be completely bounded [111], and consequently, ϕ will also be completely bounded, which is false since the transpose map is not completely bounded when $\dim \mathcal{H} = \infty$.

It is worth pointing out that if we replace $A = B(\mathcal{H})$ in the above example by the direct sum of the $n \times n$ matrix algebra $M_n(\mathbb{C})$ ($n = 1, 2, \ldots$), it is an open problem whether the dilation space Z can be taken as a Hilbert space. In fact, a positive answer would yield a counterexample to the similarity problem of Kadison [147]. In 1955, Kadison asked the following problem: Is any bounded homomorphism π from a C^*-algebra A to $B(\mathcal{H})$ is similar to a $*$-homomorphism. i.e., is there an invertible operator $T \in B(\mathcal{H})$ such that

$a \mapsto T\pi(a)T^{-1}$ is a $*$-homomorphism π from $\mathcal{A}$ to $B(\mathcal{H})$? This Similarity problem is equivalent to several long-standing problems, including the derivation problem and hyper-reflexivity problem in the theory of operator algebras and group representations [176]. It is well-known that a bounded homomorphism is similar to a $*$-homomorphism if and only if it is completely bounded [111], then this problem boils down to the implication: Does every bounded homomorphism π of a C^*-algebra $\mathcal{A}$ into $B(\mathcal{H})$ is completely bounded? While this problem remains open, it is true for some C^*-algebras: nuclear C^*-algebra [38, 49], or $\mathcal{A} = B(\mathcal{H})$, or more general, like C^*-algebras with no tracial states, commutative C^*-algebras, II_1-factor with Murry and von Neumann's property Γ, or π is cyclic [111]. So if $\mathcal{A}$ belongs to any of the above-mentioned classes and $\varphi : \mathcal{A} \to B(\mathcal{H})$ is a bounded but not completely bounded linear map, then its dilation space Z in Theorem 4.15 can never be a Hilbert space. On the other hand, if there is a non completely bounded map ϕ from a C^*-algebra to $B(\mathcal{H})$ that has a Hilbertian dilation: $\pi : \mathcal{A} \to B(\mathcal{K})$ (i.e., where the dilation space $\mathcal{K}$ is a Hilbert space), then it would be a counterexample to the Kadison's similarity problem. So we ask the following:

Problem 4.4 Is there a non-cb map that admits a Hilbert space dilation to a bounded homomorphism?

4.3.2   Commutative von Neumann Algebra Case

Since normality (i.e., ultraweakly or w^*-continuity) is often required when dealing with linear maps between von Neumann algebras, it is natural to ask whether the normality of the linear map can also be preserved via dilation, such a dilation will be called a *normal dilation*. In this subsection, we show that the answer to the above question is affirmative for normal linear maps acting on commutative von Neumann algebras.

4.3.2.1 Purely Atomic Case

This is the case where we can connect the normal dilation to the dilation of framings. The dilation of Hilbert space framings in the Banach space setting has been established in [52]. With the help of this result, we will show that for every ultraweakly continuous map ϕ from a purely atomic abelian von Neumann algebra $\mathcal{A}$ into $B(\mathcal{H})$, we can find a Banach space Z, an ultraweakly continuous unital homomorphism π from $\mathcal{A}$ into $B(Z)$, and bounded linear operators T and S such that for all $a \in \mathcal{A}$,

$$\phi(a) = S\pi(a)T.$$

We remark that this result differs from Stinespring's dilation as ϕ is not necessarily completely bounded and consequently, the dilation space is not necessarily a Hilbert space.

We already know that every frame in a Hilbert space or a framing in a Banach space induces an operator-valued measure: Let $\{x_n\}_{n=1}^\infty$ be a frame for a separable Hilbert space

$\mathcal{H}$. Let Σ be the σ-algebra of all subsets of $\mathbb{N}$. Define the map

$$E : \Sigma \to B(\mathcal{H}), \quad E(B) = \sum_{n \in B} x_n \otimes x_n, \quad B \in \Sigma,$$

where $x \otimes y$ is the mapping on $\mathcal{H}$ defined by $(x \otimes y)(u) = \langle u, y \rangle x$. Then E is a regular, positive $B(\mathcal{H})$-valued measure. Similarly, let $\{x_n, y_n\}_{n=1}^{\infty}$ be a nonzero framing for a separable Banach space X and define the map

$$E : \Sigma \to B(X), \quad E(B) = \sum_{n \in B} x_n \otimes y_n$$

for all $B \in \Sigma$. Then E is a $B(X)$-valued measure.

Now we focus on bounded linear maps on an abelian von Neumann algebra. We first prove that any framing $\{x_n, y_n\}_{n \in \Lambda}$ for a Hilbert space $\mathcal{H}$ induces a normal map from $\ell^{\infty}(\Lambda)$ to $B(\mathcal{H})$.

Proposition 4.6 ([125]) *Let $\mathcal{H}$ be a separable Hilbert space and let $\{x_i, y_i\}_{i \in \Lambda}$ be a framing of $\mathcal{H}$. Then the map ϕ from $\ell^{\infty}(\Lambda)$ into $B(\mathcal{H})$ defined by*

$$\phi : \ell^{\infty}(\Lambda) \to B(\mathcal{H}), \quad (a_i) \to \overset{sot}{\sum_{i \in \Lambda}} a_i x_i \otimes y_i$$

is well-defined, unital, linear and ultraweakly continuous.

Proof Since $\{x_i, y_i\}_{i=1}^{\infty}$ is a framing of $\mathcal{H}$, for any $x \in \mathcal{H}$, then

$$x = \sum_{i=1}^{\infty} \langle x, y_i \rangle x_i$$

converges unconditionally. For any $(a_i) \in c_{00}$, define a bounded operator $U_{(a_i)}$ by

$$U_{(a_i)} : \mathcal{H} \to \mathcal{H}, \quad U_{(a_i)}(x) = \sum a_i \langle x, y_i \rangle x_i.$$

Let B_X be the unital ball of the Banach space X. For any $x \in \mathcal{H}$, since $\sum_i \langle x, y_i \rangle x_i$ converges unconditionally, by [66, Theorem 1.9], we know that

$$\sup_{(b_i) \in B_{\ell^{\infty}}} \left\| \sum b_i \langle x, y_i \rangle x_i \right\| < +\infty.$$

Thus,

$$\sup_{(a_i)\in B_{c_{00}}} \|U_{(a_i)}(x)\| \leq \sup_{(b_i)\in B_{\ell^\infty}} \left\| \sum b_i \langle x, y_i \rangle x_i \right\| < +\infty.$$

Then by the Uniform Boundedness Principle,

$$\sup_{(a_i)\in B_{c_{00}}} \|U_{(a_i)}\| < +\infty.$$

It follows that

$$K_u = \sup_{x\in B_{\mathcal{H}}} \sup_{(\sigma_i)\subset\mathbb{D}} \left\| \sum \sigma_i \langle x, y_i \rangle x_i \right\| < +\infty.$$

Thus, for all $(a_i) \in \ell_\infty$ and $x \in \mathcal{H}$,

$$\left\| \sum a_i \langle x, y_i \rangle x_i \right\| \leq K_u \|a_i\|_{\ell^\infty} \|x\|.$$

Hence, F is well-defined, unital, linear and bounded with

$$\|F\|_{B(\ell^\infty, B(\mathcal{H}))} \leq K_u.$$

Now, we prove that F is ultraweakly continuous. If there is a net (a_i^λ) converging to 0 in the ultraweakly topology, then for any $(\gamma_i) \in \ell_1$, $\sum a_i^\lambda \gamma_i \to 0$. Let T belong to the trace class $\mathcal{S}_1(\mathcal{H})$. By the polar decomposition, $T = U|T|$ where U is a partial isometry. Moreover, recall that $\mathcal{S}_1(\mathcal{H})$ is the subset of the compact operators $K(\mathcal{H})$, $|T|$ is a self-adjoint compact operator. Thus, there is an orthonormal basis $\{e_i\}$ and a sequence $\lambda_i \geq 0$ so that

$$|T| = \sum \lambda_i e_i \otimes e_i$$

with $\|T\|_{\mathcal{S}_1} = \mathrm{tr}(|T|) = \sum_i \lambda_i < \infty$. Then for all $(a_i) \in \ell_\infty$, $(\gamma_j) \in \ell_1$ and $\{u_j\}, \{v_j\} \subset B_{\mathcal{H}}$, we have

$$\sum_i \sum_j |a_i \gamma_j \langle u_j, y_i \rangle \langle x_i, v_j \rangle| \leq \|(a_i)\|_\infty \sum_j |\gamma_j| \sum_i |\langle u_j, y_i \rangle \langle x_i, v_j \rangle|$$

$$= \|(a_i)\|_\infty \sum_j |\gamma_j| \sum_i \theta_{i,j} \langle u_j, y_i \rangle \langle x_i, v_j \rangle$$

$$= \|(a_i)\|_\infty \sum_j |\gamma_j| \left\langle \sum_i \theta_{i,j} \langle u_j, y_i \rangle x_i, v_j \right\rangle$$

$$\leq \|(a_i)\|_\infty \sum_j |\gamma_j| \sup_j \left\| \sum_i \theta_{i,j} \langle u_j, y_i \rangle x_i \right\|$$

$$\leq \|(a_i)\|_\infty \sum_j |\gamma_j| K_u < \infty,$$

where $\overline{\theta_{i,j}} = \mathrm{sgn}\{\langle u_j, y_i\rangle \langle x_i, v_j\rangle\}$. So we have

$$\sum_i \sum_j |a_i^\lambda \lambda_j \langle U e_j, y_i\rangle \langle x_i, e_j\rangle| < \infty$$

and

$$\sum_i \left|\sum_j \lambda_j \langle U e_j, y_i\rangle \langle x_i, e_j\rangle\right| < \infty.$$

Therefore,

$$\mathrm{tr}(\phi(a_i^\lambda)T) = \sum_j \langle \phi(a_i^\lambda) T e_j, e_j\rangle = \sum_j \left\langle \phi(a_i^\lambda) U\left(\sum_k \lambda_k e_k \otimes e_k\right) e_j, e_j\right\rangle$$

$$= \sum_j \langle \phi(a_i^\lambda) U \lambda_j e_j, e_j\rangle = \sum_j \lambda_j \left\langle \sum_i a_i^\lambda x_i \otimes y_i U e_j, e_j\right\rangle$$

$$= \sum_j \lambda_j \sum_i a_i^\lambda \langle U e_j, y_i\rangle \langle x_i, e_j\rangle = \sum_i a_i^\lambda \sum_j \lambda_j \langle U e_j, y_i\rangle \langle x_i, e_j\rangle,$$

which converges to 0, as claimed.   □

For a Banach space X, the weak or strong-operator topology (*wot* or *sot*) of $B(X)$ can be defined similarly as the *wot* or *sot* of $B(\mathcal{H})$ for a Hilbert space $\mathcal{H}$. While the ultraweak (or equivalently, weak*) topology of $B(\mathcal{H})$ is well-understood, we define the ultraweak topology of $B(X)$ through the natural embedding $B(X) \hookrightarrow B(X, X^{**})$, which can be identified with the dual space of $X \widehat{\otimes}_\pi X^*$, namely, $B(X, X^{**}) = (X \widehat{\otimes}_\pi X^*)^*$. While the term $X \widehat{\otimes}_\pi Y$ is the projective tensor product of X and Y [188], which is defined by endowing the tensor product $X \otimes Y$ with the projective norm $\|\cdot\|_\wedge$ by

$$\|u\|_\wedge = \inf\left\{\sum_{i=1}^n \|x_i\| \|y_i\| : u = \sum_{i=1}^n x_i \otimes y_i\right\}.$$

Viewing $X \subset X^{**}$, we define the ultraweak topology on $B(X)$ to be the weak* topology induced by the predual $X \widehat{\otimes}_\pi X^*$. Then the following lemma from [125] generalizes Proposition 4.6.

Lemma 4.2 *Let X be a Banach space and let $E : 2^{\mathbb{N}} \to B(X)$ be an operator-valued measure on $(\mathbb{N}, 2^{\mathbb{N}})$. Denote $E(\{i\})$ by E_i for all $i \in \mathbb{N}$. Then the map ϕ from ℓ^∞ into*

$B(X)$ defined by

$$\phi : \ell^\infty \to B(X), \quad (a_i) \mapsto \overset{sot}{\sum} a_i E_i$$

is well-defined, linear and ultraweakly continuous.

Proposition 4.7 ([125]) *Let $\mathcal{H}$ be a separable Hilbert space and $\phi : \ell^\infty(\mathbb{N}) \to B(\mathcal{H})$ such that $\phi(1) = I$ (i.e., unital) and $\phi(e_n)$ is at most rank one for all $n \in \mathbb{N}$, where $e_n = \chi_{\{n\}}$ and 1 is the function 1 in ℓ^∞. Then the following are equivalent:*

  (i) ϕ is ultraweakly continuous,
  (ii) the induced measure E defined by

$$E(B) = \phi\left(\sum_{n \in B} e_n\right), \quad \forall\, B \subseteq \mathbb{N}$$

  is an operator-valued measure,
(iii) ϕ is induced by a framing (x_n, y_n) for $\mathcal{H}$, i.e,

$$\phi\left(\sum_{n \in \mathbb{N}} a_n e_n\right) = \overset{sot}{\sum} a_n x_n \otimes y_n.$$

Thus, for a countable index set Λ, there is a 1-1 correspondence between the following triple: the set of (discrete) framings on a Hilbert space $\mathcal{H}$ indexed by Λ, the set of ultraweakly continuous unital linear maps from $\ell^\infty(\Lambda)$ into $B(\mathcal{H})$ and the set of purely atomic probability operator-valued measures on the σ-algebra of all subsets of Λ with rank-1 atoms in $B(\mathcal{H})$. Thus, there is a consistent dilation theory for the discrete framing, the operator-valued measure, and the ultraweakly continuous unital linear map, all have the same dilation space and the dilation procedure commutes with the correspondences in the triple. Therefore, by the dilation of framing in [52] we have the following corollary.

Corollary 4.2 *Let $\mathcal{H}$ be a separable Hilbert space and $\phi : \ell^\infty \to B(\mathcal{H})$ such that $\phi(1) = I$ and $\phi(e_n)$ is at most rank one for all $n \in \mathbb{N}$. Then there exist a separable Banach space Z, an ultraweakly continuous unital homomorphism $\pi : \ell^\infty \to B(Z)$, and bounded linear operators $T : \mathcal{H} \to Z$ and $S : Z \to \mathcal{H}$ such that*

$$\phi(a) = S\pi(a)T$$

for all $a \in \ell^\infty$, and $\pi(e_n)$ is rank one for all $n \in \mathbb{N}$.

For arbitrary ultraweakly continuous operator-valued map on ℓ^∞, Theorem 4.4 implies the following:

Theorem 4.16 ([125]) *Let $\phi : \ell^\infty \to B(\mathcal{H})$ be an ultraweakly continuous linear map. Then there exists a Banach space Z, an ultraweakly continuous unital homomorphism $\pi : \ell^\infty \to B(Z)$, and bounded linear operators $T : \mathcal{H} \to Z$ and $S : Z \to \mathcal{H}$ such that*

$$\phi(a) = S\pi(a)T$$

for all $a \in \ell^\infty$.

As every separably acting purely atomic abelian von Neumann algebra is equivalent to some ℓ^∞ via an ultraweakly continuous $*$-isomorphism, we immediately get the following result.

Theorem 4.17 ([125]) *If $\mathcal{A}$ is a purely atomic abelian von Neumann algebra acting on a separable Hilbert space, then for every ultraweakly continuous linear map $\phi : \mathcal{A} \to B(\mathcal{H})$, there exists a Banach space Z, an ultraweakly continuous unital homomorphism $\pi : \mathcal{A} \to B(Z)$, and bounded linear operators $T : \mathcal{H} \to Z$ and $S : Z \to \mathcal{H}$ such that*

$$\phi(a) = S\pi(a)T$$

for all $a \in \mathcal{A}$.

4.3.2.2 The General Abelian Case

Recall that if $\mathcal{M}$ is an abelian von Neumann algebra, then $\mathcal{M}$ is isomorphic to $L^\infty(\Omega, \Sigma, \mu)$ (the space of essentially bounded μ-measurable functions over Ω where Ω is a local compact space with a positive Radon measure μ) [178]. So in what follows, we may always assume $\mathcal{M} = L^\infty(\Omega, \mu)$. Similar to the purely atomic case, we can also associate an operator-valued measure on Ω with a linear map on the abelian von Neumann algebra $\mathcal{M}$.

Let $F : \Sigma \to B(X)$ be an operator-valued measure, where (Ω, Σ) is a measurable space, and X is a Banach space. For $x \in X$ and $x^* \in X^*$, define the vector measure $F_x : \Sigma \to X$ and the complex measure $F_{x,x^*} : \Sigma \to \mathbb{C}$ respectively by

$$F_x(B) = F(B)x, \quad \text{and} \quad F_{x,x^*}(B) = x^*(F(B)x)$$

for all $B \in \Sigma$.

We turn to the variation $|F_{x,x^*}|$ of the complex measure F_{x,x^*}. For each $B \in \Sigma$, let $|F_{x,x^*}|(B)$ be the supremum of the numbers $\sum_{j=1}^n |F_{x,x^*}(B_j)|$, where $\{B_j\}_{j=1}^n$ ranges over all finite partitions of B into Σ-measurable sets. The semivariation of F_x is the nonnegative function $\|F_x\|$ whose value on a set $B \in \Sigma$ is given by

$$\|F_x\|(B) = \sup\{|F_{x,x^*}|(B) : x^* \in X^*, \|x^*\| \leq 1\},$$

which is finite by the Uniform Boundedness Principle. Then for every $B \in \Sigma$,

$$\|F_x\|(B) = \sup \left\| \sum_n \varepsilon_n F(B_n)x \right\|,$$

where the supremum is taken over all partitions B_n's of B into finitely many disjoint members of Σ, and all finite collections ε_n's satisfying $|\varepsilon_n| \leq 1$. The semivariation of F is the nonnegative function $\|F\|$ whose value on a set $B \in \Sigma$ is given by

$$\|F\|(B) = \sup \left\{ |F_{x,x^*}|(B) : \|x\| \leq 1, \|x^*\| \leq 1 \right\},$$

which again is finite by the Uniform Boundedness Principle. This implies that

$$\|F\|(B) = \sup \left\| \sum_n \varepsilon_n F(B_n) \right\|,$$

where the supremum is taken over all partitions B_n's of B into finitely many disjoint members of Σ, and all finite collections ε_n's satisfying $|\varepsilon_n| \leq 1$.

The following shows that any operator-valued measure ϕ is associated with a positive scalar-valued measure μ. This allows us to identify the operator-valued measure with a countably additive map on $L^\infty(\mu)$.

Proposition 4.8 *Let X be a separable Banach space, and $\varphi : \Sigma \to B(X)$ be an OVM. Then there exists a nonnegative real-valued measure μ on Σ such that φ vanishes on sets of μ-measure zero. Moreover, we have $0 \leq \mu(E) \leq \|\varphi\|(E)$ for all $E \in \Sigma$.*

According to [71, Lemma 2.2], which states that if $F : \Sigma \to X$ is a countably additive vector measure and μ is a nonnegative real-valued measure on Σ. Then F vanishes on sets of μ-measure zero if and only if F is μ-continuous, i.e.,

$$\lim_{\mu(E) \to 0} F(E) = 0.$$

Thus, by Definition 4.1 of operator-valued measure or see [125], we have the following lemma.

Lemma 4.3 ([124]) *Let $F : \Sigma \to B(X)$ be a countably additive operator-valued measure and μ be a nonnegative real-valued measure on Σ. Then the following are equivalent:*

(a) F is μ-continuous, that is, F vanishes on sets of μ-measure zero.

(b) $sot\text{-}\lim_{\mu(E)\to 0} F(E) = 0$.

(c) $wot\text{-}\lim_{\mu(E)\to 0} F(E) = 0$.

As discussed before, we use the term *normal* to denote an ultraweakly continuous linear map and define the ultraweak topology on $B(X)$ to be the weak* topology of $B(X, X^{**})$ induced by the predual $X \widehat{\otimes}_\pi X^*$. Let J be the canonical embedding from $B(X)$ into $B(X, X^{**}) = (X \widehat{\otimes}_\pi X^*)^*$.

Proposition 4.9 ([124]) *Let X be a Banach space and (Ω, Σ, μ) be a σ-finite measure space. If $T : L^\infty(\mu) \to B(X)$ is a linear map, then T is ultraweak-wot continuous if and only if JT is normal.*

Let (Ω, Σ) be a measurable space and $F : \Sigma \to B(X)$ be a μ-continuous OVM, that is, $\lim_{\mu(E)\to 0} F(E) = 0$, (see Proposition 4.8 for the existence of μ). If f is a scalar-valued simple function on Ω, say $f = \sum_{i=1}^n \alpha_i \chi_{E_i}$ where α_i are nonzero scalars and E_i are pairwise disjoint members of Σ, define

$$T_F(f) = \sum_{i=1}^n \alpha_i F(E_i).$$

Then T_F is a linear map from the space of simple functions of the above form into $B(X)$. Moreover, we have

$$\|T_F(f)\| \le \|F\|(\Omega)\|f\|.$$

Since the simple functions are dense in $L^\infty(\mu)$, T_F can be uniquely extended to the entire space $L^\infty(\mu)$, still denoted by T_F. It is easy to show that $\|T_F\| = \|F\|(\Omega)$. For each $f \in L^\infty(\mu)$, we use the notation $\int f \, dF$ for $T_F(f)$.

If $Q \in B(X, Y)$ and $R \in B(Y, X)$, then $QF(\cdot)R$ is a μ-continuous OVM from Σ to $B(Y)$, denoted by E. Then we have

$$\int f \, dE = T_E(f) = QT_F(f)R = Q\left(\int f \, dF\right)R$$

for all $f \in L^\infty(\mu)$. In particular, if $x \in X$ and $x^* \in X^*$, then the following useful equality holds

$$x^*\left(\left(\int f \, dF\right)x\right) = \int f \, dF_{x,x^*}$$

Indeed, for simple functions f the above equalities are trivial and the density of simple functions in $L^\infty(\mu)$ proves the identity for all $f \in L^\infty(\mu)$.

Conversely, any bounded linear functional T on $L^\infty(\mu)$ induces a measure by $E(B) = T(\chi_B)$. Moreover, the following lemma from [71] will be useful.

Lemma 4.4 ([124]) *Let μ be a nonnegative measure on Σ, and X be a Banach space. If $T : L^\infty(\mu) \to X$ is a bounded linear map, then the following are equivalent:*

(a) T is ultraweak-weakly continuous.
(b) The representing measure of T is countably additive.
(c) The representing measure of T is μ-continuous.

Thus, we have established the relation between operator-valued measures and linear maps on $L^\infty(\mu)$ via the following theorem.

Theorem 4.18 ([124]) *Let X be a Banach space and (Ω, Σ, μ) be a finite non-negative measure. Then there is a one-to-one linear correspondence between the space of an ultraweak-wot continuous linear map (resp. homomorphism) T from $L^\infty(\mu)$ to $B(X)$ and the space of a μ-continuous operator-valued measure (resp. projection-valued measure) F from Σ to $B(X)$ defined by*

$$F \leftrightarrow T_F \quad \text{if} \quad T_F f = \int f \, dF$$

for all $f \in L^\infty(\mu)$. Moreover, $\|T_F\| = \|F\|(\Omega)$.

Combining the above theorem with Theorem 4.4, we get the following theorem which settles the problem of normal dilation for any abelian von Neumann algebra.

Theorem 4.19 ([124]) *Let (Ω, Σ, μ) be a finite measure space. If X is a Banach space, then for every ultraweak-wot continuous linear map $\phi : L^\infty(\mu) \to B(X)$, there exist a Banach space Z, a unital ultraweak-wot continuous linear homomorphism $\pi : L^\infty(\mu) \to B(Z)$, and bounded linear maps $T : X \to Z$ and $S : Z \to X$ with*

$$\|\pi\| = \|S\| = 1 \quad \text{and} \quad \|T\| = \|\phi\|$$

such that

$$\phi(f) = S\pi(f)T$$

for all $f \in L^\infty(\mu)$. Moreover, if $L^1(\mu)$ and X are separable, then the dilated space Z can be separable.

4.3.3 Noncommutative von Neumann Algebra Case

Recall from [202] that if ϕ is a bounded linear functional on a von Neumann algebra $\mathcal{M}$, then ϕ is normal if and only if ϕ is completely additive. Then the following lemma generalizes the Proposition 4.9 and shows the connection between the completely additivity on the projection lattices and the normality of linear maps.

Lemma 4.5 ([120]) *Let $\varphi : \mathcal{M} \to B(X)$ be a bounded linear map from a von Neumann algebra $\mathcal{M}$ to $B(X)$. Then the following are equivalent:*

1. *φ is completely additive,*
2. *φ is ultraweak-wot continuous,*
3. *$J\varphi$ is w^*-continuous, where J is the canonical embedding from $B(X)$ to $B(X, X^{**})$.*

Proof The equivalence (1) $\Longleftrightarrow$ (2) is clear since for any $x \in X$ and $x^* \in X^*$, the bounded linear functional $x^*(\Phi(\cdot)x)$ on $\mathcal{M}$ is completely additive if and only if $x^*(\Phi(\cdot)x)$ is normal.

(2) $\Longrightarrow$ (3) It is sufficient to prove that, for any $u = \sum_{k=1}^{\infty} x_k \otimes x_k^* \in X\widehat{\otimes}_{\pi} X^*$ with $\sum_{k=1}^{\infty} \|x_k\| \|x_k^*\| < +\infty$, $(J\Phi(\cdot))(u)$ is a normal functional on $\mathcal{M}$. Let $\phi_k = x_k^*(\Phi(\cdot)x_k)$ for each $k \in \mathbb{N}$. Since Φ is ultraweakly-wot continuous, we have $\phi_k \in \mathcal{M}_*$. Besides,

$$\|\phi_k\| \le \|\Phi\| \|x_k\| \|x_k^*\|, \quad \sum_{k=1}^{\infty} \|\phi_k\| \le \sum_{k=1}^{\infty} \|\Phi\| \|x_k\| \|x_k^*\| < +\infty,$$

we have $\sum_{k=1}^{\infty} \phi_k \in \mathcal{M}_*$. Meanwhile,

$$\sum_{k=1}^{\infty} \phi_k(\cdot) = \sum_{k=1}^{\infty} x_k^*(\Phi(\cdot)x_k) = \sum_{k=1}^{\infty} (J\Phi(\cdot))(x_k \otimes x_k^*)$$

$$= (J\Phi(\cdot))\left(\sum_{k-1}^{\infty} x_k \otimes x_k^*\right) = (J\Phi(\cdot))(u).$$

Thus, $(J\Phi(\cdot))(u)$ is a normal functional on $\mathcal{M}$.

(3) $\Longrightarrow$ (2) is obvious, since $x^*(\Phi(\cdot))x = (J\Phi(\cdot))(x \otimes x^*)$ is normal for any $x \in X$ and $x^* \in X^*$. Thus, the proof is completed. $\square$

Recall that a Jordan homomorphism $\Phi : \mathcal{M} \to \mathcal{N}$ from von Neumann algebras $\mathcal{M}$ to an algebra N is a linear map satisfying the condition $\Phi(a^2) = \Phi(a)^2$ for all $a \in \mathcal{M}$ (equivalently, $\Phi(ab + ba) = \Phi(a)\Phi(b) + \Phi(b)\Phi(a)$ for all $a, b \in \mathcal{M}$).

Theorem 4.20 *Let $\mathcal{M}$ be a σ-finite von Neumann algebra without a Type I_2 direct summand and X be a Banach space. Assume that either X has the Schur property or X is ℓ^p for some $p \in (1, 2)$. If $\varphi : \mathcal{M} \to B(X)$ is an ultraweakly-wot continuous bounded linear map, then there is a Banach space Z such that φ can be dilated to an ultraweakly-wot continuous Jordan homomorphism ϕ from $\mathcal{M}$ to $B(Z)$.*

Proof Since φ is an ultraweakly-wot bounded linear map from $\mathcal{M}$ to $B(X)$ and $\mathcal{M}$ is σ-finite, by Proposition 4.5, the restriction of φ on the projection lattice reduces to a countably additive operator-valued quantum measure $U : \mathcal{P}(\mathcal{M}) \to B(X)$. By Theorem 4.14, there exists a dilation space Z and bounded operators $S : Z \to X$ and $T : X \to Z$, a countably additive quantum measure $V : \mathcal{P}(\mathcal{M}) \to B(Z)$ such that

$$U(P) = SV(P)T, \ \forall \ P \in \mathcal{P}(\mathcal{M})$$

and V is a projection-valued dilation of U. By the Generalized Gleason Theorem (see Lemma 4.1), we can extend V uniquely to a bounded linear map ϕ from $\mathcal{M}$ to $B(Z)$. Furthermore, ϕ is countably additive on $\mathcal{P}(\mathcal{M})$. Applying Proposition 4.5 again, we get that ϕ is ultraweakly-wot continuous. It remains to show that ϕ is a Jordan homomorphism.

Since V is projection-valued, we know that ϕ is idempotent on $\mathcal{P}(\mathcal{M})$ (i.e. $\phi(P)^2 = \phi(P)$ for any $P \in \mathcal{P}(\mathcal{M})$). Let $P_\alpha, P_\beta \in \mathcal{P}(\mathcal{M})$. If $P_\alpha \perp P_\beta$, then $P_\alpha + P_\beta = P_\alpha \vee P_\beta \in \mathcal{P}(\mathcal{M})$. Since ϕ is idempotent on $\mathcal{P}(\mathcal{M})$, we get $\phi(P_\alpha \vee P_\beta) = (\phi(P_\alpha \vee P_\beta))^2$. By linearity, we have

$$\phi(P_\alpha) + \phi(P_\beta) = \phi(P_\alpha)^2 + \phi(P_\beta)^2 + \phi(P_\alpha)\phi(P_\beta) + \phi(P_\beta)\phi(P_\alpha).$$

Thus

$$\phi(P_\alpha)\phi(P_\beta) + \phi(P_\beta)\phi(P_\alpha) = 0.$$

Now let $x = \sum_{i=1}^{N} \lambda_i P_i \in \mathcal{M}$, where $P_1, \ldots, P_n \in \mathcal{P}(\mathcal{M})$ are mutually orthogonal and $\lambda_i \in \mathbb{R}$. Then $x = x^*$. Upon computing $\phi(x^2)$, we have

$$\phi(x^2) = \phi\left(\sum_{i=1}^{n} \lambda_i^2 P_i\right) = \sum_{i=1}^{n} \lambda_i^2 \phi(P_i)$$

and

$$\phi(x)^2 = \sum_{i=1}^{N} \lambda_i^2 \phi(P_i) + \sum_{1 \leq i < j \leq N} \lambda_i \lambda_j (\phi(P_i)\phi(P_j) + \phi(P_j)\phi(P_i)) = \phi(x^2).$$

Since any self-adjoint element $a \in \mathcal{M}_{sa}$ can be approximated by a finite real linear combination of mutually orthogonal projections [151], we get $\phi(a^2) = \phi(a)^2$ for all $a \in \mathcal{M}_{sa}$.

For any two projections $P_1, P_2 \in \mathcal{P}(\mathcal{M})$. Then $y = P_1 + P_2 \in \mathcal{M}_{sa}$, and so $\phi(y^2) = \phi(y)^2$, which implies that

$$\phi(P_1)\phi(P_2) + \phi(P_2)\phi(P_1) = \phi(P_1 P_2 + P_2 P_1).$$

Therefore, for any $z = \sum_{j=1}^{M} \kappa_j P_j$ with $\{\kappa_j\}_{1 \leq j \leq M} \subset \mathbb{C}$ and $\{P_j\}_{1 \leq j \leq M} \subset \mathcal{P}(\mathcal{M})$, we get

$$\phi(z^2) = \sum_{j=1}^{M} \kappa_j^2 \phi(P_j) + \sum_{1 \leq i < j \leq M} \kappa_i \kappa_j \phi(P_i P_j + P_j P_i)$$

$$= \sum_{j=1}^{M} \kappa_j^2 \phi(P_j)^2 + \sum_{1 \leq i < j \leq M} \kappa_i \kappa_j (\phi(P_1)\phi(P_2) + \phi(P_2)\phi(P_1))$$

$$= \phi(z)^2.$$

Since the set of finite linear combinations of projections in $\mathcal{M}$ is norm dense in $\mathcal{M}$, we obtain that ϕ is a Jordan homomorphism.   □

Algebraic and Generalized Choi-Kraus Dilations

Inspired by the results about projection-valued dilation for operator-valued measures or more generally bounded homomorphism dilation for bounded linear maps on Banach algebras in previous chapters, we now explore a pure algebraic structure of the dilation theory. We first consider the general linear systems acting on unital algebras and vector spaces. By introducing two natural dilation structures, namely the canonical and the universal dilation systems, we get that every linearly minimal dilation is equivalent to a reduced homomorphism dilation of the universal dilation, and all the linearly minimal homomorphism dilations can be classified by the associated reduced subspaces. We then focus on the ∗-homomorphism dilation for linear maps between two matrix algebras. By using the generalized Choi-Kraus representation of a linear map, we obtain induced ∗-homomorphism dilations which will be called the generalized Choi-Kraus dilation [119]. We also examine the connections between the generalized Choi-Kraus dilations with other well-established dilations including the universal dilation. In particular, we show that any linearly minimal ∗-homomorphism dilation is equivalent to a linearly minimal generalized Choi-Kraus dilation. Moreover, we present a necessary and sufficient condition for the equivalence of two linearly minimal generalized Choi-Kraus dilations and then classify them by the generalized Choi-ranks.

5.1 Algebraic Dilation for Linear Systems

In the dilation theorems for general operator-valued measures or general bounded linear maps, the dilation space was built on the elementary dilation equipped with a proper dilation norm so that the involved homomorphisms and linear maps are continuous with respect to the dilation norm. To better understand the topological nature of the dilation theory for continuous maps, a good understanding of the purely algebraic aspects of the dilation theory for linear maps is naturally needed. In this section, we present several structural results involving the classification of algebraic homomorphism dilations for linear maps acting on arbitrary vector spaces.

5.1.1 Principle and Universal Dilations

A linear system is a triple $(\varphi, \mathcal{A}, V)$ such that φ is a unital linear map from a unital algebra $\mathcal{A}$ to $L(V)$, where V is a vector space and $L(V)$ denotes the space of all linear maps from V to V.

Definition 5.1 A homomorphism dilation system of a linear system (φ, V) is a unital homomorphism π from $\mathcal{A}$ to a linear operator space $L(W)$ for some vector space W such that there exist an injective linear map $T : V \to W$ and a surjective linear map $S : W \to V$ with the property that for all $a \in \mathcal{A}$, the following diagram commutes:

$$
\begin{array}{ccc}
W & \xrightarrow{\ \pi(a)\ } & W \\
{\scriptstyle T}\big\uparrow & & \big\downarrow{\scriptstyle S} \\
V & \xrightarrow{\ \varphi(a)\ } & V.
\end{array}
$$

That is

$$\varphi(a) = S\pi(a)T, \quad \forall a \in \mathcal{A}.$$

We will use (π, S, T, W) to denote this homomorphism dilation system (also called dilation for short), and the dimension of W is called the dilation dimension. For convenience, we call T the analysis operator and S the synthesis operator, and W the dilation space. If $\ker(S)$ contains a nonzero π-invariant subspace, then we say that (π, S, T, W) is reducible, and otherwise it is called irreducible.

We first point out that any dilation can be reduced to an irreducible dilation. Suppose that K is a π-invariant nonzero subspace of $\ker(S)$. Define $\widetilde{W} = W/K$, and let $\widetilde{S} : \widetilde{W} \to V$, $\widetilde{T} : V \to \widetilde{W}$ and $\widetilde{\pi} : \mathcal{A} \to L(\widetilde{W})$ be the induced linear maps. Then we have for any $a \in \mathcal{A}$ and any $v \in V$ that

$$\widetilde{S}\widetilde{\pi}(a)\widetilde{T}(v) = \varphi(a)v.$$

Thus $(\widetilde{\pi}, \widetilde{S}, \widetilde{T}, \widetilde{W})$ is a homomorphism dilation of (φ, V) and we call it a reduced homomorphism dilation of (π, S, T, W) associated with K. If K is a maximal π-invariant subspace contained in $\ker(S)$, then $\ker(\widetilde{S})$ does not contain any nonzero

$\widetilde{\pi}$-invariant subspace, and hence the reduced dilation homomorphism system $(\widetilde{\pi}, \widetilde{S}, \widetilde{T}, \widetilde{W})$ is irreducible.

Definition 5.2 A homomorphism dilation system $(\pi, S, T, W,)$ of a linear system (φ, V) is called *linearly minimal* if span$\{\pi(\mathcal{A})TV\} = W$, and it is called a *principle dilation* if it is both linearly minimal and irreducible.

Let (π, S, T, W) be a dilation for (φ, V). By replacing W with span$\{\pi(\mathcal{A})TV\}$, we get a linearly minimal dilation. Then, the reduced dilation system of the new linearly minimal dilation corresponding to a maximal invariant subspace is irreducible. Therefore, any dilation leads to a principle one. In what follows, we will focus only on linearly minimal dilations.

We construct two very special but important dilations, namely the canonical dilation and the universal dilation, where the canonical dilation serves as the "smallest" dilation system, the universal one indeed serves as the "largest" one for a given linear system. Both are essential for establishing a structural theory for dilations.

We first introduce the canonical dilation (cf. [125, Section 4.1]). Let $(\varphi, \mathcal{A}, V)$ be a linear system. For $a \in \mathcal{A}, x \in V$, define $\alpha_{a,x} \in L(\mathcal{A}, V)$ by

$$\alpha_{a,x}(\cdot) := \varphi(\cdot a)x.$$

Let

$$W_c := \text{span}\{\alpha_{a,x} : a \in \mathcal{A}, x \in V\} \subset L(\mathcal{A}, V).$$

Define $\pi_c : \mathcal{A} \to L(W_c)$ by $\pi_c(a)(\alpha_{b,x}) := \alpha_{ab,x}$. It is easy to see that π_c is a unital homomorphism. For $x \in V$, we define

$$T_c : V \to L(\mathcal{A}, V) \quad T_c x := \alpha_{I,x} = \phi(\cdot I)x = \phi(\cdot)x.$$

Define

$$S_c : W \to W, \quad S_c\left(\sum_i c_i \alpha_{a,x}\right) := \sum_i c_i \phi(a)x.$$

For arbitrary elements $a \in \mathcal{A}, x \in V$, we have

$$S_c\pi_c(a)T_c x = S_c\pi_c(a)\alpha_{I,x} = S_c\alpha_{a,x} = \phi(a)x.$$

Hence $\varphi(a) = S\pi_c(a)T$ for all $a \in \mathcal{A}$. This implies that (π_c, S_c, T_c, W_c) is a dilation homomorphism of (φ, V), and we will call it the *canonical dilation* of (φ, V). It follows that (π_c, S_c, T_c, W_c) is a linearly minimal dilation.

For irreducibility, note that

$$\ker S_c = \left\{ \sum_i c_i \alpha_{a_i, x_i} \in W_c : \sum_i c_i \varphi(a_i) x_i = 0 \right\}.$$

Let $w = \sum_i c_i \alpha_{a_i, x_i} \in \ker S_c$. Then we have $\pi_c(a)w \in \ker S_c$ for all $a \in \mathcal{A}$ if and only if $\sum_i c_i \varphi(aa_i)(x_i) = 0$ for all $a \in \mathcal{A}$, which in turn is equivalent to the condition that $w = \sum_i c_i \alpha_{a_i, x_i} = 0$ as an element in $L(\mathcal{A}, V)$. Theorefore $\ker(S)$ does not contain any nontrivial π_c-invariant subspaces and consequently we obtain the following statemnet:

Proposition 5.1 *The canonical dilation of a linear system $(\varphi, \mathcal{A}, V)$ is a principle dilation.*

Remark 5.1 In the case that φ is already a unital homomorphism, it can be verified that the canonical dilation π_c must be φ. With this identification, it is easy to see that S and T constructed in the canonical dilation are inverse to each other.

We will see in the next subsubsection that the canonical dilation is the one that has the "smallest" dilation dimension and all the principle dilations are equivalent. Note that for any linearly minimal dilation (π, S, T, W) for a finite-dimensional system $(\varphi, \mathcal{A}, V)$, we always have $\dim W \leq \dim \mathcal{A} \cdot \dim V$. Next, we construct a linearly minimal dilation which has the maximal dilation dimension $\dim \mathcal{A} \cdot \dim V$, and we will show later that any linearly minimal dilation system is equivalent to a reduced dilation system of this dilation.

Let $W_u = \mathcal{A} \otimes V$. For $\sum_i c_i a_i \otimes x_i \in W_u$ where $\{c_i\} \subset \mathbb{C}, \{a_i\} \subset \mathcal{A}, \{x_i\} \subset V$, and for all $b \in \mathcal{A}$, we define the following maps

$$\pi_u : \mathcal{A} \to L(W_u) \quad \pi_u(b)\left(\sum_i c_i a_i \otimes x_i \right) = \sum_i c_i (ba_i) \otimes x_i.$$

$$T_u : V \to W_u \quad T_u x = 1 \otimes x, \ \forall x \in V.$$

$$S_u : W_u \to V \quad S_u\left(\sum_i c_i a_i \otimes x_i \right) = \sum_i c_i \varphi(a_i) x_i.$$

It can be checked that all the maps are well-defined and π_u is a homomorphism and

$$S_u \pi_u(a) T_u x = \varphi(a)x, \ \forall a \in \mathcal{A}, x \in V.$$

Thus, (π_u, S_u, T_u, W_u) is a homomorphism dilation system.

Definition 5.3 The above dilation (π_u, S_u, T_u, W_u) is called the *universal dilation* for the linear system $(\varphi, \mathcal{A}, V)$.

As $\mathrm{span}\{\pi_u(a)Tx : a \in \mathcal{A}, x \in V\} = W$, it follows that (π_u, S, T, W) is a linearly minimal dilation system with the property that $\dim W = \dim \mathcal{A} \cdot \dim V$.

5.1.2   Structural Theorems

In this subsection, we present some results on the classifications of all linearly minimal homomorphism dilation systems.

Definition 5.4 Let (π_1, S_1, T_1, W_1) and (π_2, S_2, T_2, W_2) be two linearly minimal homomorphism dilation systems for a linear system $(\varphi, \mathcal{A}, V)$. We say that they are equivalent if there exists a bijective linear map $R : W_1 \to W_2$ such that

$$RT_1 = T_2, \quad S_2 R = S_1, \quad \pi_1(a) = R^{-1}\pi_2(a)R, \ \forall\, a \in \mathcal{A}.$$

Remark 5.2 It is worth pointing out that $S_2 R = S_1$ automatically follows from the other two conditions by

$$S_2 R w = S_2\left(\sum_i c_i R\pi_1(a_i)T_1 x_i\right) = S_2\left(\sum_i c_i \pi_2(a_i)RT_1 x_i\right)$$

$$= \sum_i c_i S_2\pi_2(a_i)T_2 x_i = \sum_i c_i \varphi(a_i)x_i$$

$$= \sum_i c_i S_1\pi_1(a_i)T_1 x_i = S_1 w.$$

The following theorem shows that all the principle dilations are equivalent.

Theorem 5.1 *If (π_1, S_1, T_1, W_1) and (π_2, S_2, T_2, W_2) are two principle homomorphism dilation systems for $(\varphi, \mathcal{A}, V)$, then (π_1, S_1, T_1, W_1) and (π_2, S_2, T_2, W_2) are equivalent, hence they are all equivalent to canonical dilation.*

Proof Since both dilations are linearly minimal, we have that $W_i = \mathrm{span}\,\pi(\mathcal{A})T_i(V)$ for $i = 1, 2$. Define $R : W_1 \to W_2$ by

$$R(w) = \sum_i c_i \pi_2(a_i)T_2(v_i),$$

where $w = \sum_i c_i \pi_1(a_i)T_1(v_i)$. In order for T to be well-defined that induces the equivalence between π_1 and π_2, it suffices to show that

$$w = \sum_i c_i \pi_1(a_i)T_1(v_i) = 0 \quad \Longleftrightarrow \quad \sum_i c_i \pi_2(a_i)T_2(v_i) = 0.$$

Assume to the contrary that $w \neq 0$. Since

$$S_1 w = \sum_i c_i \varphi(a_i) v_i = S_2 \sum_i c_i \pi_2(a_i) T_2(v_i) = S_2(0) = 0,$$

we get that $w \in \ker S_1$. Moreover,

$$S_1 \pi_1(a) w = \sum_i c_i S_1 \pi_1(a a_i) T_1(v_i) = \sum_i c_i S_1 \varphi(a a_i)(v_i)$$

$$= \sum_i c_i S_2 \pi_2(a a_i) T_2(v_i) = S_2 \pi_2(a) \sum_i c_i \pi_2(a_i) T_2(v_i)$$

$$= S_2 \pi_2(a)(0) = 0.$$

Thus, $\pi_1(a)w \in \ker S_1$ for all $a \in \mathcal{A}$. So $M = \{\pi_1(a)w : a \in \mathcal{A}\}$ is a nonzero π_1-invariant subspace inside $\ker S_1$, which leads to a contradiction since the dilation (π_1, S_1, T_1, W_1) is irreducible. The argument for the other direction is the same. By the definition of R, we clearly have for any $w = \sum_i c_i \pi_1(a_i) T_1(v_i) \in W_1$ that

$$R \pi_1(a) w = R\left(\sum_i c_i \pi_1(a a_i) T_1(v_i) \right) = \sum_i c_i \pi_2(a a_i) T_2(v_i)$$

$$= \pi_2(a)\left(\sum_i c_i \pi_2(a_i) T_2(v_i) \right) = \pi_2(a) R w$$

and $R T_1(v) = T_2 v$ for any $v \in V$. Thus, we get $\pi_1(a) = R^{-1} \pi_2(a) R$ and $R T_1 = T_2$, and therefore, by Remark 5.2, we get that (π_1, S_1, T_1, W_1) and (π_2, S_2, T_2, W_2) are equivalent. $\qquad\square$

Intuitively, the principle dilation has the "smallest" dilation dimension when we deal with the finite-dimensional case (i.e., $\dim V < \infty$ and $\dim \mathcal{A} < \infty$). Furthermore, we deduce the following corollaries from the perspective of dilation dimensions.

Corollary 5.1 *Let $(\varphi, \mathcal{A}, V)$ be a linear system such that both $\mathcal{A}$ and V are finite-dimensional.*

1. *Assume that (π, S, T, W) is a principle dilation of $(\varphi, \mathcal{A}, V)$ such that $\dim W = \dim \mathcal{A} \cdot \dim V$. Then any linearly minimal dilation of $(\varphi, \mathcal{A}, V)$ is irreducible, and hence a principle dilation.*
2. *Assume that (π, S, T, W) is a principle dilation of $(\varphi, \mathcal{A}, V)$. If (π_1, S_1, T_1, W_1) is a linearly minimal dilation of $(\varphi, \mathcal{A}, V)$ such that $\dim W_1 \leq \dim W$, then it is irreducible.*

Corollary 5.2 *Let (π_1, S_1, T_1, W_1) be a linearly minimal dilation of $(\varphi, \mathcal{A}, V)$.*

1. *If* $\ker S_1$ *does not contain any nonzero* π*-invariant subspaces, then* π_1 *is equivalent to the canonical homomorphism* π_c.
2. *Assume that* $\dim W_1 < \infty$. *If* π_1 *is equivalent to the canonical homomorphism* π_c, *then* $\ker S_1$ *does not contain any nonzero* π_1*-invariant subspaces.*

Remark 5.3 We do not know if (2) is still true when $\dim W_1$ is infinite-dimensional.

The term "universal dilation" is justified by the following theorem.

Theorem 5.2 *For a linear system* $(\varphi, \mathcal{A}, V)$, *any linearly minimal homomorphism dilation system is equivalent to a reduced homomorphism dilation system of its universal dilation.*

Proof Let (π_1, S_1, T_1, W_1) be a linearly minimal dilation system. Define K by

$$K = \left\{ w = \sum_i c_i a_i \otimes x_i : \sum_i c_i \pi_1(a_i) T_1 x_i = 0 \right\}.$$

We claim that K is a π_u-invariant subspace contained in $\ker S$. In fact, if $w = \sum_i c_i a_i \otimes x_i \in K$, then

$$S_u w = \sum_i c_i S_u(a_i \otimes x_i) = \sum_i c_i S_u \pi_u(a_i) T_u x = \sum_i c_i \varphi(a_i) x_i$$

$$= \sum_i c_i S_1 \pi_1(a_i) T_1 x_i = S_1 \sum_i c_i \pi_1(a_i) T_1 x_i = S(0) = 0.$$

Thus $K \subseteq \ker S_u$. Moreover, for any $a \in \mathcal{A}$ and $w = \sum_i c_i a_i \otimes x_i \in K$, we have

$$\sum_i c_i \pi_1(a a_i) T_1 x_i = \pi_1(a) \left(\sum_i c_i \pi_1(a_i) T_1 x_i \right) = 0.$$

Thus $\pi_u(a) w = \sum_i c_i (a a_i) \otimes x_i \in K$. Therefore, K is a π_u-invariant subspace contained in $\ker S_u$.

Let $(\tilde{\pi}_u, \tilde{S}_u, \tilde{T}_u, W_u/K)$ be the reduced dilation homomorphism. Define $R : W/K \to W_1$ by

$$R[w] = \sum_i c_i \pi_1(a_i) T_1 x_i$$

for any $[w] \in W_u/K$ represented by $w = \sum_i c_i a_i \otimes x_i$. Then, by the definition of K, we have $R[w] = \sum_i c_i a_i \otimes x_i = 0$ if and only if $w \in K$. Hence, R is a well-defined injective linear map. Clearly, it is also surjective since $\pi_1(\mathcal{A}) T_1 V = W_1$. Moreover, for any $w = \sum_i c_i a_i \otimes x_i \in W$, we have

$$\pi_1(a)R([w]) = \pi_1(a)\left(\sum_i c_i \pi_1(a_i)T_1 x_i\right) = \sum_i c_i \pi_1(aa_i)T_1 x_i$$

$$= R\left[\sum_i c_i(aa_i) \otimes x_i\right] = R\tilde{\pi}_u(a)([w]).$$

So $\pi_1(a) = R\tilde{\pi}_u(a)R^{-1}$ for any $a \in \mathcal{A}$. Moreover, for any $x \in V$, we have

$$R\tilde{T}_u x = R[Tx] = R[I \otimes x] = \pi_1(I)T_1 x.$$

Hence $R\tilde{T}_u = T_1$. Thus, (π_1, S_1, T_1, W_1) and $(\tilde{\pi}_u, \tilde{S}_u, \tilde{T}_u, W_u/K)$ are equivalent.   □

In order to classify the linearly minimal homomorphism dilation systems we introduce the following definition.

Definition 5.5 Let (π_u, S_u, T_u, W_u) be the universal dilation and (π_1, S_1, T_1, W_1) be a linearly minimal homomorphism dilation system for a linear system $(\varphi, \mathcal{A}, V)$. Then the π_u-invariant subspace K_1 introduced in the above proof

$$K_1 = \left\{w = \sum_i c_i a_i \otimes x_i : \sum_i c_i \pi_1(a_i)T_1 x_i = 0\right\}$$

will be called the *reduced subspace* associated with (π_1, S_1, T_1, W_1).

Remark 5.4 We point out that the reduced subspace K of a linearly minimal dilation (π_1, S_1, T_1, W_1) is different from the maximal π_1-invariant subspace M contained in $\ker S_1$, which is used to reduce (π_1, S_1, T_1, W_1) to the "smallest" dilation - the principle dilation, while K is a π_u-invariant subspace contained in W_u (i.e., $\mathcal{A} \otimes V$) that is used to reduce the universal dilation to (π_1, S_1, T_1, W_1).

The following result presents us with a classification for all linearly minimal dilations for a given linear system.

Theorem 5.3 *Let K_1 and K_2 be the reduced subspaces associated with the linearly minimal dilations (π_1, S_1, T_1, W_1) and (π_2, S_2, T_2, W_2), respectively. Then (π_1, S_1, T_1, W_1) and (π_2, S_2, T_2, W_2) are equivalent if and only if $K_1 = K_2$.*

Proof By Theorem 5.2, we only need to prove that if (π_1, S_1, T_1, W_1) and (π_2, S_2, T_2, W_2) are equivalent, then $K_1 = K_2$. Let $R : W_1 \to W_2$ be a bijective linear map. Then by the assumption we have that $\pi_2(a)R = R\pi_1(a)$ for all $a \in \mathcal{A}$, $S_2 R = S_1$ and $RT_1 = T_2$.

Let $w = \sum_i c_i a_i \otimes x_i \in W_u$. Since

$$\sum_i c_i \pi_1(a_i) T_1 x_i = R^{-1} \sum_i c_i \pi_2(a_i) R T_1 x_i = R^{-1} \sum_i c_i \pi_2(a_i) T_2 x_i,$$

we get that $\sum_i c_i \pi_i(a_i) x_i = 0$ if and only if $\sum_i c_i \pi_2(a_i) T_2 x_i = 0$, i.e., $w \in K_1$ if and only if $w \in K_2$. Hence $K_1 = K_2$.   □

Theorem 5.3 shows that the equivalent class of linearly minimal dilations is uniquely determined by the reduced subspace. However, the structure of the reduced subspaces is complex and we will show by examples later that there could be infinitely many inequivalent linearly minimal dilations even in the finite-dimensional case.

Additionally, there is a weaker version of equivalence which seems also relevant to the dilation theory: Let (π_1, S_1, T_1, W_1) be a linearly minimal dilation for a linear system (φ, V), and π_2 be a homomorphism from $\mathcal{A}$ to $L(W_2)$ such that π_1 and π_2 are equivalent in the usual sense, i.e.

$$\pi_1(a) = R^{-1} \pi_2(a) R, \ \forall a \in \mathcal{A}$$

for some isomorphism $R : W_1 \to W_2$. It is interesting to know that under what conditions do we have that the homomorphisms π_1 and π_2 for linearly minimal dilations (π_1, S_1, T_1, W_1) and (π_2, S_2, T_2, W_2) are equivalent. For this purpose, we introduce the following concept of equivalence for the reduced invariant subspaces.

Definition 5.6 Let (π_u, S_u, T_u, W_u) be the universal dilation of a linear system $(\varphi, \mathcal{A}, V)$. Two π_u-invariant subspaces K_1 and K_2 of ker S_u are called *strongly isomorphic* if there is an isomorphism $R : W_u \to W_u$ such that $R(K_1) = K_2$ and

$$\pi_u(a) R w - R \pi_u(a) w \in K_2, \ \forall a \in \mathcal{A}, w \in W_u,$$

i.e., the quotient maps of $\pi_u(a)$ and R on W_u / K_2 commute for all $a \in \mathcal{A}$.

Theorem 5.4 *Let K_1 and K_2 be the reduced subspaces associated with the linearly minimal dilation (π_1, S_1, T_1, W_1) and (π_2, S_2, T_2, W_2), respectively. Then π_1 and π_2 are equivalent if and only if K_1 and K_2 are strongly isomorphic.*

Proof By Theorem 5.2, we can assume that (π_i, S_i, T_i, W_i) is the reduced dilation of the universal dilation associated with $K_i (i = 1, 2)$.

"⟸". Assume that K_1 and K_2 are strongly isomorphic. Then there is an isomorphism $R : W_u \to W_u$ such that $R(K_1) = K_2$ and

$$\pi_u(a) R w - R \pi_u(a) w \in K_2, \ \forall a \in \mathcal{A}, w \in W_u.$$

Let $\widetilde{R} : W_1 = W_u/K_1 \to W_u/K_2 = W_2$ be defined by

$$\widetilde{R}[w]_1 = [Rw]_2, \quad w \in W_u,$$

where we use $[\cdot]_i, i = 1, 2$ to denote the element in the corresponding quotient space. Then $\widetilde{R}$ is a bijective linear transformation. Note that π_2 is the reduced homomorphism of π_u on W_u/K_2, we have that

$$\pi_2(a)\widetilde{R}([w]_1) = \pi_2(a)[Rw] = [\pi_u(a)Rw]_2,$$

and

$$\widetilde{R}\pi_1(a)[w]_1 = \widetilde{R}[\pi_u(a)w] = [R\pi_u(a)w]_2.$$

Thus, from $\pi_u(a)Rw - R\pi_u(a)w \in K_2$, we obtain that $\pi_2(a)\widetilde{R}[w]_1 = \widetilde{R}\pi_1(a)[w]_1$, which implies that π_1 and π_2 are equivalent.

" $\Longrightarrow$ ". Assume that π_1 and π_2 are equivalent. Then there is bijective linear map $L : W_u/K_1 \to W_u/K_2$ such that $\pi_2(a)L = L\pi_1(a)$ for all $a \in \mathcal{A}$. Since $\dim(K_1) = \dim(K_2)$, we obtain that there exists a bijective linear map $R : W_u \to W_u$ such that $R(K_1) = K_2$ and the induced quotient map $\widetilde{R}$ is L. Moreover, from $\pi_2(a)L = L\pi_1(a)$ we have that $\pi_2(a)\widetilde{R} = \widetilde{R}\pi_1(a)$, which is equivalent to the condition that

$$\pi_u(a)Rw - R\pi_u(a)w \in K_2, \quad \forall a \in \mathcal{A}, w \in W_u.$$

Thus, K_1 and K_2 are strongly isomorphic.   □

Theorems 5.3 and 5.4 provide us with two classifications for linearly minimal dilations based on the reduced subspaces in $\ker S_u$ where S_u is defined by $S_u(a \otimes x) = \varphi(a)x$. These lead to many interesting questions, especially in the finite-dimensional case. For example,

1. Under what condition on $(\varphi, \mathcal{A}, V)$ do we have the property that for every k between the dimensions of V and $\mathcal{A} \otimes V$ there exists a linearly minimal dilation with dilation dimension k.
2. When do we have only finite many inequivalent linearly minimal homomorphism dilations? Examples 5.2.2 and 5.2.5 show that we could have infinitely many inequivalent classes even if both $\mathcal{A}$ and V are finite-dimensional.)
3. Under what condition do we have that the principle and universal dilations are the only two classes of linearly minimal dilations?
4. We will construct an example showing that there exist reduced subspaces K_1 and K_2 that are strongly isomorphic and $K_1 \neq K_2$. However, it would be interesting to know

that if the condition $\dim K_1 = \dim K_2$ automatically implies that they are strongly isomorphic.

In what follows, we will answer some of these questions and at the same time construct some examples showing the complexity of these problems.

Let

$$M = \left\{ \sum_i c_i a_i \otimes x_i : \sum_i c_i \varphi(a a_i) x_i = 0, \forall a \in \mathcal{A} \right\}.$$

Then M is the largest π_u-invariant subspace contained in $\ker S_u$. Hence, by Theorem 5.1, the universal homomorphism dilation is equivalent to the principle dilation if and only if $M = \{0\}$. Moreover, we have the following proposition.

Proposition 5.2 *A linear system (φ, V) has only one equivalent class of linearly minimal homomorphism dilations if and only if $M = \{0\}$.*

And if $M \neq \{0\}$, then the universal dilation is not equivalent to the principle dilation. Note that if a is a nonzero element in the left ideal of $\ker(\varphi)$. Then for any $x \in V$ and any $b \in \mathcal{A}$ we have $\varphi(ba)x = 0$, which implies that $a \otimes x \in M$.

Corollary 5.3 *Let $(\varphi, \mathcal{A}, V)$ be a linear system. If $\ker(\varphi)$ contains a proper left ideal, then the universal dilation is not equivalent to its principle dilation.*

5.2   Generalized Choi-Kraus Dilations

In this section, we explore the dilations for linear maps between two matrix algebras [119]. Since every such a map is completely bounded [192] and has a generalized Choi-Kraus representation $\varphi(X) = \sum_{k=1}^{L} A_k X B_k^*$, it automatically induces a $*$-homomorphism dilation by the representation matrix system $\{A_k, B_k\}$, which we call it the generalized Choi-Kraus dilation for φ, and the Choi-Kraus dilation when $A_k = B_k$ for a completely positive (CP) map. We examine the connections between the generalized Choi-Kraus dilations with the several well-established dilations including the universal dilation, and present the conditions for the equivalence of two linearly minimal generalized Choi-Kraus dilations and then classify them by the generalized Choi rank. For convenience, we use the following standard notations:

- A vector $v \in \mathbb{C}^n$ will be treated as a column vector and its adjoint gives a row vector. We use $\{e_i\}_{i=1}^{n}$ for the standard basis of $\mathbb{C}^n$, still as column vectors.
- M_n^+—the set of positive semidefinite $n \times m$ complex matrices, and we write $X \geq 0$ if $X \in M_n^+$.

- $E_{i,j}$—the standard matrix units and I_n is the identity matrix in M_n; Use $A(i,j)$ to denote the (i,j)-entry of a matrix A.
- $L(M_n, M_d)$—the set of all linear maps from M_n to M_d and $CP(M_n, M_d)$—the set of all completely positive maps from M_n to M_d.
- $\ker A$—the kernel of the matrix A; $\operatorname{Im} A$—the image space of A; $\operatorname{Rank}(A)$—the rank of A. A^*—the adjoint of matrix A, A^t—the transpose of matrix A.
- $A \otimes B$—tensor product of the matrices A and B. Suppose $A = (a_{i,j})$, for convenience, we set $A \otimes B$ in block form by

$$\begin{pmatrix} a_{1,1}B & \cdots & a_{1,n}B \\ \vdots & \ddots & \vdots \\ a_{n,1}B & \cdots & a_{n,n}B \end{pmatrix},$$

which is also referred to as the Kronecker product of A and B.

5.2.1 Generalized Choi-Kraus Dilations

The following is the celebrated Choi-Kraus representation theorem [48] for CP maps between matrix algebras.

Theorem 5.5 *Let* $\Phi : M_n \to M_d$ *be linear. The following conditions on* Φ *are equivalent:*

1. Φ *is completely positive (i.e., CP).*
2. Φ *is n-positive.*
3. $P_\Phi = (\Phi(E_{i,j})) \in M_n(M_d)^+$ *where* $E_{i,j}$ *are the standard matrix units of* M_n.
4. $\Phi(X) = \sum_{i=1}^r A_i X A_i^*$ *for some* r, d *and* $A_i \in M_{d \times n}$ *(Choi-Kraus representation).*

Similar to the Choi matrix $(\phi(E_{i,j}))$ for $\phi \in CP(M_n, M_d)$, the generalized Choi matrix G_φ of $\varphi \in L(M_n, M_d)$ is defined by

$$G_\varphi = (\varphi(E_{i,j})) \in M_n(M_d).$$

If we write G_φ as $\sum_{i=1}^m v_i w_i^*$, where for $1 \le i \le m$, $v_i = \begin{pmatrix} \alpha_1^i \\ \vdots \\ \alpha_n^i \end{pmatrix}$, $w_i = \begin{pmatrix} \beta_1^i \\ \vdots \\ \beta_n^i \end{pmatrix} \in$

$\mathbb{C}^{nd}$, and $\alpha_j^i, \beta_j^i \in \mathbb{C}^d$, $1 \le j \le n$. Let $A_i = [\alpha_1^i, \vdots, \cdots, \vdots, \alpha_n^i]_{d,n}$ and $B_i :=$

$[\beta_1^i, \vdots, \cdots, \vdots, \beta_n^i]_{d,n}$, then it can be verified that

$$\varphi(X) = \sum_{i=1}^{m} A_i X B_i^*, \tag{5.1}$$

which is called a *generalized Choi-Kraus representation* of φ [173], where $\{A_i, B_i^*\}_{i=1}^{m}$ is called the Kraus-matrix pairs

Conversely, if $\varphi(X) = \sum_{i=1}^{m} A_i X B_i^*$ for some $A_i := \left[\alpha_1^i, \vdots, \cdots, \vdots, \alpha_n^i\right]_{d,n}$ and

$B_i := \left[\beta_1^i, \vdots, \cdots, \vdots, \beta_n^i\right]_{d,n}$, and set $v_i = \begin{pmatrix} \alpha_1^i \\ \vdots \\ \alpha_n^i \end{pmatrix}$, $w_i = \begin{pmatrix} \beta_1^i \\ \vdots \\ \beta_n^i \end{pmatrix} \in \mathbb{C}^{nd}$, then $G_\varphi =$

$\sum_{i=1}^{m} v_i w_i^*$. In particular, if we define

$$A_\varphi = \sum_{i=1}^{m} v_i e_i^*, \quad B_\varphi = \sum_{i=1}^{m} w_i e_i^*, \tag{5.2}$$

then

$$G_\varphi = A_\varphi B_\varphi^*. \tag{5.3}$$

Similar to the Choi rank $\mathrm{Cr}(\phi)$ of $\phi \in CP(M_n, M_d)$ to $\varphi \in L(M_n, M_d)$, the generalized Choi rank of φ is given by

$$\mathrm{Gr}(\varphi) := \min \left\{ k : \varphi(X) = \sum_{i=1}^{k} A_i X B_i^* \right\},$$

and we have

$$\mathrm{Gr}(\varphi) = \mathrm{Rank}(G_\varphi). \tag{5.4}$$

For $\varphi \in L(M_n, M_d)$ with a generalized Choi-Kraus representation $\varphi(X) = \sum_{i=1}^{m} A_i X B_i^*$ where $A_i, B_i \in M_{d,n}, 1 \leq i \leq m$. Define $\pi : M_n \to M_{mn}$, and $A, B : \mathbb{C}^{mn} \to \mathbb{C}^d$ by

$$A = (A_1, \cdots, A_m), \quad \pi_1(X) = \begin{pmatrix} X & \cdots & 0 \\ \vdots & \ddots & \vdots \\ 0 & \cdots & X \end{pmatrix} = I_m \otimes X, \quad B = (B_1, \cdots, B_m).$$

It is easy to see that π is a unital $*$-homomorphism and for all $X \in M_n, x \in \mathbb{C}^d$,

$$A\pi(X)B^*x = \sum_{i=1}^{m} A_i X B_i^* x = \varphi(X)x.$$

In the case that $\varphi \in CP(M_n, M_d)$ with a Choi-Kraus representation $\varphi(X) = \sum_{i=1}^{m} A_i X A_i^*$. Define $\pi : M_n \to M_{mn}$, and $A : \mathbb{C}^{mn} \to \mathbb{C}^d$ by

$$A = (A_1, \cdots, A_m), \quad \pi(X) = \begin{pmatrix} X & \cdots & 0 \\ \vdots & \ddots & \vdots \\ 0 & \cdots & X \end{pmatrix} = I_m \otimes X.$$

Then π is a unital $*$-homomorphism and $\varphi(X) = A\pi(X)A^*$ for all $X \in M_n$.

Definition 5.7 For $\varphi \in L(M_n, M_d)$, the above constructed dilation $(\pi, A, B^*, \mathbb{C}^{mn})$ is called the *generalized Choi-Kraus dilation* induced by $\varphi(X) = \sum_{i=1}^{m} A_i X B_i^*$. For $\varphi \in CP(M_n, M_d)$, $(\pi, A, A^*, \mathbb{C}^{mn})$ is called the *Choi-Kraus dilation* of φ induced by $\varphi(X) = \sum_{i=1}^{m} A_i X A_i^*$.

Remark 5.5 (i) A CP map can have a generalized Choi-Kraus representation $\varphi(X) = \sum_{i=1}^{m} A_i X B_i^*$, where A_i's are not necessarily the same as B_i's. In this case the induced generalized Choi-Kraus dilation could be quite different (inequivalent) from the Choi-Kraus dilation induced by $\varphi(X) = \sum_{i=1}^{m} C_i X C_i^*$ (see Proposition 5.5 or Example 5.2.6).

(ii) A natural question concerning the generalized Choi-Kraus dilation $(\pi, A, B^*, \mathbb{C}^{mn})$ for φ is whether $\mathrm{span}\{\pi(X)B^*x, X \in M_n, x \in \mathbb{C}^d\}$ is $\mathbb{C}^{mn}$. This is not true in general. So we need to consider a subspace of $\mathbb{C}^{mn}$ as the dilation space. Thus, it is natural to only consider the linear minimal generalized Choi-Kraus dilations in the above definition.

Next, we discuss some basic properties of generalized Choi-Kraus dilations. The first result tells us that it is enough to work with linearly minimal generalized Choi-Kraus dilations.

Theorem 5.6 *For $\varphi \subset L(M_n, M_d)$, any linearly minimal $*$-homomorphism dilation of φ is equivalent to a linearly minimal generalized Choi-Kraus dilation*

Proof Suppose that (π, S, T, W_1) is a linearly minimal $*$-homomorphism dilation. Then, by Definition 5.5, the reduced subspace associated with it is

$$K = \left\{ w = \sum_i c_i X_i \otimes x_i \in W_u : \sum_i c_i \pi(a_i) T x_i = 0 \right\}.$$

By Theorem 5.2, it remains to show that there is a linearly minimal generalized Choi-Kraus dilation such that the corresponding reduced subspace is K as well.

We first recall the universal dilation of $\varphi \in L(M_n, M_d)$, as a special linear system $(\varphi, M_n, \mathbb{C}^d)$, the dilation space is $W_u = M_n \otimes \mathbb{C}^d$, in the light of the one-to-one correspondence between W_u and $\mathbb{C}^{n^2 d}$. That is, here we choose $\mathbb{C}^{n^2 d}$ as the dilation space of the universal dilation. Then the maps $T_u : \mathbb{C}^d \to W_u$, $\pi_u : M_n \to L(W_u)$ are realized concretely as $T_u : \mathbb{C}^d \to \mathbb{C}^{n^2 d}$, $\pi_u : M_n \to M_{n^2 d}$ and formulated by

$$T_u = \begin{pmatrix} E_{1,1} \\ \vdots \\ E_{1,d} \\ E_{2,1} \\ \vdots \\ E_{n,d} \end{pmatrix} \qquad \pi_u(X) = \begin{pmatrix} X & \cdots & 0 \\ \vdots & \ddots & \vdots \\ 0 & \cdots & X \end{pmatrix} = I_{nd} \otimes X, \ \forall \, X \in M_n. \tag{5.5}$$

Write $G_\varphi = \big(\varphi(E_{i,j})\big) = \sum_{i=1}^{nd} v_i e_i^*$, where v_i is the column vector of G_φ and $\{e_i\}_{i=1}^{nd}$ is the standard basis of $\mathbb{C}^{nd}$. Similar to the construction of Choi-Kraus representation matrix systems, we arrange v_i into $V_i \in M_{d,n}$ and set $S_u = (V_1, \cdots, V_{nd})$. Represent $K \subset W_u = \mathbb{C}^{n^2 d} = \mathbb{C}^{nd} \otimes \mathbb{C}^n$, for arbitrary $w \in K$ as $\sum_{i=1}^n w_i \otimes e_i$, where $\{w_i\}_{1=i}^n \subset \mathbb{C}^{nd}$ and $\{e_i\}_{1=i}^n$ is standard basis of $\mathbb{C}^n$. Since K is π_u-invariant subspace in $\ker S_u$, that is, for all $X \in M_n$, $\pi_u(X)w \in K \subset \ker S_u$, it follows that

$$\pi_u(E_{i,j})w \in K \subset \ker S_u, \ 1 \le i, j \le n.$$

On the one hand, fix j, for $1 \le i \le n$, as

$$\pi_u(E_{i,j})w = (I_{nd} \otimes E_{i,j})w = w_j \otimes e_i,$$

then the condition $\{w_j \otimes e_i\}_{i=1}^n \subset \ker S_u$ implies that $w_j \in \ker G_\varphi$. On the other hand, for each i, for $1 \le j \le n$, $\pi(E_{i,j})w \in K$, that is, $\{w_j \otimes e_i\}_{j=1}^n \subset K$. Thus we conclude that $K = M \otimes \mathbb{C}^n$, where M is a subspace of $\ker G_\varphi$.

We construct a matrix D such that $\ker D = M$, a slight of variation of Douglas' Factorization Theorem in [68], there exists C such that $CD = G_\varphi$, as we can write $C = \sum_{i=1}^m v_i e_i^*$, $D = \sum_{i=1}^m e_i w_i^*$ for some m, where $\{e_i\}_{i=1}^m$ is the standard basis of $\mathbb{C}^m$ and $\{v_i, w_i\}_{i=1}^m \subset \mathbb{C}^{nd}$. Besides, we can rearrange vectors $\{v_i, w_i\}_{i=1}^m$ into the Kraus matrix pairs $\{A_i, B_i\}_{i=1}^m \subset M_{d,n}$ such that $\varphi(X) = \sum_{i=1}^m A_i X B_i^*$. Then we can show that the reduced subspace associated with the linearly minimal dilation (π, A, B^*, W_2) induced by $\varphi(X) = \sum_{i=1}^m A_i X B_i^*$ is K as well. Therefore, (π, S, T, W_1) is equivalent to the linearly minimal generalized Choi-Kraus dilation (π, A, B^*, W_2). □

Proposition 5.3 *Let* $\varphi \in L(M_n, M_d)$ *with* $Gr(\varphi) = r$. *Then the generalized Choi-Kraus dilation* $(\pi, A, B^*, \mathbb{C}^{nr})$ *of* φ *induced by its generalized Choi-Kraus representation* $\varphi(X) = \sum_{i=1}^{r} A_i X B_i^*$ *is a principle dilation.*

Proof The generalized Choi-Kraus dilation $(\pi, A, B^*, \mathbb{C}^{nr})$ of φ is given by

$$A = (A_1, \cdots, A_r), \quad B = (B_1, \cdots, B_r), \quad \pi(X) = \begin{pmatrix} X & \cdots & 0 \\ \vdots & \ddots & \vdots \\ 0 & \cdots & X \end{pmatrix} = I_r \otimes X,$$

for all $X \in M_n$.

We first show that it is irreducible, namely, $\ker A$ does not contain any nontrivial π-invariant subspaces. Suppose K is π-invariant subspace in $\ker A$, that is, if $v \in K$, then $\pi(X)v \in K \subset \ker A$ for all $X \in M_n$. Note that $K \subset W \subset \mathbb{C}^{nr}$ and set

$$v = \begin{pmatrix} \alpha_1 \\ \vdots \\ \alpha_r \end{pmatrix} \in \mathbb{C}^{nr} \quad \text{where} \quad \alpha_i = \begin{pmatrix} \alpha_i(1) \\ \vdots \\ \alpha_i(n) \end{pmatrix} \in \mathbb{C}^n.$$

We first choose $E_{1,1}$. Then $\pi_1(E_{1,1})v \in K \subset \ker A$ and thus

$$A\pi_1(E_{1,1})v = (A_1, \cdots, A_r) \begin{pmatrix} E_{1,1}\alpha_1 \\ \vdots \\ E_{1,1}\alpha_r \end{pmatrix} = \mathbf{0} \text{ (zero vector)}.$$

That is,

$$\alpha_1(1) \begin{pmatrix} A_1(1, 1) \\ \vdots \\ A_1(d, 1) \end{pmatrix} + \alpha_2(1) \begin{pmatrix} A_2(1, 1) \\ \vdots \\ A_2(d, 1) \end{pmatrix} + \cdots + \alpha_r(1) \begin{pmatrix} A_r(1, 1) \\ \vdots \\ A_r(d, 1) \end{pmatrix} = \mathbf{0} \text{ (zero vector)}.$$

Continuing this way, we take $\{E_{i,1}\}_{i=2}^{n}$ and arrange $\{\pi_1(E_{i,1})v\}_{i=1}^{n} \subset \ker A$ into

$$\alpha_1(1)A_1 + \alpha_2(1)A_2 + \cdots + \alpha_r(1)A_r = \mathbf{0} \text{ (zero matrix)}.$$

As $Gr(\varphi) = r$ and $\varphi(X) = \sum_{i=1}^{r} A_i X B_i^*$, then we claim $\{A_i\}_{i=1}^{r}$ is linearly independent, otherwise, we can recombine the Kraus matrix pairs $\{A_i, B_i^*\}_{i=1}^{r}$ into $\{C_i, D_i^*\}_{i=1}^{k}$ where $k < r$, which contradicts the minimality of the generalized Choi rank. Consequently,

$$\alpha_1(1) = \alpha_2(1) = \cdots = \alpha_r(1) = 0.$$

For each $2 \leq j \leq n$, we repeat the above argument, that is, we take $E_{1,j}, E_{2,j}, \cdots, E_{n,j}$ such that $\pi(E_{i,j})v \in \ker A$ for all $1 \leq i \leq n$. Then it follows that

$$\alpha_1(j)A_1 + \alpha_2(j)A_2 + \cdots + \alpha_r(j)A_r = \mathbf{0} \text{ (zero matrix)}.$$

Similarly, the linear independence of $\{A_i\}_{i=1}^r$ implies $\alpha_1(j) = \cdots = \alpha_r(j) = 0$. To summarise,

$$\alpha_i(j) = 0, \ 1 \leq i \leq r, \ 1 \leq j \leq n,$$

that is, $v = 0$, meaning $\{0\}$ is the only π-invariant subspace in $\ker A$.

Next, suppose W is the linearly minimal dilation space, that is, $W = \mathrm{span}\{\pi(X)B^*x : X \in M_n, x \in \mathbb{C}^d\}$. Using the notations defined in Eq. (5.2) and (5.5), we have,

$$\sum_k c_k \pi(X_k)B^*x_k = (B_\varphi^* \otimes I_n)\left(\sum_k c_k X_k \otimes x_k\right). \tag{5.6}$$

Hence

$$W = (B_\varphi^* \otimes I_n)(W_u) = (B_\varphi^* \otimes I_n)(\mathbb{C}^{n^2 d}).$$

That is, the linearly minimal dilation space of $(\pi, A, B^*, \mathbb{C}^{nr})$ is $\mathrm{Im}(B_\varphi^* \otimes I_n)$.

A similar argument gives that $\{B_i\}_{i=1}^r$ is also linearly independent, and thus $\mathrm{Rank}(B_\varphi^*) = r$. Therefore $\dim W = \mathrm{Rank}(B_\varphi^* \otimes I_n) = nr$. Moreover, $W \subset \mathbb{C}^{nr}$, then it follows that $W = \mathbb{C}^{nr}$. Therefore, we have proved that $(\pi, A, B, \mathbb{C}^{nr})$ is irreducible and linearly minimal, and hence it is a principle dilation. $\qquad\square$

Remark 5.6 With the notations as above, if $\varphi(X) = \sum_{i=1}^r V_i X W_i^*$ is another generalized Choi-Kraus representation of φ, then, by Theorem 5.1, $(\pi, V, W^*, \mathbb{C}^{nr})$ induced by $\varphi(X) = \sum_{i=1}^r V_i X W_i^*$ is equivalent to $(\pi, A, B^*, \mathbb{C}^{nr})$. Moreover, the reduced subspace associated with $(\pi, A, B^*, \mathbb{C}^{nr})$ is $K = \ker(B_\varphi^* \otimes I_n)$. A direct application of the Sylvester Theorem [213], which states that for $P \in M_{m,n}, Q \in M_{n,l}$, then $\mathrm{Rank}(PQ) = \mathrm{Rank}(Q) - \dim(\mathrm{Im}\, Q \cap \ker P)$, gives that if $\mathrm{Rank}(PQ) = \mathrm{Rank}(Q)$, then $\ker PQ = \ker Q$. As $\mathrm{Rank}(B_\varphi^*) = r = \mathrm{Gr}(\varphi) = \mathrm{Rank}(G_\varphi)$, then

$$K = \ker(B_\varphi^* \otimes I_n) = \ker(A_\varphi \otimes I_n)(B_\varphi^* \otimes I_n) = \ker(G_\varphi \otimes I_n).$$

Meanwhile, by Corollary 5.2, in the finite-dimensional case, a linearly minimal homomorphism dilation system is equivalent to a principle dilation. Thus, by Theorem 5.1 and 5.2, a generalized Choi-Kraus dilation of $\varphi \in L(M_n, M_d)$ is a principle dilation if and only if the reduced subspace is $\ker(G_\varphi \otimes I_n)$.

From Theorem 5.2, any linearly minimal dilation is equivalent to a reduced dilation of the universal dilation. Meanwhile, any linearly minimal dilation can be reduced to a principle one with respect to a maximal invariant subspace by [128]. Next, we quantify the relation between the generalized Choi-Kraus dilations with universal dilation defined in Eq. (5.5) and the principle dilations from Proposition 5.3.

Theorem 5.7 *Let $\varphi \in L(M_n, M_d)$ with $Gr(\varphi) = r$. Let $(\pi_u, S_u, T_u, \mathbb{C}^{n^2 d})$ be the universal dilation, $(\pi_1, A, B^*, \mathbb{C}^{nr})$ be the principle dilation induced by $\varphi(X) = \sum_{i=1}^{r} A_i X B_i^*$ and $(\pi_2, E, F^*, \mathrm{Im}(F_\varphi^* \otimes I_n))$ be a linearly minimal generalized Choi-Kraus dilation induced by $\varphi(X) = \sum_{i=1}^{m} E_i X F_i^*$. Then*

1. *$(\pi_2, E, F^*, \mathrm{Im}(F_\varphi^* \otimes I_n))$ is equivalent to the reduced dilation of $(\pi_u, S_u, T_u, \mathbb{C}^{n^2 d})$ with respect to $K_1 = \ker(F_\varphi^* \otimes I_n)$.*
2. *$(\pi_1, A, B^*, \mathbb{C}^{nr})$ is equivalent to the reduced dilation of $(\pi_2, E, F^*, \mathrm{Im}(F_\varphi^* \otimes I_n))$ with respect to $K_2 = \ker(E_\varphi \otimes I) \bigcap \mathrm{Im}(F_\varphi^* \otimes I_n)$.*

Proof Similar to Eq. (5.6), we have

$$\sum_k c_k \pi_2(X_k) F^* x_k = (F_\varphi^* \otimes I_n)\left(\sum_k c_k X_k \otimes x_k\right).$$

Then the reduced subspace for $(\pi_2, E, F^*, \mathrm{Im}(F_\varphi^* \otimes I_n))$ can be simplified as $\ker(B_\varphi^* \otimes I_n)$. Thus if we define the following induced map

$$\widetilde{F_\varphi^* \otimes I_n} : \mathbb{C}^{n^2 d}/K_1 \to \mathrm{Im}(F_\varphi^* \otimes I_n), \quad \left[\sum_k c_k X_k \otimes x_k\right] \mapsto \sum_k c_k \pi_2(X_k) F^* x_k,$$

then it is easy to verify that $\widetilde{F_\varphi^* \otimes I_n}$ is a well-defined bijection from the quotient space $\mathbb{C}^{n^2 d}/K_1$ into $\mathrm{Im}(F_\varphi^* \otimes I_n)$ and

$$\widetilde{F_\varphi^* \otimes I_n}\widetilde{T_u} = F, \quad \pi_2(X)\widetilde{F_\varphi^* \otimes I_n} = \widetilde{F_\varphi^* \otimes I_n}\widetilde{\pi_u(X)}, \forall X \in M_n,$$

where $\widetilde{S_u}, \widetilde{T_u}$ and $\widetilde{\pi_u(X)}$ are corresponding induced maps. Thus, it follows from $E\widetilde{F_\varphi^* \otimes I_n} = \widetilde{S_u}$ that $(\pi_2, E, F^*, \mathrm{Im}(F_\varphi^* \otimes I_n))$ is equivalent to the reduced dilation of $(\pi_u, S_u, T_u, \mathbb{C}^{n^2 d})$ with respect to $K_1 = \ker(F_\varphi^* \otimes I_n)$.

Let $w = \sum_k c_k \pi(X_k) F^* x_k \in \ker(E_\varphi \otimes I) \bigcap \mathrm{Im}(F_\varphi^* \otimes I_n)$. Then we have

$$Ew = \Big(E_{1,1}, \cdots, E_{d,1}, E_{1,2}, \cdots, \cdots, E_{d,n}\Big)(E_\varphi \otimes I_n)w = 0.$$

and

$$(E_\varphi \otimes I_n)\pi_2(X)w = (E_\varphi \otimes I_n)(I_m \otimes X) = (I_{nd} \otimes X)(E_\varphi \otimes I_n)w.$$

Thus $K_2 = \ker(E_\varphi \otimes I) \bigcap \mathrm{Im}(F_\varphi^* \otimes I_n)$ is a π-invariant subspace in $\ker E_\varphi$. By Eq. (5.2) and the minimality of number of the Kraus-matrix pairs $\{A_i, B_i^*\}$, it is easy to show that $\ker B_\varphi^* = \ker G_\varphi$. Meanwhile $G_\varphi = E_\varphi F_\varphi^*$. Thus $\ker F_\varphi^* \subset \ker G_\varphi = \ker B_\varphi^*$. Due to the fact that $(\ker T)^\perp = \mathrm{Im}(T^*)$, for all $T \in M_{i,j}$ where $\perp$ denotes the orthogonal complement, we have $\mathrm{Im}(B_\varphi) \subset \mathrm{Im}(F_\varphi)$. Thus, by Douglas' Factorization Theorem [68], there exists $V \in M_{m,r}$ such that $B_\varphi = F_\varphi V$. Write V as $(v_{i,j})$. By the correspondence between the Kraus matrix pairs and the decomposition of the generalized Choi matrix, we get $B_i = \sum_{j=1}^m \overline{v_{j,i}} F_j^*$ for all $1 \leq i \leq r$.

Setting $R = V^* \otimes I_n : \mathrm{Im}(F_\varphi^* \otimes I_n) \to \mathbb{C}^{nr}$ we get

$$R\Big(\sum c_k \pi_2(X_k) F^* x_k\Big) = \Big(\sum c_k \pi_1(X_k) B^* x_k\Big).$$

It can be checked that $K_2 = \ker R$. Define the quotient map $\overline{R}$ by

$$\overline{R} : \mathrm{Im}(F_\varphi^* \otimes I_n)/K_2 \to \mathbb{C}^{nr}, \quad \Big[\sum c_k \pi_2(X_k) F^* x_k\Big] \mapsto \sum c_k \pi_1(X_k) B^* x_k.$$

Then $\overline{R}$ is a well-defined bijection, and moreover

$$\overline{R}\pi_2(X) = \pi_1(X)\overline{R}, \quad \overline{R}\overline{F^*} = B, \quad A\overline{R} = \overline{E}, \quad \forall X \in M_n,$$

where $\overline{E}, \overline{F^*}, \overline{\pi_2(X)}$ are the corresponding induced maps. Thus, the principle dilation $(\pi_1, A, B^*, \mathbb{C}^{nr})$ is equivalent to the reduced dilation of $(\pi_2, E, F^*, \mathrm{Im}(F_\varphi^* \otimes I_n))$ with respect to $K_2 = \ker(E_\varphi \otimes I) \bigcap \mathrm{Im}(F_\varphi^* \otimes I_n)$. $\square$

Combining the above concrete reduced subspace for generalized Choi-Kraus dilations and the classification theorems 5.2 and 5.3, we can quickly identify their dilation type. With the same notation as above, if $\mathrm{Rank}(F_\varphi^*) = nd$, then $K_1 = \{0\}$. If $\mathrm{Rank}(F_\varphi^*) = r$, as $G_r(\varphi) = r = \mathrm{Rank}(G_\varphi) = \mathrm{Rank}(E_\varphi F_\varphi^*)$, then by the Sylvester Theorem [213] again, we have $\ker E_\varphi \cap \mathrm{Im} F_\varphi^* = \{0\}$, and thus $K_2 = \{0\}$. Consequently, we have the following:

Corollary 5.4 *Let $\varphi \in L(M_n, M_d)$ with $Gr(\varphi) = r$. If $(\pi, E, F^*, \mathrm{Im}(F_\varphi^* \otimes I_n))$ is the linearly minimal generalized Choi-Kraus dilation induced by $\varphi(X) = \sum_{i=1}^m E_i X F_i^*$. Then.*

1. *If* $\text{Rank}(F_\varphi^*) = nd$, *then* $(\pi, E, F^*, \text{Im}(F_\varphi^* \otimes I_n))$ *is equivalent to the universal dilation.*
2. *If* $\text{Rank}(F_\varphi^*) = r$, *then* $(\pi, E, F^*, \text{Im}(F_\varphi^* \otimes I_n))$ *is a principle dilation.*

Suppose $(\pi, A, B, \mathbb{C}^{nm})$ is the generalized Choi-Kraus dilation induced by $\varphi(X) = \sum_{i=1}^m A_i X B_i^*$. As the linearly minimal dilation space is $\text{Im}(B_\varphi^* \otimes I_n)$, we have the following:

Corollary 5.5 *Let* $\varphi \in L(M_n, M_d)$ *with generalized Choi-Kraus representation* $\varphi(X) = \sum_{i=1}^m A_i X B_i^*$. *If* $\text{Rank } B_\varphi = m$, *then* $(\pi, A, B, \mathbb{C}^{mn})$ *is a linearly minimal dilation.*

By Theorem 5.2, two linearly minimal dilations are equivalent if and only if the reduced spaces associated with them are the same. Thus, by Theorem 6.8, we get

Corollary 5.6 *For* $\varphi \in L(M_n, M_d)$ *with* $\varphi(X) = \sum_{i=1}^m A_i X B_i^* = \sum_{j=1}^n E_j X F_j^*$. *Then the linearly minimal generalized Choi-Kraus dilations* $(\pi_1, A, B^*, \text{Im}(B_\varphi^* \otimes I_n))$ *and* $(\pi_2, E, F^*, \text{Im}(F_\varphi^* \otimes I_n))$ *induced by its representations respectively are equivalent if and only if* $\ker B_\varphi^* = \ker F_\varphi^*$.

5.2.2   Classifications for Linear Maps on Matrix Algebras M_n

For a CP map, all the Choi-Kraus dilations are Stinespring dilations. By Theorem 5.2, the reduced subspace associated with a linearly minimal Choi-Kraus dilation $(\pi, A, A^*, \text{Im}(A_\varphi^* \otimes I_n))$ induced by $\varphi(X) = \sum_{i=1}^m A_i X A_i^*$ is $\ker(A_\varphi^* \otimes I_n)$. Using the fact that $\ker T = \ker T^*T$ get that the reduced space is $\ker(C_\varphi \otimes I_n)$, where $A_\varphi A_\varphi^* = C_\varphi$. Thus, by Remark 5.6, all the linearly minimal Choi-Kraus dilations are principle dilations, hence are equivalent. The following result shows that they are actually also unitarily equivalent.

Proposition 5.4 *For every* $\phi \in CP(M_n, M_d)$, *all the linearly minimal Choi-Kraus dilations of* ϕ *are unitarily equivalent.*

Proof Suppose that $\text{Cr}(\varphi) = r$ and $\varphi(X) = \sum_{i=1}^r A_i X A_i^* = \sum_{j=1}^m B_j X B_j^*$ has two Choi-Kraus representations of φ with induced dilations $(\pi_1, A, A^*, \mathbb{C}^{nr})$ and $(\pi_2, B, B^*, \mathbb{C}^{nm})$, respectively. By Proposition 5.3, we have that $(\pi_1, A, A^*, \mathbb{C}^{nr})$ is a principle dilation, and $(\pi_2, B, B^*, \text{Im}(B_\varphi^* \otimes I_n))$ is a linearly minimal Choi-Kraus dilation.

By the Choi's second representation theorem [48, 98], there exists a unique matrix $U = (u_{i,j}) \in M_{m,r}$ such that $U^*U = I_r$, $B_i = \sum_{j=1}^r u_{i,j} A_j$. Define $\mathcal{U} = (\overline{u_{i,j}}) \otimes I_n :$ $\mathbb{C}^{nr} \to \mathbb{C}^{nm}$. Then it can be checked that $B^* = \mathcal{U} A^*$, $B\mathcal{U} = A$ and $\pi_2(X)\mathcal{U} = \mathcal{U}\pi_1(X)$. Meanwhile, for any $\sum_k c_k \pi(X_k) A^* x_k \in \mathbb{C}^{nr}$, we have

$$\mathcal{U}\left(\sum_k c_k \pi_1(X_k) A^* x_k\right) = \sum_k c_k \pi_2(X_k) B^* x_k.$$

Furthermore, as $\mathcal{U}^*\mathcal{U} = I_{nr}$ and $\mathcal{U} : \mathbb{C}^{nr} \to W_B$ is a surjective isometry, we get that $\mathcal{U}$ is unitary. Hence $(\pi_1, A, A^*, \mathbb{C}^{nr})$ and (π_2, B, B^*, W_B) are unitarily equivalent. By the transitivity of equivalence, we get that all the linearly minimal Choi-Kraus dilations are unitary equivalent.   □

For $\varphi \in L(M_n, M_d)$ with $\mathrm{Gr}(\varphi) = r$, a basis fact about the dimension k of the linearly minimal dilation space of a generalized Choi-Kraus dilation of φ is nm for some $m : r \le m \le nd$. This leads to some natural questions. For example,

(i) Under what condition is there only one (two, three, $\cdots$) equivalent class of linearly minimal generalized Choi-Kraus dilations?
(ii) Clearly equivalent linearly minimal dilations have the same dilation dimension. Is the converse also true? For question (i), we have

Proposition 5.5 *Let $\varphi \in L(M_n, M_d)$ with $\mathrm{Gr}(\varphi) = r$.*

1. *There is only one equivalent class of linearly minimal generalized Choi-Kraus dilations for φ if and only if $r = nd$.*
2. *There are two equivalent classes of linearly minimal generalized Choi-Kraus dilations for φ if and only if $r = nd - 1$.*
3. *There are infinite many inequivalent classes of linearly minimal generalized Choi-Kraus dilations for φ if and only if $r \le nd - 2$.*

Proof For $\varphi \in L(M_n, M_d)$, by Theorem 5.2, the reduced subspace associated with the linearly minimal generalized Choi-Kraus dilation $(\pi, A, B^*, \mathrm{Im}(B_\varphi^* \otimes I_n))$ induced by $\varphi(X) = \sum_{i=1}^m A_i X B_i^*$ is $\ker(B_\varphi^* \otimes I_n)$, that is, $\ker B_\varphi^* \otimes \mathbb{C}^n$. Since the reduced subspace associated with any principle generalized Choi-Kraus dilation is $\ker(G_\varphi \otimes I_n)$, which can be verified to be the largest reduced subspace, we get that

$$\ker(B_\varphi^* \otimes I_n) \subset \ker(G_\varphi \otimes I_n) = \ker G_\varphi \otimes \mathbb{C}^n.$$

Thus $\ker B_\varphi^* \subset \ker G_\varphi$. Moreover, by Theorem 5.2, each reduced subspace determines an equivalent class of linearly minimal dilations, and then the classification of all linearly minimal generalized Choi-Kraus dilations of φ can be reduced to the structure of the linear subspaces of $\ker G_\varphi$. Thus, we have

- There is only one equivalent class of linearly minimal generalized Choi-Kraus dilations if and only if $\ker G_\varphi$ has only one linear subspace if and only if $\ker G_\varphi = \{0\}$, that is, dim $\ker G_\varphi = 0$

- There are two equivalent classes linearly minimal generalized Choi-Kraus dilations if and only if $\ker G_\varphi$ has only two different linear subspaces if and only if $\dim \ker G_\varphi = 1$.
- There are infinite equivalent classes of linearly minimal generalized Choi-Kraus dilations if and only if $\ker G_\varphi$ has infinite different linear subspaces if and only if $\dim \ker G_\varphi \geq 2$.

As seen in Eq. (5.4) that $\mathrm{Rank}(G_\varphi) = \mathrm{Gr}(\varphi)$ and $\dim \ker G_\varphi + \dim \mathrm{Im}(G_\varphi) = nd$, that is,

$$\dim \ker G_\varphi = nd - \mathrm{Gr}(\varphi).$$

Thus, those statements are a direct consequence of the above characterizations.   □

Example 5.2.1 Let $\mathcal{A} = \mathcal{T}_n$ be the algebra of all the $n \times n$ upper triangular matrices, $\mathcal{T}_{n,0}$ be the algebra of all the $n \times n$ strictly upper triangular matrices, and $\varphi(a) = \frac{1}{n}\mathrm{tr}(a)$. Then $\mathcal{T}_{n,0}$ is a proper ideal contained in $\ker(\varphi)$. Thus, the universal dilation system is not equivalent to its canonical dilation system.

(i) For $n = 2$, set

$$A = \begin{pmatrix} a & b \\ 0 & c \end{pmatrix},$$

we have $\varphi(A) = \frac{1}{2}(a + c)$.

Then the universal dilation system $(\pi_u, S_u, T_u, \mathbb{C}^3)$ and the canonical dilation system $(\pi_c, S_c, T_c, \mathbb{C}^2)$ are given by

$$S_u = \begin{pmatrix} \frac{1}{2} & 0 & \frac{1}{2} \end{pmatrix}, \quad \pi_u(A) = \begin{pmatrix} a & 0 & 0 \\ 0 & a & b \\ 0 & 0 & c \end{pmatrix}, \quad T_u = \begin{pmatrix} 1 & 0 & 1 \end{pmatrix}^t$$

and

$$S_c = \begin{pmatrix} \frac{1}{2} & \frac{1}{2} \end{pmatrix}, \quad \pi_c(A) = \begin{pmatrix} a & 0 \\ 0 & c \end{pmatrix}, \quad T_c = \begin{pmatrix} 1 & 1 \end{pmatrix}^t.$$

These are the only two linearly minimal homomorphism dilations.

(ii) For the case $n = 3$, we have $\varphi(A) = \frac{1}{3}(a + d + f)$ where

$$A = \begin{pmatrix} a & b & c \\ 0 & d & e \\ 0 & 0 & f \end{pmatrix}.$$

In this case, the universal dilation system $(\pi_u, S_u, T_u, \mathbb{C}^6)$ and the canonical dilation system $(\pi_c, S_c, T_c, \mathbb{C}^3)$ are given by

$$S_u = \left(\tfrac{1}{3}\ 0\ \tfrac{1}{3}\ 0\ 0\ \tfrac{1}{3} \right), \quad \pi_u(A) = \begin{pmatrix} a\ 0\ 0\ 0\ 0\ 0 \\ 0\ a\ b\ 0\ 0\ 0 \\ 0\ 0\ d\ 0\ 0\ 0 \\ 0\ 0\ 0\ a\ b\ c \\ 0\ 0\ 0\ 0\ d\ e \\ 0\ 0\ 0\ 0\ 0\ f \end{pmatrix} \quad T_u = \begin{pmatrix} 1\ 0\ 1\ 0\ 0\ 1 \end{pmatrix}^t$$

and

$$S_c = \left(\tfrac{1}{3}\ \tfrac{1}{3}\ \tfrac{1}{3} \right), \quad \pi_c(A) = \begin{pmatrix} a\ 0\ 0 \\ 0\ d\ 0 \\ 0\ 0\ f \end{pmatrix} \quad T_c = \begin{pmatrix} 1\ 1\ 1 \end{pmatrix}^t.$$

In order to identify the rest of the equivalent classes of homomorphism dilations, we need to identify all the π_u-invariant subspaces in $\ker S_u$. Note that $\ker S_u = \operatorname{span}\{e_2, e_4, e_5\}$, and it is easy to verify that the maximal π_u-invariant subspace is $\operatorname{span}\{e_2, e_4\}$, and any one dimensional subspace of $\operatorname{span}\{e_2, e_4\}$ is also π_u-invariant. Hence, by Theorem 5.3, we only have one equivalent class of 4-dimensional homomorphism dilation systems and infinitely many inequivalent classes of 5-dimensional homomorphism dilation systems.

The 4-dimensional equivalent class of homomorphism dilation systems is represented by $(\pi_4, S_4, T_4, \mathbb{C}^4)$:

$$S_4 = \left(\tfrac{1}{3}\ \tfrac{1}{3}\ 0\ \tfrac{1}{3} \right), \quad \pi_4(A) = \begin{pmatrix} a\ 0\ 0\ 0 \\ 0\ d\ 0\ 0 \\ 0\ 0\ d\ e \\ 0\ 0\ 0\ f \end{pmatrix}, \quad T_4 = \begin{pmatrix} 1\ 1\ 0\ 1 \end{pmatrix}^t.$$

Two classes of homomorphism dilations systems represented by $(\pi_{5,1}, S_{5,1}, T_{5,1}, \mathbb{C}^5)$ and $(\pi_{5,2}, S_{5,2}, T_{5,2}, \mathbb{C}^5)$ associate with π_u-invariant subspaces $K_1 = \operatorname{span}\{e_2\}$ and $K_2 = \operatorname{span}\{e_4\}$, respectively, are given by

$$S_{5,1} = \left(\tfrac{1}{3}\ \tfrac{1}{3}\ 0\ 0\ \tfrac{1}{3} \right), \quad \pi_{5,1}(A) = \begin{pmatrix} a\ 0\ 0\ 0\ 0 \\ 0\ d\ 0\ 0\ 0 \\ 0\ 0\ a\ b\ c \\ 0\ 0\ 0\ d\ e \\ 0\ 0\ 0\ 0\ f \end{pmatrix}, \quad T_5 = \begin{pmatrix} 1\ 1\ 0\ 0\ 1 \end{pmatrix}^t$$

and

$$S_{5,2} = \begin{pmatrix} \frac{1}{3} & 0 & \frac{1}{3} & 0 & \frac{1}{3} \end{pmatrix}, \quad \pi_{5,2}(A) = \begin{pmatrix} a & 0 & 0 & 0 & 0 \\ 0 & a & b & 0 & 0 \\ 0 & 0 & d & 0 & 0 \\ 0 & 0 & 0 & d & e \\ 0 & 0 & 0 & 0 & f \end{pmatrix}, \quad T_{5,2} = \begin{pmatrix} 1 & 0 & 1 & 0 & 1 \end{pmatrix}^{t}.$$

We leave the construction of the homomorphism dilation associated with the π_u-invariant subspace $K_{\alpha,\beta} = \mathrm{span}\{\alpha e_2 + \beta e_4\}$ for the interested readers. □

However, if we consider the map on the whole matrix algebra, for example, let t be the transpose map on M_2 defined by $\tau(A) = A^t$. Then it can be shown that there is no nontrivial π_u-invariant subspaces in $\ker S_u$, and thus the above formula also gives us the canonical dilation. But the situation becomes quite different from triangular matrices. For simplicity, let us examine the transpose map on $\mathcal{T}_2$ and $\mathcal{T}_3$.

Example 5.2.2 (i) Let $\tau : \mathcal{T}_2 \to M_2$ be the transpose map. Then, the universal dilation system is given by

$$S_u = \begin{pmatrix} 1 & 0 & 0 & 0 & 0 & 0 \\ 0 & 1 & 0 & 0 & 0 & 1 \end{pmatrix}, \quad \pi_u\left(\begin{pmatrix} a & b \\ 0 & c \end{pmatrix}\right) = \begin{pmatrix} a & 0 & 0 & 0 & 0 & 0 \\ 0 & a & b & 0 & 0 & 0 \\ 0 & 0 & c & 0 & 0 & 0 \\ 0 & 0 & 0 & a & 0 & 0 \\ 0 & 0 & 0 & 0 & a & b \\ 0 & 0 & 0 & 0 & 0 & c \end{pmatrix}, \quad T_u = \begin{pmatrix} 1 & 0 & 1 & 0 & 0 & 0 \\ 0 & 0 & 0 & 1 & 0 & 1 \end{pmatrix}^{t}.$$

The canonical dilation system is given by

$$S_c = \begin{pmatrix} 1 & 0 & 0 & 0 \\ 0 & 1 & 1 & 0 \end{pmatrix}, \quad \pi_c\left(\begin{pmatrix} a & b \\ 0 & c \end{pmatrix}\right) = \begin{pmatrix} a & 0 & 0 & 0 \\ 0 & c & 0 & 0 \\ 0 & 0 & a & b \\ 0 & 0 & 0 & c \end{pmatrix}, \quad T_c = \begin{pmatrix} 1 & 0 & 0 & 1 \\ 0 & 1 & 0 & 0 \end{pmatrix}^{t}.$$

Furthermore, we have

$$\ker S_u = \mathrm{span}\{e_2 - e_6, e_3, e_4, e_5\}.$$

In $\ker S_u$, the maximal π_u-invariant subspace is $M = \mathrm{span}\{e_4, e_5\}$, and for any given α, β, the one-dimensional subspace $K_{\alpha,\beta} = \mathrm{span}\{\alpha e_4 + \beta e_5\}$ is π_u-invariant. So, again, there are infinitely many inequivalent classes of 5-dimensional dilations. The two special ones corresponding to $K_{1,0}$ and $K_{0,1}$ are represented by

$$\begin{pmatrix} 1 & 0 & 0 & 0 & 0 \\ 0 & 1 & 0 & 0 & 1 \end{pmatrix} \begin{pmatrix} a & 0 & 0 & 0 & 0 \\ 0 & a & b & 0 & 0 \\ 0 & 0 & c & 0 & 0 \\ 0 & 0 & 0 & a & b \\ 0 & 0 & 0 & 0 & c \end{pmatrix} \begin{pmatrix} 1 & 0 \\ 0 & 0 \\ 1 & 0 \\ 0 & 0 \\ 0 & 1 \end{pmatrix} \quad \text{and} \quad \begin{pmatrix} 1 & 0 & 0 & 0 & 0 \\ 0 & 1 & 0 & 0 & 1 \end{pmatrix} \begin{pmatrix} a & 0 & 0 & 0 & 0 \\ 0 & a & b & 0 & 0 \\ 0 & 0 & c & 0 & 0 \\ 0 & 0 & 0 & a & 0 \\ 0 & 0 & 0 & 0 & c \end{pmatrix} \begin{pmatrix} 1 & 0 \\ 0 & 0 \\ 1 & 0 \\ 0 & 1 \\ 0 & 1 \end{pmatrix}.$$

(ii) Let $\tau : \mathcal{T}_3 \to \mathbb{M}_3$ be the transpose map

$$\tau \left(\begin{pmatrix} a & b & c \\ 0 & d & e \\ 0 & 0 & f \end{pmatrix} \right) = \begin{pmatrix} a & 0 & 0 \\ b & d & 0 \\ c & e & f \end{pmatrix}.$$

Then we have the canonical dilation $\pi_c : \mathcal{T}_3 \to M_{10}$ by

$$S_c = \begin{pmatrix} 1 & 0 & 0 & 0 & 0 & 0 & 0 & 0 & 0 & 0 \\ 0 & 1 & 0 & 0 & 0 & 0 & 0 & 1 & 0 & 0 \\ 0 & 0 & 1 & 1 & 0 & 1 & 0 & 0 & 0 & 0 \end{pmatrix}$$

and

$$\pi_c \left(\begin{pmatrix} a & b & c \\ 0 & d & e \\ 0 & 0 & f \end{pmatrix} \right) = \begin{pmatrix} a & 0 & 0 & 0 & 0 & 0 & 0 & 0 & 0 & 0 \\ 0 & d & 0 & 0 & 0 & 0 & 0 & 0 & 0 & 0 \\ 0 & 0 & f & 0 & 0 & 0 & 0 & 0 & 0 & 0 \\ 0 & 0 & 0 & a & b & 0 & 0 & 0 & 0 & c \\ 0 & 0 & 0 & 0 & d & 0 & 0 & 0 & 0 & e \\ 0 & 0 & 0 & 0 & 0 & d & e & 0 & 0 & 0 \\ 0 & 0 & 0 & 0 & 0 & 0 & f & 0 & 0 & 0 \\ 0 & 0 & 0 & 0 & 0 & 0 & 0 & a & b & 0 \\ 0 & 0 & 0 & 0 & 0 & 0 & 0 & 0 & d & 0 \\ 0 & 0 & 0 & 0 & 0 & 0 & 0 & 0 & 0 & f \end{pmatrix} \quad \text{and} \quad T_c = \begin{pmatrix} 1 & 0 & 0 \\ 0 & 1 & 0 \\ 0 & 0 & 1 \\ 0 & 0 & 0 \\ 0 & 0 & 0 \\ 0 & 0 & 0 \\ 0 & 1 & 0 \\ 0 & 0 & 0 \\ 1 & 0 & 0 \\ 1 & 0 & 0 \end{pmatrix}.$$

$\square$

In what follows, we present some examples of linear maps on the matrix algebras. We first give an example with full rank, as mentioned before, the transpose map.

Example 5.2.3 Let $\varphi : M_n \rightarrow M_n$ be the transpose map, i.e. $\varphi(X) = X^t$, for all $X \in M_n$. One of the generalized Choi-Kraus representations is $\varphi(X) = \sum_{1 \le i,j \le n} E_{i,j} X E_{i,j}$.
Define

$$\pi(X) = \begin{pmatrix} X & \cdots & 0 \\ \vdots & \ddots & \vdots \\ 0 & \cdots & X \end{pmatrix} = I_{n^2} \otimes X, \ \forall \, X \in M_n$$

and

$$A = (E_{1,1}, \cdots, E_{1,n}, E_{2,1}, \cdots, \cdots, E_{n,n}),$$

$$B = (E_{1,1}, \cdots, E_{n,1}, E_{1,2}, \cdots, \cdots, E_{n,n}).$$

Then $(\pi, A, B^*, \mathbb{C}^{n^3})$ is a linearly minimal generalized Choi-Kraus dilation. It can be checked that $(\pi, A, B^*, \mathbb{C}^{n^3})$ is a universal dilation as well as a principle dilation, by Theorem 5.2, the largest reduced subspace is $\{0\}$. Thus, any linearly minimal generalized Choi-Kraus dilation is equivalent to the universal dilation. This also follows from Proposition 5.5 since its generalized Choi matrix $G_\varphi = (\varphi(E_{i,j})) = (E_{j,i})$ is a unitary and hence has full rank. Therefore, all the linearly minimal dilations of φ are equivalent.

□

Here is an example with exactly two equivalent classes of linearly minimal generalized Choi-Kraus dilations.

Example 5.2.4 Let $\varphi : M_2 \rightarrow M_2$ be defined by

$$\varphi \left(\begin{pmatrix} a_{1,1} & a_{1,2} \\ a_{2,1} & a_{2,2} \end{pmatrix} \right) = \begin{pmatrix} a_{1,1} + a_{2,2} & 0 \\ 0 & -a_{1,1} \end{pmatrix}.$$

We can decompose the generalized Choi-matrix G_φ into the sum of rank-one matrices and rearrange them into Kraus matrix pairs. Define

$$\pi_1(X) = \begin{pmatrix} X & 0 & 0 \\ 0 & X & 0 \\ 0 & 0 & X \end{pmatrix} = I_3 \otimes X, \ \forall \, X \in M_2$$

and

$$A = \begin{pmatrix} 1 & 0 & 0 & 0 & 0 & 1 \\ 0 & 0 & -1 & 0 & 0 & 0 \end{pmatrix} \quad B = \begin{pmatrix} 1 & 0 & 0 & 0 & 0 & 1 \\ 0 & 0 & 1 & 0 & 0 & 0 \end{pmatrix}.$$

Then $(\pi_1, A, B^*, \mathbb{C}^6)$ is a generalized principle Choi-Kraus dilation. And the universal dilation $\pi_u : M_2 \to M_8$ is given by

$$\pi_u(X) = \begin{pmatrix} X & \cdots & 0 \\ \vdots & \ddots & \vdots \\ 0 & \cdots & X \end{pmatrix} = I_4 \otimes X, \ \forall\, X \in M_2$$

with

$$S_u = \begin{pmatrix} 1\,0\ \ 0\ \ 0\,0\,1\,0\,0 \\ 0\,0\,-1\,0\,0\,0\,0\,0 \end{pmatrix}, \quad T_u = \begin{pmatrix} 1\,0\,0\,0\,0\,1\,0\,0 \\ 0\,0\,1\,0\,0\,0\,0\,1 \end{pmatrix}^t.$$

Clearly, $(\pi_1, A, B^*, \mathbb{C}^6)$ is not equivalent to $(\pi_u, S_u, T_u, \mathbb{C}^8)$. Moreover, it can be verified that $\{0\}$ and $\ker(G_\varphi \otimes I_2)$ are the only two reduced subspaces and hence there are only two equivalent classes of linearly minimal generalized Choi-Kraus dilations.   □

The next example shows that there are infinitely many inequivalent linearly minimal dilations even with the same dilation dimension. This answers question (ii) negatively.

Example 5.2.5 Let $A = \begin{pmatrix} 1 & 2 \\ 3 & 4 \end{pmatrix}$. Define $\varphi \in L(M_2, M_2)$ by $\varphi(X) = A \circ X$, for all $X \in M_2$, where $A \circ B$ denotes the Hadamard (entrywise) product of A and B (also called Schur product). Specifically,

$$\varphi\left(\begin{pmatrix} a_{1,1} & a_{1,2} \\ a_{2,1} & a_{2,2} \end{pmatrix} \right) = \begin{pmatrix} a_{1,1} & 2a_{1,2} \\ 3a_{2,1} & 4a_{2,2} \end{pmatrix}.$$

Set

$$A = \begin{pmatrix} 1\,0\,2\,0 \\ 0\,3\,0\,4 \end{pmatrix}, \quad B = \begin{pmatrix} 1\,0\,0\,0 \\ 0\,0\,0\,1 \end{pmatrix}, \quad \pi_1(X) = \begin{pmatrix} X & 0 \\ 0 & X \end{pmatrix}, \ \forall X \in M_2.$$

Then $(\pi_1, A, B^*, \mathbb{C}^4)$ is a principle dilation.
Define

$$\pi_u(X) = \begin{pmatrix} X & 0 & 0 & 0 \\ 0 & X & 0 & 0 \\ 0 & 0 & X & 0 \\ 0 & 0 & 0 & X \end{pmatrix}, \ \forall X \in M_2$$

with

$$S_u = \begin{pmatrix} 1\,0\,0\,0\,0\,0\,2\,0 \\ 0\,3\,0\,0\,0\,0\,0\,4 \end{pmatrix}, \quad T_u = \begin{pmatrix} 1\,0\,0\,0\,0\,1\,0\,0 \\ 0\,0\,1\,0\,0\,0\,0\,1 \end{pmatrix}^t.$$

Then $(\pi_u, S_u, T_u, \mathbb{C}^8)$ is the universal dilation.

Besides, if we define

$$\pi_3(X) = \begin{pmatrix} X & 0 & 0 \\ 0 & X & 0 \\ 0 & 0 & X \end{pmatrix}, \quad \forall X \in M_2$$

with

$$C = \begin{pmatrix} 1\,0\,1\,0 & -2 & 0 \\ 0\,3\,0\,1 & 0 & -4 \end{pmatrix}, \quad D_\lambda = \begin{pmatrix} 1 & \lambda & 0 & \lambda & 0 & \lambda \\ 1-\lambda & 0 & 1-\lambda & 0 & 1-\lambda & -1 \end{pmatrix},$$

then $(\pi_3, C, D_\lambda^*, \mathbb{C}^6)$ is a linearly minimal generalized Choi-Kraus dilation.

Let

$$R_\lambda = \begin{pmatrix} I_2 & (1-\lambda)I_2 & \lambda I_2 & 0 \\ 0 & (1-\lambda)I_2 & \lambda I_2 & 0 \\ 0 & (1-\lambda)I_2 & \lambda I_2 & -I_2 \end{pmatrix}.$$

The reduced subspace associated with $(\pi_3, C, D_\lambda^*, \mathbb{C}^6)$ is the $\ker R_\lambda$. Taking $\lambda_1 \neq \lambda_2$, clearly $\ker R_{\lambda_1} \neq \ker R_{\lambda_2}$. Then $(\pi_3, C, D_{\lambda_1}^*, \mathbb{C}^6)$ and $(\pi_3, C, D_{\lambda_2}^*, \mathbb{C}^6)$ are not equivalent. In summary, there are infinitely many inequivalent classes of 6-dimensional dilations. □

Although none of the above three examples is completely positive. It is also a trivial exercise to construct similar examples for quantum channels (i.e., trace-preserving CP maps). Here is one that admits infinitely many equivalent classes of linearly minimal generalized dilations.

Example 5.2.6 Let $\phi \in CP(M_2, M_2)$ be

$$\phi\left(\begin{pmatrix} a_{1,1} & a_{1,2} \\ a_{2,1} & a_{2,2} \end{pmatrix}\right) = \begin{pmatrix} a_{1,1} & 0 \\ 0 & a_{2,2} \end{pmatrix}.$$

Define

$$A = \begin{pmatrix} 1 & 0 & 0 & 0 \\ 0 & 0 & 0 & 1 \end{pmatrix}, \quad \pi_1(X) = \begin{pmatrix} X & 0 \\ 0 & X \end{pmatrix} \quad \forall X \in M_2,$$

and

$$B = \frac{1}{\sqrt{2}} \begin{pmatrix} 1 & 0 & 1 & 0 \\ 0 & 1 & 0 & -1 \end{pmatrix}, \quad \pi_2(X) = \begin{pmatrix} X & 0 \\ 0 & X \end{pmatrix}, \quad \forall X \in M_2.$$

Then $(\pi_1, A, A^*, \mathbb{C}^4)$ and $(\pi_2, B, B^*, \mathbb{C}^4)$ are linearly minimal Choi-Kraus dilations hence unitarily equivalent by Proposition 5.4.

The universal dilation $(\pi_u, S_u, T_u, \mathbb{C}^8)$ is given by

$$\pi_u(X) = \begin{pmatrix} X & \cdots & 0 \\ \vdots & \ddots & \vdots \\ 0 & \cdots & X \end{pmatrix} = I_4 \otimes X, \ \forall \, X \in M_2$$

with

$$S_u = \begin{pmatrix} 1 & 0 & 0 & 0 & 0 & 0 & 0 & 0 \\ 0 & 0 & 0 & 0 & 0 & 0 & 0 & 1 \end{pmatrix}, \quad T_u = \begin{pmatrix} 1 & 0 & 0 & 0 & 0 & 1 & 0 & 0 \\ 0 & 0 & 1 & 0 & 0 & 0 & 0 & 1 \end{pmatrix}^t.$$

Besides, if we set

$$\pi(X) = \begin{pmatrix} X & 0 & 0 \\ 0 & X & 0 \\ 0 & 0 & X \end{pmatrix} = I_3 \otimes X, \ \forall \, X \in M_2$$

with

$$E = \begin{pmatrix} \frac{3}{5} & 0 & \frac{4}{5} & 0 & 0 & 0 \\ 0 & 0 & 0 & 0 & 0 & 1 \end{pmatrix}, \quad F_\alpha = \begin{pmatrix} \frac{3}{5} & \frac{4\alpha}{5} & \frac{4}{5} & -\frac{3\alpha}{5} & 0 & 0 \\ \frac{4(1-\alpha)}{5} & 0 & -\frac{3(1-\alpha)}{5} & 0 & 0 & 1 \end{pmatrix},$$

it follows that $(\pi, E, F_\alpha^*, \mathbb{C}^6)$ is a linearly minimal generalized Choi-Kraus dilation.

By Corollary 5.6, if $\alpha_1 \neq \alpha_2$, then $(\pi, E, F_{\alpha_1}^*, \mathbb{C}^6)$ and $(\pi, E, F_{\alpha_2}^*, \mathbb{C}^6)$ are not equivalent. Thus, there are infinitely many equivalent classes of linearly minimal generalized Choi-Kraus dilations. □

Quantum Detection and Phase Retrieval Frames

6

Positive operator valued measures (POVMs) have an important physical significance, as they represent the generalized observables in the quantum theory of measurement and in its applications, such as quantum optics or theory of measurement in phase-space, which are called generalized measurements in quantum mechanics. So POVMs provide some basic mathematical tools in quantum information theory. A special class of POVMs, namely frame POVMs, is induced by Hilbert space frames. Such a POVM, due to its simplicity, plays special roles in applications including the quantum detection problem in quantum information theory and the phase retrieval problem in engineering. In this chapter, we review some basics and present some of our recent work related to quantum detection and phase retrievable frames.

6.1   Quantum States and Observables

We begin with a brief review of the mathematical framework of quantum states and present them as the following Postulates [175].

(P1) To each isolated physical system, there corresponds a Hilbert space $\mathcal{H}$, called the state space, and each unit vector in $\mathcal{H}$ represents a possible state, called the state vector or pure state.

When $\dim \mathcal{H} = 2$, the system is called a qubit system. Usually, we assume that $\mathcal{H}$ is finite-dimensional unless specified and take the bra-ket notation, that is, for $\psi \in \mathcal{H}$, $A \in B(\mathcal{H})$, we view $|\psi\rangle \in \mathcal{H}$ as a vector and $\langle\psi| \in \mathcal{H}^*$ as a functional. Thus, we have $A|h\rangle \in \mathcal{H}$, $\langle h|A \in \mathcal{H}^*$ and $\langle h|A|h\rangle \in \mathbb{C}$. Due to Postulate (2) on measurement, it turns out that measurements do not distinguish between different unit vectors ψ from the one-dimensional subspace spanned by the given state vector ψ and hence states should really refer to one-dimensional subspace and not just a unit vector. This means that the probabilities of outcomes really depend on the one-dimensional subspace generated by a vector. Thus, we can replace states by rank one projections and lengths by trace by the following extended postulate:

A state vector $h \in \mathcal{H}$ can be replaced by an operator $|h\rangle\langle h|$ acting on $\mathcal{H}$, which we call a *pure state*. And a "state" of a system is described by a mixed state

$$\rho = \sum_{i \geq 1} p_i |h_i\rangle\langle h_i|, \quad \sum_i p_i = 1, \quad p_i \geq 0$$

which we interpret as the pure states $|h_i\rangle\langle h_i|$ being "mixed" with "probability" p_i. Mathematically speaking, the state ρ is a positive matrix with trace one, which we call it a *density matrix*.

When we want to observe a system, we need the observables, i.e., quantum measurement.

(P2) Quantum measurements are always described by a class of operators $\{M_i\}$ where i is one of the outcomes.

The probability that we observe the outcome i, for a state ψ before we measure, is given by $p_i = \|M_i\psi\|^2$ and the state changes to $\frac{M_i\psi}{\|M_i\psi\|}$. As quantum mechanics is inherently probabilistic. So the sum of the probabilities of all possible outcomes must equal 1. Thus, for every state vector $\psi \in \mathcal{H}_s$ we have

$$1 = \sum_{i=1}^{k} p_i = \sum_{i=1}^{k} \|M_i\psi\|^2 = \sum_{i=1}^{k} \langle M_i\psi, M_i\psi \rangle = \sum_{i=1}^{k} \langle \psi, M_i^* M_i\psi \rangle.$$

Since the above equality holds for every $\psi \in \mathcal{H}$ with $\|\psi\| = 1$, it forces $\sum_{i=1}^{k} M_i^* M_i = I$. In other words, a quantum measurement is a family of operators $\{M_i\}_{i \in I} \subseteq B(\mathcal{H})$ acting on the density matrix satisfying the condition $\sum_{i \in I} M_i^* M_i = I_{\mathcal{H}}$ where $\{P_i = M_i^* M_i\}_{i \in I}$ is a positive operator valued measure (POVM).

(P3) If a quantum event occurs transforming pure states in $\mathcal{H}_1$ to $\mathcal{H}_2$, (then there exists a linear map $\Phi : L(\mathcal{H}_1) \longrightarrow L(\mathcal{H}_2)$ such that Φ maps density matrices to density matrices.

Arising from the measurement system, the above postulate is generalized as quantum channels (completely positive and trace-preserving map), which play an essential role in quantum information theory [98].

As shown in the above postulate, positive operator-valued measures (POVMs) are called generalized measurements in quantum mechanics and are the basic mathematical tools in quantum information theory. There is an extensive literature on POVMs and we refer the reader to some standard references c.f. [17, 34, 125, 131] for more details. More generally, given a positive operator-valued measure V acting on a Hilbert space $\mathcal{H}$ over a σ-algebra Σ of subsets. The probability outcome of the quantum measurement on the state ρ is formulated by

$$P_B = \text{tr}(V(B)\rho), \forall B \in \Sigma,$$

where $\text{tr}(T)$ is the trace of the operator given by $\text{tr}(T) = \sum_{i \in I} \langle Te_i, e_i \rangle$ for an orthonormal basis $\{e_i\}_{i \in I}$ of $\mathcal{H}$.

6.2   Quantum Detection from Frame POVMs

In quantum information theory, quantum state tomography asks to recover states from the probability of observing outcomes from measurements of the system on these states, and retrieving data from quantum systems is carried out according to quantum measurement theory. Let $\mathcal{S}(\mathcal{H})$ be the set of *states* or *density operators* on $\mathcal{H}$ consisting of ρ in the trace class operators on $\mathcal{H}$ such that $\rho \geq 0$ and $\text{tr}(\rho) = 1$. Mathematically speaking, the quantum detection problem asks to characterize the quantum measurement or a positive operator-valued measure (POVM) $V : \Sigma \to B(\mathcal{H})$ such that

$$\text{tr}(\rho_1 V(E)) = \text{tr}(\rho_2 V(E)), \forall E \in \Sigma.$$

implies that $\rho_1 = \rho_2$. Such a POVM is said to be *informationally complete*.

A well-examined class of POVMs is the one that is induced by a frame which we will call a frame POVM. Given a Parseval frame $\{x_i\}_{i \in I}$ (the index I could be finite or infinite) for $\mathcal{H}$, define

$$V : 2^I \to B(X) \quad V(B) = \sum_{i \in B} x_i \otimes x_i$$

where $x_i \otimes x_i(x) = \langle x, x_i \rangle x_i$. It can be verified that V is a POVM. A frame-induced quantum measurement is given by the function

$$p : 2^I \to \mathbb{R}, \quad p(B) = \sum_{k \in B} \text{tr}(\rho x_k x_k^*) = \sum_{k \in B} \langle \rho x_k, x_k \rangle.$$

In this special case, the quantum detection problem asks to characterize all the Parseval frames $\{x_i\}_{i \in I}$ such that the map $M \colon \mathcal{S}(\mathbb{H}) \to L(2^I, \mathbb{R})$ defined by $M(\rho)(B) = p(B) = \sum_{k \in B} \langle \rho x_k, x_k \rangle$ for $B \subset I$ is injective?

Similarly, given a continuous Parseval (Ω, μ)-frame $\mathcal{F}$. Let Σ be the σ-algebra of subsets over Ω. Define

$$U : \Sigma \to B(\mathcal{H}), \quad U(E) = \int_E \mathcal{F}(\omega) \otimes \mathcal{F}(\omega) \mathrm{d}\mu(\omega).$$

in the sense of

$$\langle U(E)x, y \rangle = \int_E \langle x, \mathcal{F}(\omega) \rangle \langle \mathcal{F}(\omega), y \rangle \mathrm{d}\mu(\omega), \forall x, y \in \mathcal{H}.$$

Then U is a frame POVM induced by the frame $\mathcal{F}$. In this case, the quantum detection problem asks to determine the injectivity of the map $\mathbb{P} : \mathcal{S}(\mathcal{H}) \to B(\Sigma, \mathbb{R})$, where

$$\mathbb{P}(\rho)(E) = \mathrm{tr}(\rho U(E)) = \mathrm{tr}\left(\rho\left(\int_E \mathcal{F}(\omega) \otimes \mathcal{F}(\omega) \mathrm{d}\mu(\omega) \right) \right) \forall E \in \Sigma. \tag{6.1}$$

In what follows, we say that a frame $\mathcal{F}$ is *quantum injective*, or simply *injective* if $\mathbb{P}$ is injective.

6.2.1   Injective Discrete Frames

In 2008, J. Benedetto et al. [33] used concepts from classical mechanics and orthogonal group techniques to convert the frame quantum detection problem into a set of ordinary differential equations to give a general method and construct a tight frame that minimizes an error term. In 2017, Botelho-Andrade et al. [18] used discrete frame measurements to solve the quantum injectivity problem in both real and complex cases and in both finite-dimensional and infinite-dimensional cases. Additionally, they also worked on the problem in a much more general setting including:

1. Self-adjoint operators which may not be positive.
2. Operators which are not trace one but are Hilbert-Schmidt.
3. Frames which are not Parseval.

As for discrete frames, the frame quantum detection problem can be formulated as follows.

Definition 6.1 A family of vectors $\mathcal{X} = \{x_k\}_{k \in I}$ in a Hilbert space $\mathbb{H}$ is said to be injective if whenever a Hilbert-Schmidt self-adjoint operator T satisfies

$$\langle Tx_k, x_k \rangle = 0, \ \forall \, k \in I, \tag{6.2}$$

then $T = 0$.

Remark 6.1 Note that in the above definition, we do not require the frame to be Parseval as it is easy to see that the injectivity of a frame is preserved under linear isomorphisms so that if we have a frame $\mathcal{X} = \{x_k\}_{k \in I}$ giving injectivity, then its canonical Parseval frame $\{S^{-1/2}x_k\}_{k \in I}$ is injective. And every operator can be presented as a linear combination of self-adjoint operators, even positive operators, thus, we relax, when possible, the usual assumption of positivity in the quantum information theory literature and deal with self-adjoint operators which may not be positive.

6.2.1.1 Finite-dimensional Case

In 2017, Botelho-Andrade et al. [18] presented the solution to the finite-dimensional injectivity problem for both real and complex cases. We take the real case $\mathbb{R}^n$ as an example.

Note that, in finite-dimensional real space, given a self-adjoint operator $T = (a_{ij})_{i,j=1}^n$ on $\mathbb{R}^n$ and a vector $x = (x_1, x_2, \ldots, x_n) \in \mathbb{R}^n$, we have

$$\langle Tx, x \rangle = \sum_{i=1}^n \sum_{j=1}^n a_{ij} x_i x_j = \sum_{i=1}^n a_{ii} x_i^2 + 2 \sum_{i=1}^n \sum_{j=i+1}^n a_{ij} x_i x_j.$$

And we can also turn the inner product in $\mathbb{R}^n$ or $\mathbb{C}^n$ into the Hilbert-Schmidt inner product $(\langle A, B \rangle_{HS} = \mathrm{tr}(A^*B))$ on M_n. Then we have

$$\langle Tx, x \rangle = \langle T, x^*x \rangle_{HS}.$$

Let $\mathcal{E} = \{e_j\}_{j=1}^n$ be an eigenbasis for T with respect to the eigenvalues $\{\lambda_j\}_{j=1}^n$. Then for every $k = 1, 2, \ldots, m$ we have

$$\langle Tx, x \rangle = \sum_{j=1}^n \lambda_j |\langle x_k, e_j \rangle|^2 = 0$$

By the above basic computation and the fact that the orthogonal complement space of a spanning set is 0, Botelho-Andrade et al. [18] obtained various classifications of the frame $\mathcal{X}$ that give injectivity for the quantum detection problem.

Definition 6.2 For a self-adjoint operator $T = (a_{ij})_{i,j=1}^n$ on $\mathbb{R}^n$, the vectorization $\tilde{T}$ of T in $\mathbb{R}^{\frac{n(n+1)}{2}}$ is defined by:

$$\tilde{T} = (a_{11}, 2a_{12}, \ldots, 2a_{1n}; a_{22}, 2a_{23}, \ldots, 2a_{2n}; \ldots; a_{(n-1)(n-1)}, 2a_{(n-1)n}; a_{nn})$$

For a vector $x = (x_1, x_2, \ldots, x_n) \in \mathbb{R}^n$, we associate a vector $\tilde{x}$ in $\mathbb{R}^{\frac{n(n+1)}{2}}$ by

$$\tilde{x} = (x_1 x_1, \ldots, x_1 x_n; x_2 x_2, \ldots, x_2 x_n; \ldots; x_{n-1} x_{n-1}, x_{n-1} x_n; x_n x_n)$$

Various results in [18] are summarized as follows.

Theorem 6.1 *Let $\mathcal{X} = \{x_k\}_{k=1}^m$ be a frame for $\mathbb{R}^n$. The following are equivalent:*

1. *$\mathcal{X}$ gives injectivity.*
2. *$\{\tilde{x}_k\}_{k=1}^m$ spans $\mathcal{K} := \mathbb{R}^{n(n+1)/2}$.*
3. *For every orthonormal basis $\mathcal{E} = \{e_j\}_{j=1}^n$ for $\mathbb{H}^n$ we have:*

$$H(\mathcal{E}) =: \mathrm{span}\{(|\langle x_k, e_1 \rangle|^2, |\langle x_k, e_2 \rangle|^2, \ldots, |\langle x_k, e_n \rangle|^2) : k = 1, 2, \ldots, m\} = \mathbb{R}^n$$

Next is a simple example of an injective frame.

Example 6.2.1 The frame

$$\{e_i\}_{i=1}^n \cup \{e_i + e_j : i < j\}_{i,j=1}^n$$

gives injectivity.

For the complex case, we can slightly modify some of the definitions to have similar results, we omit it and leave it to the interested readers.

Next, we will show that the family of m-element frames which solve the quantum detection injectivity problem is open and dense in the family of all m-element frames. For this, we need to measure the distance between m-element frames.

Definition 6.3 Given frames $\mathcal{X} = \{x_k\}_{k=1}^m$ and $\mathcal{Y} = \{y_k\}_{k=1}^m$ for a real Hilbert space $\mathbb{R}^n$, or complex one $\mathbb{C}^n$, the distance between them is

$$d(\mathcal{X}, \mathcal{Y})^2 = \sum_{k=1}^m \|x_k - y_k\|^2.$$

Theorem 6.2 ([20]) *The set of all m-element frames on $\mathbb{R}^n$, or $\mathbb{C}^n$ that give injectivity in the frame quantum detection problem is open and dense in the space of all m-element frames on $\mathbb{H}^n$. In particular, the set of all m-element Parseval frames which give injectivity is dense in the set of all m-element Parseval frames.*

6.2.1.2 Infinite-dimensional Case

In the infinite-dimensional case, we work with a general case: the Hilbert-Schmidt operators. i.e. operators $T = (a_{ij})_{i,j=1}^{\infty}$ with $\sum_{i,j=1}^{\infty} |a_{ij}|^2 < \infty$. The injectivity problem asks to find the frames $\{x_k\}_{k=1}^{\infty}$ in infinite dimensional Hilbert space $\mathcal{H}$ such that T, S are Hilbert Schmidt positive self-adjoint operators on $\mathcal{H}$ and

$$\langle Tx_k, x_k \rangle = \langle Sx_k, x_k \rangle, \ \forall k \in \mathbb{N},$$

then $T = S$.

Similar to the finite case, the classification in [18] of injective frames for Hilbert-Schmidt operators is shown as follows.

Theorem 6.3 *Let* $\mathcal{X} = \{x_k\}_{k=1}^{\infty}$ *be a frame for an infinite-dimensional real or complex Hilbert space* $\mathcal{H}$. *The following are equivalent:*

1. $\mathcal{X}$ *is injective.*
2. *For every orthonormal basis* $\mathcal{E} = \{e_j\}_{j=1}^{\infty}$ *for* $\mathcal{H}$ *we have:*

$$H(\mathcal{E}) =: \overline{\text{span}}\{(|\langle x_k, e_1 \rangle|^2, |\langle x_k, e_2 \rangle|^2, \ldots) : k = 1, 2, \ldots\} = \ell_2.$$

3. $\overline{\text{span}}\{\tilde{x}_k\}_{k=1}^{\infty} = \tilde{\mathcal{H}}$, *where* $\tilde{\mathcal{H}}$ *is the direct sum of the real Hilbert spaces* ℓ_2:

$$\tilde{\mathcal{H}} = \left(\sum_{i=1}^{\infty} \oplus \ell^2 \right)_{\ell_2}$$

and for $x = \{x_i\}_{i=1}^{\infty}$ *in real* ℓ^2, $\tilde{x}$ *is defined to be* $(\overline{x}_1, \overline{x}_2, \ldots, \overline{x}_n, \ldots) \in \tilde{\mathcal{H}}$ *with*

$$\overline{x}_1 = (x_1 x_1, x_1 x_2, \ldots); \ \ldots; \ \overline{x}_n = (x_n x_n, x_n x_{n+1}, \ldots); \ \ldots$$

For $x = \{x_i\}_{i=1}^{\infty}$ *in complex* ℓ_2, $\tilde{x}$ *is defined to be* $(\overline{x}_1, \overline{x}_2, \ldots, \overline{x}_n, \ldots) \in \tilde{\mathcal{H}}$ *with*

$$\overline{x}_1 = (|x_1|^2, \text{Re}(\tilde{x}_1 x_2), \text{Im}(\tilde{x}_1 x_2), \text{Re}(\tilde{x}_1 x_3), \text{Im}(\tilde{x}_1 x_3), \ldots)$$

$$\overline{x}_n = (|x_n|^2, \text{Re}(\tilde{x}_n x_{n+1}), \text{Im}(\tilde{x}_n x_{n+1}), \text{Re}(\tilde{x}_n x_{n+2}), \text{Im}(\tilde{x}_n x_{n+2}), \ldots); \ \ldots$$

The following provides some concrete examples of injective frames.

Theorem 6.4 ([20]) *Let* $\{e_i\}_{i=1}^{\infty}$ *be the canonical basis for the real Hilbert space* ℓ_2 *and let* $a_i \neq 0$ *for* $i = 1, 2, \ldots$ *be such that* $\sum_{i=1}^{\infty} a_i^2 < \infty$. *Define*

$$x_k = a_k(e_1 + e_{k+1}), \ \text{for } k = 1, 2, \ldots$$

Let L be the right shift operator on ℓ_2. Then the family

$$\{e_i\}_{i=1}^{\infty} \cup \{\frac{1}{2^i} L^i x_k\}_{i=0,k=1}^{\infty,\infty}$$

is a frame for ℓ_2 which gives injectivity.

We have seen in Theorem 6.2, that the injective frames are open and dense in finite-dimensional cases. However, the solutions to the injectivity problem in infinite dimensions are neither open nor dense in the class of frames. Similarly, Definition 6.3 is generalized to the infinite-dimensional case, and it follows that:

Theorem 6.5 ([20]) *Let $\mathcal{X} = \{e_i\}_{i=1}^{\infty} \cup \{\frac{1}{2^i} L^i x_k\}_{i=0,k=1}^{\infty}$ be the injective frame for the real space ℓ_2 as in Theorem 6.3. Then for any $\varepsilon > 0$, there is a frame $\mathcal{Y}$ such that $d(\mathcal{X}, \mathcal{Y}) < \varepsilon$, and $\mathcal{Y}$ is not injective.*

6.2.2   Injective Continuous Frames

The characterization of Botelho-Andrade et al. [18, 20] for the quantum injective discrete frames was given in terms of the spanning properties of some derived sequences from the frame vectors. While discrete frames might be much better suited for computations, we would also naturally wonder how much of the results are still valid for other types of frames, for instance, continuous frames. On the one hand, it is reasonable to expect similar results in terms of a continuous frame. It is impractical, on the other hand, to verify the injectivity by performing uncountably many operations. So when dealing with continuous frames, researchers often sample the continuous frame into a discrete frame and then use the discrete frame for computations instead of the entire continuous frame. For example, Freeman, et al. [88] solved the discretization problem for the continuous frame and showed when a continuous frame for a Hilbert space may be sampled to obtain a frame. In this subsection, we present a few similar results for injective continuous frames [118] in terms of their discrete representations that were introduced in [93]. Constructions of injective continuous frames and their stability will also be discussed. The following list of notations will be used in this subsection.

- $B_{sa}(\mathcal{H})$ – the real linear space of self-adjoint bounded operators on a separable Hilbert space $\mathcal{H}$;
- $B(\mathcal{H})_+$ – for the real cone of positive operators on $\mathcal{H}$;
- $S_1(\mathcal{H})$ – the space of trace class operators on $\mathcal{H}$;
- $S_2(\mathcal{H})$ – the Hilbert space of Hilbert-Schmidt operators endowed with the inner product$\langle T, S \rangle_{HS} = \mathrm{tr}(T S^*)$;

Let (Ω, Σ, μ) be a measure space, where Σ is the σ-algebra over Ω and μ is σ-finite positive measure. It is not hard to see from Proposition 6.2 that quantum injectivity of a continuous (Ω, μ)-frame $\mathcal{F}$ is equivalent to the condition that if $\langle T\mathcal{F}(\omega), \mathcal{F}(\omega)\rangle = 0$, for a.e $\omega \in \Omega$ for a self-adjoint trace class operator T with trace zero, then $T = 0$. Similarly, we say that $\mathcal{F}$ is $\mathcal{S}_2$-*injective* (resp, $\mathcal{S}_1$-*injective*) if whenever a self-adjoint Hilbert-Schmidt (resp, self-adjoint trace-class) operator T satisfies

$$\langle T\mathcal{F}(\omega), \mathcal{F}(\omega)\rangle = 0, \text{ for a.e. } \omega \in \Omega. \tag{6.3}$$

then $T = 0$.

Clearly, we have "$\mathcal{S}_2$-injectivity $\Rightarrow$ $\mathcal{S}_1$-injectivity $\Rightarrow$ quantum injectivity." In the case that $\mathcal{F}$ is a Parseval frame, quantum injectivity also implies the $\mathcal{S}_1$-injectivity (see Corollary 6.1). We point out that the Parseval frame requirement is not too much to ask since every frame is similar to a Parseval frame, and similar frames preserve $\mathcal{S}_j$-injectivity, here we say that two continuous frames $\mathcal{F}$ and $\mathcal{G}$ are *similar* if there is bounded invertible operator S such that $\mathcal{F}(\omega) = S\mathcal{G}(\omega)$ (a.e. $\omega \in \Omega$).

As every operator can be decomposed into the sum of two self-adjoint operators, and any self-adjoint operator can be decomposed into the sum of two positive operators, then we have the following two elementary facts:

Proposition 6.1 *Given a measure space* (Ω, Σ, μ) *and a* (Ω, μ)-*frame* $\mathcal{F}$ *for* $\mathcal{H}$. *For* $j = 1$ *or* 2, *the following statements are equivalent:*

1. *If* $T, S \in \mathcal{S}_j(\mathcal{H})$ *are positive operators, and*

$$\langle T\mathcal{F}(\omega), \mathcal{F}(\omega)\rangle = \langle S\mathcal{F}(\omega), \mathcal{F}(\omega)\rangle, \text{ for a.e. } \omega \in \Omega,$$

then $T = S$.
2. *If* $T, S \in \mathcal{S}_j(\mathcal{H})$ *are self-adjoint operators, and*

$$\langle T\mathcal{F}(\omega), \mathcal{F}(\omega)\rangle = \langle S\mathcal{F}(\omega), \mathcal{F}(\omega)\rangle, \text{ for a.e. } \omega \in \Omega,$$

then $T = S$.
3. $\mathcal{F}$ *is* $\mathcal{S}_j$-*injective.*
4. *For any* $T \in \mathcal{S}_j(\mathcal{H})$, *the condition* $\langle T\mathcal{F}(\omega), \mathcal{F}(\omega)\rangle = 0$ $(a.e.\omega \in \Omega)$ *implies that* $T = 0$.

Taking the trace condition into consideration, we get the following generalization of Proposition 6.1.

Proposition 6.2 *Given a (Ω, μ)-frame $\mathcal{F}$ for Hilbert space $\mathcal{H}$. Then the following statements are equivalent:*

1. *$\mathcal{F}$ is quantum injective.*
2. *If T, S are self-adjoint trace class operators and trace one, and*

$$\langle T\mathcal{F}(\omega), \mathcal{F}(\omega) \rangle = \langle S\mathcal{F}(\omega), \mathcal{F}(\omega) \rangle, \; \text{for a.e } \omega \in \Omega,$$

then $T = S$.
3. *If T is a self-adjoint trace class operator and trace zero, and*

$$\langle T\mathcal{F}(\omega), \mathcal{F}(\omega) \rangle = 0, \; \text{for a.e } \omega \in \Omega,$$

then $T = 0$.
4. *If T is a trace class operator and trace zero, and*

$$\langle T\mathcal{F}(\omega), \mathcal{F}(\omega) \rangle = 0, \; \text{for a.e } \omega \in \Omega,$$

then $T = 0$.

Given a Parseval frame, a self-adjoint trace-class operator T, since

$$I = \int_\Omega \mathcal{F}(\omega) \otimes \mathcal{F}(\omega) d\mu(\omega),$$

we have

$$tr(T) = \int_\Omega \langle T\mathcal{F}(\omega), \mathcal{F}(\omega) \rangle d\mu(\omega).$$

Thus, the condition $\langle T\mathcal{F}(\omega), \mathcal{F}(\omega) \rangle = 0$, for a.e. $\omega \in \Omega$ implies $tr(T) = 0$. By the equivalence of (1) and (3) in Proposition 6.2, we have the following corollary.

Corollary 6.1 *If $\mathcal{F}$ is a Parseval frame, then $\mathcal{F}$ is $\mathcal{S}_1$-injective if and only if it is quantum injective.*

6.2.2.1 Characterizations of Injective Frames

To obtain characterizations of quantum injective continuous frames, we use the following discrete representations for continuous frames proposed in [93]. Let $\mathbb{I}$ denote the finite index $\{1, 2, \cdots, N\}$ or the infinite index $\mathbb{N}$.

Proposition 6.3 *Let $\{e_i\}_{i \in \mathbb{I}}$ be an orthonormal basis for $\mathcal{H}$. Then the following are equivalent:*

1. $\mathcal{F}$ is a Parseval (Ω, μ)-frame for $\mathcal{H}$.
2. There exists an orthonormal set $\{\varphi_i\}_{i \in \mathbb{I}}$ in $L^2(\Omega, \mu)$ having the property that $\sum_{i \in \mathbb{I}} |\varphi_i(\omega)|^2 < \infty$ for a.e. $\omega \in \Omega$ and $\mathcal{F}(\omega) = \sum_{i \in \mathbb{I}} \varphi_i(\omega) e_i$ holds for a.e. $\omega \in \Omega$.

For general continuous frames, we also have

Proposition 6.4 *The following statements are equivalent:*

1. $\mathcal{F}$ is a (Ω, μ)-frame for $\mathcal{H}$.
2. $\mathcal{F}(\omega) = \sum_{i \in \mathbb{I}} \varphi_i(\omega) e_i$ for some orthonormal basis $\{e_i\}_{i \in \mathbb{I}}$ of $\mathcal{H}$ and some family $\{\varphi_i\}_{i \in \mathbb{I}}$ in $L^2(\Omega, \mu)$ with the properties that $\{\varphi_i\}_{i \in \mathbb{I}}$ is a Riesz basis for $\overline{span}\{\varphi_i\}_{i \in \mathbb{I}}$ and that for a.e. $\omega \in \Omega$, $\sum_{i \in \mathbb{I}} |\varphi_i(\omega)|^2 < \infty$.
3. $\mathcal{F}(\omega) = \sum_{i \in \mathbb{I}} \varphi_i(\omega) e_i$ for some Riesz basis $\{e_i\}_{i \in \mathbb{I}}$ of $\mathcal{H}$ and some set $\{\varphi_i\}_{i \in \mathbb{I}}$ orthonormal in $L^2(\Omega, \mu)$ with $\sum_{i \in \mathbb{I}} |\varphi_i(\omega)|^2 < \infty$ for a.e. $\omega \in \Omega$,
4. $\mathcal{F}(\omega) = \sum_{i \in \mathbb{I}} \varphi_i(\omega) e_i$ for some Riesz basis $\{e_i\}_{i \in \mathbb{I}}$ of $\mathcal{H}$ and some family $\{\varphi_i\}_{i \in \mathbb{I}}$ in $L^2(\Omega, \mu)$ with the properties that $\{\varphi_i\}_{i \in \mathbb{I}}$ is a Riesz basis for $\overline{span}\{\varphi_i\}_{i \in \mathbb{I}}$ and that $\sum_{i \in \mathbb{I}} |\varphi_i(\omega)|^2 < \infty$ for a.e. $\omega \in \Omega$.

By this discrete representation, we investigate the analogous spanning properties to characterize the injectivity. We first present several characterizations for $\mathcal{S}_2$-injective frames.

Theorem 6.6 *Let $\mathcal{F}$ be a (Ω, μ)-frame for $\mathcal{H}$ with a representation $\mathcal{F}(\omega) = \sum_{i \in \mathbb{I}} \varphi_i(\omega) e_i$ for some orthonormal basis $\{e_i\}_{i \in \mathbb{I}}$ of $\mathcal{H}$. Then the following are equivalent:*

1. $\mathcal{F}$ is $\mathcal{S}_2$-injective.
2. There exists a set $E \subseteq \Omega$ with $\mu(E) = 0$, if we set $M(\omega) = (\varphi_i(\omega)\overline{\varphi_j(\omega)})_{i,j}$ for all $\omega \in E^C$, then $\mathrm{span}_{\omega \in E^C} M(\omega)$ is dense in $S_2(\ell^2(\mathbb{I}))_{sa}$.

Proof Our proof starts with two observations: First, let $\mathcal{F}$ be a (Ω, μ)-frame for $\mathcal{H}$ and $\mathcal{F}(\omega) = \sum_{i \in \mathbb{I}} \varphi_i(\omega) e_i$ for some orthonormal basis $\{e_i\}_{i \in \mathbb{I}}$ of $\mathcal{H}$ and some family $\{\varphi_i\}_{i \in \mathbb{I}}$ in $L^2(\Omega, \mu)$ with the properties that $\{\varphi_i\}_{i \in \mathbb{I}}$ is a Riesz basis for $\overline{span}\{\varphi_i\}_{i \in \mathbb{I}}$ and $\sum_{i \in \mathbb{I}} |\varphi_i(\omega)|^2 < \infty$ a.e. Ω, that is, there exists a set $E \subseteq \Omega$ with $\mu(E) = 0$ such that $\sum_{i \in \mathbb{I}} |\varphi_i(\omega)|^2 < \infty$ for all $\omega \in E^C$. We define the following matrix for each $\omega \in E^C$ by

$$M(\omega) = \left(\varphi_i(\omega)\overline{\varphi_j(\omega)}\right)_{i,j}.$$

We claim $M(\omega)$ is a self-adjoint Hilbert-Schmidt operator on $\ell^2(\mathbb{I})$ for $\omega \in E^C$. Obviously it is self-adjoint. Meanwhile, its Hilbert-Schmidt norm is finite for all $\omega \in E^C$ based on the following computation

$$\left\| (\varphi_i(\omega)\overline{\varphi_j(\omega)})_{i,j} \right\|_2 = \left(\sum_{i\in\mathbb{I}, j\in\mathbb{I}} |\varphi_i(\omega)\overline{\varphi_j(\omega)}|^2 \right)^{1/2} = \sum_{i\in\mathbb{I}} |\varphi_i(\omega)|^2 < \infty.$$

Secondly, let T be a self-adjoint Hilbert-Schmidt operator. We set $a_{i,j} = \langle Te_i, e_j \rangle$ and define the matrix $(a_{i,j})_{i,j}$ on $\ell^2(\mathbb{I})$. We propose to prove an analogous result for $(a_{i,j})_{i,j}$, namely, it is a self-adjoint Hilbert-Schmidt operator on Hilbert space. Since T is self-adjoint, then

$$a_{i,j} = \langle Te_i, e_j \rangle = \langle e_i, Te_j \rangle = \overline{\langle Te_j, e_i \rangle} = \overline{a_{j,i}},$$

hence the matrix $(a_{i,j})_{i,j}$ is self-adjoint. Moreover,

$$\|(a_{i,j})_{i,j}\|_2 = \left(\sum_{i,j\in\mathbb{I}} |\langle Te_i, e_j \rangle|^2 \right)^{1/2} = \left(\sum_{i\in\mathbb{I}} \|Te_i\|^2 \right)^{1/2} = \|T\|_2.$$

Now we use the representation of (Ω, μ)-frame and we obtain

$$\langle T\mathcal{F}(\omega), \mathcal{F}(\omega) \rangle = \langle T(\sum_{i\in\mathbb{I}} \varphi_i(\omega)e_i), (\sum_{j\in\mathbb{I}} \varphi_j(\omega)e_j) \rangle$$

$$= \sum_{i\in\mathbb{I}} \sum_{j\in\mathbb{I}} \varphi_i(\omega)\overline{\varphi_j(\omega)}\langle Te_i, e_j \rangle$$

$$= \left\langle (\varphi_i(\omega)\overline{\varphi_j(\omega)})_{i,j}, (\overline{a_{i,j}})_{i,j} \right\rangle_{HS}.$$

Recall that the set of self-adjoint Hilbert-Schmidt operators under the inner product $\langle T, S \rangle_{HS} = \operatorname{tr}(TS^*)$ is the real Hilbert space. By orthogonality, we conclude that if $\mathcal{F}$ is $\mathcal{S}_2$-injective, then the orthogonal complement space of span $\{(\varphi_i(\omega)\overline{\varphi_j(\omega)})_{i,j}\}$ for all $\omega \in E^C$ is 0, hence that $\operatorname{span}_{\omega\in E^C}$ is dense in $S_2(\ell^2(\mathbb{I}))_{sa}$.

Conversely, for any self-adjoint Hilbert-Schmidt operator T and if there exists a set $E \subseteq \Omega$ with $\mu(E) = 0$, if $\langle T\mathcal{F}(\omega), \mathcal{F}(\omega) \rangle = 0$ for all $\omega \in E^C$, then

$$\left\langle (\varphi_i(\omega)\overline{\varphi_j(\omega)})_{i,j}, (\overline{a_{i,j}})_{i,j} \right\rangle_{HS} = 0, \ \forall \omega \in E^C.$$

Since $\operatorname{span}_{\omega\in E^C}\{M(\omega)\}$ is dense in $S_2(\ell^2(\mathbb{I}))_{sa}$, we have $(\overline{a_{i,j}})_{i,j} = 0$, hence $(a_{i,j})_{i,j} = 0$ as well as $T = 0$. Thus $\mathcal{F}$ is $\mathcal{S}_2$-injective. $\qquad\square$

Similar to the discrete case, we define $\mathbb{H}$ as the direct sum of real Hilbert spaces ℓ^2 in the sense

$$\mathbb{H} = \left(\sum_{i \in \mathbb{I}} \oplus \ell^2 \right)_{\ell^2}$$

with the inner product $\langle X, Y \rangle_{\mathbb{H}} = \sum_{i \in \mathbb{I}} \langle x_i, y_i \rangle$ for $X = (x_i)_{i \in \mathbb{I}}, Y = (y_i)_{i \in \mathbb{I}} \in \mathbb{H}$.
By turning Eq. (6.3) into the inner product in $\mathbb{H}$ in the sense

$$\langle T \mathcal{F}(\omega), \mathcal{F}(\omega) \rangle = \langle A_T, S(\omega) \rangle_{\mathbb{H}}.$$

where A_T is $(A_1, A_2, A_3, \cdots, A_i, \cdots)_{i \in \mathbb{I}}$ with

$$A_i = (a_{i,i}, 2 \operatorname{Re}(a_{i,i+1}), -2 \operatorname{Im}(a_{i,i+1}), 2 \operatorname{Re}(a_{i,i+2}), -2 \operatorname{Im}(a_{i,i+2}), \cdots),$$

and $S(\omega)$ will be defined in the next theorem, we can obtain another characterization of $\mathcal{S}_2$-injectivity in the next theorem.

Theorem 6.7 *Let $\mathcal{F}$ be a (Ω, μ)-frame for an infinite dimensional Hilbert space $\mathcal{H}$ and $\mathcal{F}(\omega) = \sum_{i \in \mathbb{I}} \varphi_i(\omega) e_i$ for some orthonormal basis $\{e_i\}_{i \in \mathbb{I}}$ of $\mathcal{H}$. Then the following are equivalent:*

1. $\mathcal{F}$ is $\mathcal{S}_2$-injective,
2. There exists a set $E \subseteq \Omega$ with $\mu(E) = 0$, if we set

$$S(\omega) = (|\varphi_1(\omega)|^2, \operatorname{Re}(\varphi_1(\omega)\overline{\varphi_2(\omega)}), \operatorname{Im}(\varphi_1(\omega)\overline{\varphi_2(\omega)}) \cdots,$$

$$|\varphi_2(\omega)|^2, \operatorname{Re}(\varphi_2(\omega)\overline{\varphi_3(\omega)}), \operatorname{Im}(\varphi_2(\omega)\overline{\varphi_3(\omega)}) \cdots, \cdots) \ \forall \ \omega \in E^C.$$

then $\operatorname{span}_{\omega \in E^C}\{S(\omega)\}$ is dense in $\mathbb{H}$.

By the spectral decomposition of compact operators, for a self-adjoint Hilbert-Schmidt operator T, there is eigenbasis $\{e_i\}$ and $\lambda = (\lambda_1, \cdots, \lambda_i, \cdots) \in \ell^2(\mathbb{I})$ such that $T e_i = \lambda_i e_i$. Moreover, if we choose the representation $\mathcal{F}(\omega) = \sum_{I \in \mathbb{I}} \varphi_i(\omega) e_i$ for $\mathcal{F}$, then

$$\langle T \mathcal{F}(\omega), \mathcal{F}(\omega) \rangle = \sum_{i \in \mathbb{I}} \lambda_i |\varphi_i(\omega)|^2.$$

Based on these observations, we present the following results.

Theorem 6.8 *Let $\mathcal{F}$ be a (Ω, μ)-frame for the Hilbert spaces $\mathcal{H}$. Then the following are equivalent:*

1. $\mathcal{F}$ is $\mathcal{S}_2$-injective.

2. *Given any representation $F(\omega) = \sum_{I \in \mathbb{I}} \varphi_i(\omega) e_i$ for some orthonormal basis $\{e_i : i \in \mathbb{I}\}$. There exists a set $E \subseteq \Omega$ with $\mu(E) = 0$, if we set*

$$f(\omega) = (|\varphi_1(\omega)|^2, |\varphi_2(\omega)|^2, |\varphi_3(\omega)|^2, \cdots, |\varphi_i(\omega)|^2, \cdots) \,\forall\, \omega \in E^C,$$

then $\mathrm{span}_{\omega \in E^C}\{f(\omega)\}$ is dense in the real Hilbert space $\ell^2(\mathbb{I})$.

Replacing $\ell^2(\mathbb{I})$ by $c_0(\mathbb{I})$, we obtain the following:

Theorem 6.9 *Given a (Ω, μ)-frame for Hilbert space $\mathcal{H}$. Then the following are equivalent:*

1. *$\mathcal{F}$ is quantum injective.*
2. *Given any representation $F(\omega) = \sum_{I \in \mathbb{I}} \varphi_i(\omega) e_i$ for some orthonormal basis $\{e_i : i \in \mathbb{I}\}$. There exists a set $E \subseteq \Omega$ with $\mu(E) = 0$, if we set*

$$g(\omega) = (|\varphi_1(\omega)|^2, |\varphi_2(\omega)|^2, |\varphi_3(\omega)|^2, \cdots, |\varphi_i(\omega)|^2, \cdots) \,\omega \in E^C,$$

then $\mathrm{span}_{\omega \in E^C}\{g(\omega)\}$ is dense in the real space $c_0(\mathbb{I})$.

6.2.2.2 Constructions and Stability

Next, we construct some examples and quantify their stability. The following dimension formula for continuous Parseval frames clearly distinguishes whether the underlying Hilbert space is finite-dimensional or infinite-dimensional.

Proposition 6.5 *For any Parseval (Ω, μ)-frame $\mathcal{F}$ in a Hilbert space $\mathcal{H}$, we have*

$$\dim \mathcal{H} = \int_{\Omega} \|\mathcal{F}(\omega)\|^2 \mathrm{d}\mu(\omega).$$

Proof By Proposition 6.3, we can write $\mathcal{F}(\omega) = \sum_{i \in \mathbb{I}} \varphi_i(\omega) e_i$ for some orthonormal basis $\{e_i\}_{i \in \mathbb{I}}$ of $\mathcal{H}$ and an orthonormal set $\{\varphi_i\}_{i \in \mathbb{I}}$ in $L^2(\Omega, \mu)$. Thus

$$\int_{\Omega} \|F(\omega)\|^2 \mathrm{d}\mu(\omega) = \int_{\Omega} \sum_{i \in \mathbb{I}} |\varphi_i(\omega)|^2 \mathrm{d}\mu(\omega) = \sum_{i \in \mathbb{I}} \|\varphi_i\|^2 = \mathrm{card}(\mathbb{I}) = \dim \mathcal{H}.$$

We first present two examples demonstrating how the characterizations from Sect. 6.2.2 can be used to construct injective or non-injective continuous frames.

Example 6.2.2 Let $\mathcal{H}$ be an infinite dimensional Hilbert space, and let $\mathcal{F}$ be the following Parseval continuous frame with respect to a σ-finite measure space (Ω, μ) defined by

$$\mathcal{F}(\omega) := \sum_{i=1}^{\infty} \frac{\chi_{\Omega_i}}{\sqrt{\mu(\Omega_i)}} e_i, \qquad \text{for } \omega \in \Omega,$$

where $\{\Omega_k\}$ is a sequence of mutually disjoint, finite, positive measure subsets of Ω with union Ω. For simplicity of notation, we set $\varphi_i(\omega) = \frac{\chi_{\Omega_i}}{\sqrt{\mu(\Omega_i)}}$. Then $\mathcal{F}(\omega) := \sum_{i=1}^{\infty} \varphi_i(\omega)e_i$, for $\omega \in \Omega$. By the equivalent characterization in Theorem 6.6, the span of the operators

$$(\varphi_i(\omega)\overline{\varphi_j(\omega)})_{i,j} = \begin{pmatrix} 0 & \vdots & 0 \\ \cdots & \frac{1}{\mu(\Omega_k)} & \cdots \\ 0 & \vdots & 0 \end{pmatrix}.$$

is not dense in $S_2(\ell^2)_{sa}$. If $\omega \in \Omega$, then $\omega \in \Omega_k$ for some k. Thus,

$$\langle T\mathcal{F}(\omega), \mathcal{F}(\omega) \rangle = \frac{1}{\mu(\Omega)} \langle Te_k, e_k \rangle = 0,$$

from which we can not derive $T = 0$. Therefore, $\mathcal{F}$ is not S_2-injective. The same is true for the self-adjoint trace class operator case. Thus, $\mathcal{F}$ can neither give S_2-injectivity nor give S_1-injectivity.

Example 6.2.3 Let $\mathcal{H}$ be a Hilbert space and $\{e_i\}_{i=1}^{\infty}$ be an orthonormal basis for H. Let $\chi = \chi_{[0,1)}$, $\psi = \chi_{[0,1/2)} - i\chi_{[1/2,1)}$. Define

$$\varphi_1(t) = \chi$$

$$\varphi_2(t) = \chi(t-1) + \psi(4(t - \frac{1}{2}))$$

$$\vdots$$

$$\varphi_n(t) = \chi(t - n + 1) + \psi(2^n(t - 1 + 2^{\frac{1}{n-1}}) + \cdots + \psi(2^n(t - n + 1 + 2^{\frac{1}{n-1}}))$$

$$\vdots$$

Define $\mathcal{F} : \mathbb{R}^+ \to \mathcal{H}$ by $\mathcal{F}(t) = \sum_{i=1}^{\infty} \varphi_i(t)e_i$. By the equivalent characterization of Riesz bases [51], it can be easily verified that $\{\varphi_i\}$ is a Riesz basis for $\overline{span}\{\varphi_i\}$, and hence $\mathcal{F}(t) = \sum_{i=1}^{\infty} \varphi_i(t)e_i$ is a continuous frame with respect to the Lebesgue measure m on $\mathbb{R}^+$ for $\mathcal{H}$ by Corollary 6.4.

Now, we check the injectivity. By the definition of $\mathcal{F}$, we get

$$\mathcal{F}(t) = \begin{cases} e_1, & t \in [0, \frac{1}{2}); \\ e_1 + e_2, & t \in [\frac{1}{2}, \frac{5}{8}); \\ \vdots & \vdots \\ e_2, & t \in [1, \frac{7}{4}); \\ e_2 + e_3, & t \in [\frac{7}{4}, \frac{29}{16}); \\ \vdots & \vdots \end{cases}$$

Let T be a self-adjoint Hilbert-Schmidt operator with

$$\langle T\mathcal{F}(t), \mathcal{F}(t) \rangle = 0, \forall t \in \mathbb{R}^+.$$

For $t \in [0, \frac{3}{4}) \cup [1, \frac{7}{4})$, we get

$$\langle Te_1, e_1 \rangle = 0; \langle T(e_1 + e_2), (e_1 + e_2) \rangle = 0; \langle T(e_1 - ie_2), (e_1 - ie_2) \rangle = 0; \langle Te_2, e_2 \rangle = 0.$$

This implies that

$$\langle Te_1, e_1 \rangle = 0; \quad \langle Te_1, e_2 \rangle = \langle Te_2, e_1 \rangle = 0; \quad \langle Te_2, e_2 \rangle = 0.$$

Repeating the previous argument for all the intervals in $\mathbb{R}^+$, we will get $\langle Te_i, e_j \rangle = 0$ for all $i, j \in \mathbb{N}^+$, which implies that $T = 0$. Thus $\mathcal{F}$ is $\mathcal{S}_2$-injectivity (and hence it is also $\mathcal{S}_1$-injective and quantum injective).

Now, we examine the stability property for injective continuous frames. To do this, we need to define a metric between frames. There have been various metrics introduced in the literature (c.f. [7, 12, 19]). Here we will use the following metric to measure the distance between continuous frames.

Definition 6.4 Given two (Ω, μ)-frames $\mathcal{F}, \mathcal{G}$ for a Hilbert space $\mathcal{H}$. The distance between $\mathcal{F}, \mathcal{G}$ is defined by

$$d^2(\mathcal{F}, \mathcal{G}) = \int_\Omega \|\mathcal{F}(\omega) - \mathcal{G}(\omega)\|^2 d\mu(\omega).$$

Let $\{e_i\}_{i \in \mathbb{I}}$ is an orthonormal basis for $\mathcal{H}$ and $\mathcal{F}(\omega) = \sum_{i \in \mathbb{I}} \varphi_i(\omega)e_i$, $\mathcal{G}(\omega) = \sum_{i \in \mathbb{I}} \phi_i(\omega)e_i$. Then we have

$$d^2(\mathcal{F}, \mathcal{G}) = \int_\Omega \|\mathcal{F}(\omega) - \mathcal{G}(\omega)\|^2 d\mu(\omega). = \int_\Omega \|\sum_{i \in \mathbb{I}}(\varphi_i(\omega) - \phi_i(\omega))e_i\|^2 d\mu(\omega).$$

$$= \int_{\Omega} \sum_{i \in \mathbb{I}} |\varphi_i(\omega) - \phi_i(\omega)|^2 d\mu(\omega) = \sum_{i \in \mathbb{I}} \|\varphi_i - \phi_i\|_2^2.$$

The following establishes the stability of finite-dimensional injective Parseval frames.

Theorem 6.10 *Let $\mathcal{F}$ be an injective Parseval (Ω, μ)-frame for a finite-dimensional Hilbert space. Then there is $\delta > 0$ such that $\mathcal{G}$ also gives injectivity for any Parseval (Ω, μ)-frame $\mathcal{G}$ satisfying*

$$d(\mathcal{F}, \mathcal{G}) < \delta.$$

We can perturb Example 6.2.3 to show that quantum injective frames are unstable in infinite-dimensional cases.

Example 6.2.4 Let $\mathcal{F}(t) = \sum_{i=1}^{\infty} \varphi_i(t) e_i$ be the quantum injective frame for Hilbert space $\mathcal{H}$ as in Example 6.2.3. For any $\varepsilon > 0$, there exists a continuous frame $\mathcal{G}$ such that $d(\mathcal{F}, \mathcal{G}) < \varepsilon$, but $\mathcal{G}$ is not quantum injective.

6.2.2.3 Finite-dimensional Case

Note that when card($\mathbb{I}$) is finite, that is, $\mathcal{H}$ is a finite-dimensional Hilbert space. Then S_1-injectivity and S_2-injectivity coincide. Even if $\mathcal{H}$ is finite-dimensional, there is still considerable demand to work with continuous frames for $\mathcal{H}$ [210]. The discrete representation of a continuous frame helps us construct injective frames for finite-dimensional Hilbert spaces.

Suppose that we have an n-dimensional Hilbert space $\mathcal{H}$, and a continuous Parseval frame $\mathcal{F}$ with a representation $\mathcal{F}(\omega) = \sum_{i=1}^{n} \varphi_i(\omega) e_i$ for some orthonormal basis $\{e_i\}_{i=1}^n$ and orthonormal set $\{\varphi_i(\omega)\}_{i=1}^n$ in $L_2(\Omega, \mu)$. From Theorem 6.7, we need to find an orthonormal set $\{\varphi_i\}_i^n$ such that the sequence

$$(|\varphi_1(\omega)|^2, \cdots, \mathrm{Re}(\varphi_1(\omega)\overline{\varphi_n}(\omega)), \mathrm{Im}(\varphi_1(\omega)\overline{\varphi_n}(\omega));$$

$$|\varphi_2(\omega)|^2, \cdots, \mathrm{Re}(\varphi_2(\omega)\overline{\varphi_n}(\omega)), \mathrm{Im}(\varphi_2(\omega)\overline{\varphi_n}(\omega)); \cdots; |\varphi_n(\omega)|^2)$$

spans the real Hilbert space $\ell_2^{n^2}$. This is equivalent to the condition that pointwise multiplication vectors $\{\varphi_i \overline{\varphi_i}\}_{1 \leq i \leq n}$ and $\{\mathrm{Re}(\varphi_i \overline{\varphi_j}), \mathrm{Im}(\varphi_i \overline{\varphi_j})\}_{1 \leq i < j \leq n}$ are linearly independent. Quite often we make an appropriate partition of Ω and then define φ_i on disjoint partition subsets to ensure the linear independence of the set $\{\varphi_i \overline{\varphi_i}\}_{1 \leq i \leq n}$ and $\{\mathrm{Re}(\varphi_i \overline{\varphi_j}), \mathrm{Im}(\varphi_i \overline{\varphi_j})\}_{1 \leq i < j \leq n}$. Thus, we get a continuous frame $\mathcal{F}(\omega) = \sum_{i=1}^{n} \varphi_i(\omega) e_i$ which is injective. As a byproduct of this fact, if we consider a finite or countable set Ω with the counting measure μ to obtain a discrete frame and it gives injectivity in finite-dimensional if and only if $\{\varphi_i \overline{\varphi_i}\}_{1 \leq i \leq n}$, and $\{\mathrm{Re}(\varphi_i \overline{\varphi_j}), \mathrm{Im}(\varphi_i \overline{\varphi_j})\}_{1 \leq i < j \leq n}$ are linearly independent, which is equivalent to the condition that determinant of the matrix which is composed of those vectors, is not zero. Using this idea, we construct the following two examples of injective frames.

Example 6.2.5 Define the matrix

$$
\begin{pmatrix}
W_{1,1} & 0 & 0 & \cdots & 0 \\
W_{2,1} & W_{2,2} & 0 & \cdots & 0 \\
W_{3,1} & W_{3,2} & W_{3,3} & \cdots & 0 \\
\vdots & \vdots & \vdots & \ddots & \vdots \\
W_{n,1} & W_{n,2} & W_{n,3} & \cdots & W_{n,n}
\end{pmatrix},
$$

where for any $1 \leq l \leq n$, we choose $2l - 1$ linearly independent column vectors $\{V_{l,k}\}_{1 \leq k \leq 2l-1}$ in $\mathbb{R}^{2l-1}$, and all coordinates of $V_{l,2l-1}(1 \leq l \leq n)$ are not zero. Let i be the complex unit and set

$$
W_{l,k} = \begin{cases} V_{l,2k-1} + i V_{l,2k}, & k < l; \\[2mm] V_{l,2k-1}, & k = l. \end{cases}
$$

and $\mathbf{0}$ denotes the zero vector. Take the column vector $\{\varphi_i\}$ and define $\mathcal{F} = \sum_{i=1}^{n} \varphi_i e_i$. It can be verified that the pointwise multiplication of column vectors gives a linearly independent vector set. Thus, $\mathcal{F}$ is a frame that gives injectivity for the self-adjoint operator on $\mathbb{C}^n$.

Due to the representation of the continuous frame, if we want to construct a Parseval frame, we only need to choose an orthonormal set $\{\tilde{\varphi}_i\}_{i=1}^{n}$. Thus, we have the following example.

Example 6.2.6 Define a matrix as follows

$$
\begin{pmatrix}
\lambda_{1,1}U_{1,1} & 0 & 0 & \cdots & 0 \\
\lambda_{2,1}U_{2,1} & \lambda_{2,2}U_{2,2} & 0 & \cdots & 0 \\
\lambda_{3,1}U_{3,1} & \lambda_{3,2}U_{3,2} & \lambda_{3,3}U_{3,3} & \cdots & 0 \\
\vdots & \vdots & \vdots & \ddots & \vdots \\
\lambda_{n,1}U_{n,1} & \lambda_{n,2}U_{n,2} & \lambda_{n,3}U_{n,3} & \cdots & \lambda_{n,n}U_{n,n}
\end{pmatrix}
$$

where

$$
U_{l,k} = \begin{cases} \frac{\sqrt{2}}{2}(V_{l,2k-1} + i V_{l,2k}), & k < l; \\[2mm] V_{l,2k-1}, & k = l. \end{cases}
$$

and $\mathbf{0}$ denotes the zero vector.

The coefficients $\{\lambda_{i,j}\}_{j \le i}$ satisfy

$$\lambda_{i,j} \ne 0 \quad \text{for} \quad 1 \le j \le i \le n, \quad \text{and} \quad \sum_{j \le i} |\lambda_{i,j}|^2 = 1, \forall\ 1 \le j \le n.$$

The column vectors $\{V_{l,k}\}_{1 \le k \le 2l-1}$ form an orthonormal basis for $\mathbb{R}^{2l-1}$, and all the coordinates of $V_{l,2l-1}$ are not zero for all $1 \le l \le n$. Take the column vectors $\{\widetilde{\varphi}_i\}_{1 \le i \le n}$ and define $\mathcal{F} = \sum_{i=1}^{n} \widetilde{\varphi}_i e_i$. A similar argument shows that $\mathcal{F}$ is a Parseval frame that gives injectivity.

Remark 6.2 Obviously, by Theorem 6.7, $\mathcal{F} = \sum_{i=1}^{n} \widetilde{\varphi}_i e_i$ gives injective. Meanwhile, from the equivalent characterization of a continuous frame, it is easily seen that it is a Parseval frame. If we take the row vectors in our construction, it leads to an ordinary discrete frame. The representation of continuous frames on a finite or a countable set provides us with a different perspective on the construction of a more variety of frames with desired properties.

6.2.3   Injective Finite Multiwindow Gabor Frames

Frames with good structures, such as group representation frames [210], are of particular interest for both theoretical development and applications. Gabor frames, collections of modulations and translations of a single vector or multiple vectors, is such a typical example that has been extensively studied in Gabor analysis [106, 163] and time-frequency analysis [102]. In this subsection, we briefly discuss the quantum detection problem for multi-windowed Gabor frames [121].

We will be working in $\mathbb{C}^N$ and also list the notations that will be used.

- Vectors $x \in \mathbb{C}^n$ will be treated as column vectors and written as $x = (x(i))_{i=0}^{N-1}$.
- $\omega = e^{\frac{2\pi i}{N}}$ -the N-th root of unity.
- $\mathbb{Z}_N = \{0, \ldots, N-1\}$, the ring of integers modulo N. $\mathbb{Z}_N^2 = \mathbb{Z}_N \times \mathbb{Z}_N = \{(k, l) : k, l \in \mathbb{Z}_N\}$.
- For any set Λ, denote $\sharp \Lambda$ as the cardinality of the set.

Let $k \in \mathbb{Z}_N, l \in \mathbb{Z}_N$. The translation operator T_k and the modulation operator M_l: $\mathbb{C}^N \to \mathbb{C}^N$ are defined by

$$(T_k x)(n) = x(n-k), \quad (M_\ell x)(n) = \omega^\ell x(n)$$

We will use $\pi(k, l)$ to denote $M_l T_k$. Some facts are worth pointing out, such as the well-known commutation relations shown in [95] between translations and modulations

$$M_\ell T_k = \omega^{k\ell} T_k M_\ell, \quad \pi(m, n)\pi(k, l) = \omega^{kn-ml}\pi(k, l)\pi(m, n).$$

Then it follows that

$$\pi(m,n)\pi(k,l) = \omega^{-nk}\pi(m+k,n+l), \quad \forall(m,n),(k,l) \in \mathbb{Z}_N^2$$

which implies that π is a projective unitary representation of $\mathbb{Z}_N^2$ with the multiplier $m((m,n),(k,l)) = \omega^{-nk}$ [210]. Moreover, $\{\frac{1}{\sqrt{N}}\pi(m,n)\}_{(m,n)\in\mathbb{Z}_N^2}$ forms an orthonormal basis of M_N as

$$\langle\pi(m,n),\pi(k,l)\rangle_{HS} = N\delta_{m,k}\delta_{n,l}.$$

Furthermore, for every nonzero vector $\varphi \in \mathbb{C}^N$,

$$\sum_{(m,n)\in\mathbb{Z}_N^2} |\langle\pi(m,n)\varphi,x\rangle|^2 = \sum_{(m,n)\in\mathbb{Z}_N^2} |\langle\pi(m,n),x\varphi^*\rangle_{HS}|^2$$

$$= N\|x\varphi^*\|_{HS}^2 = (N\|\varphi\|^2)\|x\|^2,$$

that is, the collection $\{\pi(m,n)\varphi\}_{(m,n)\in\mathbb{Z}_N^2}$ is a tight frame with frame bound $N\|\varphi\|^2$ hence spans $\mathbb{C}^N$, which implies the irreducibility of π. Such a frame is called a (single window) Gabor frame.

One of the very basic questions is to characterize the window vectors φ such that $\{\pi(m,k)\varphi\}_{(m,k)\in\mathbb{Z}_N^2}$ is quantum injective? Bojarovska and Flinth [29] investigate the problem which will be discussed in the next section and give a simple characterization for these types of frames $\{f_i\}$ with the maximal span property [16], i.e., $\text{span}\{f_i \otimes f_i\}_{i=1}^N$ contains all the self-adjoint operators. This property exactly coincides with the equivalent condition in Theorem 6.1, which is a sufficient but not necessary condition for phase retrieval, see [29, Theorem 2.2]. We summarize it as the following theorem.

Theorem 6.11 *Let π be the Gabor representation of $G = \mathbb{Z}_n \times \mathbb{Z}_n$ on $\mathbb{C}^n$ and $\varphi \in \mathbb{C}^n$. Then $\{\pi(m,k)\varphi\}_{(m,k)\in G}$ has the maximal span property if and only if*

$$\langle\pi(m,k)\varphi,\varphi\rangle \neq 0$$

for every $(m,k) \in G$, if and only if it is quantum injective. Then the corresponding Gabor frame $G = \{\pi(m,k)\varphi\}_{(m,k)\in G}$ allows phase retrieval.

Recall that a multi-window Gabor frame is a Gabor frame $\{\pi(m,n)\varphi_r\}_{(m,n)\in\Lambda}^{1\leq r\leq s}$ generated by s-window vectors $\varphi_1, \ldots, \varphi_s$ with $\Lambda \subseteq \mathbb{Z}_N^2$. We always assume that all the window vectors are nonzero. Moreover, if $\Lambda = \mathbb{Z}_N^2$, then it is called a full Gabor frame. The following characterizes all injective full multi-window Gabor frames.

Theorem 6.12 Let $\{\pi(m,n)\varphi_r\}^{1\le r\le s}_{(m,n)\in\mathbb{Z}^2_N}$ be a full multi-window Gabor frame generated by window vectors $\varphi_1,\ldots,\varphi_s$ with $\mathbb{Z}^2_N$. Then, the following are equivalent.

1. $\{\pi(m,n)\varphi_r\}^{1\le r\le s}_{(m,n)\in\mathbb{Z}^2_N}$ is quantum injective;
2. $\forall (k,l)\in\mathbb{Z}^2_N,\ \langle\pi(k,l)\varphi_r,\varphi_r\rangle\ne 0$ for some $1\le r\le s$;

A conclusion given by [96] says that for arbitrary $S\subseteq\mathbb{Z}_N$ with $\sharp S>\frac{N}{2}$, we define the support of g to be the set $\mathrm{supp}(g)=\{i\,|\,g_i\ne 0\}$, then there exists a unit vector g with $\mathrm{supp}(g)=S$, that gives quantum injectivity, which immediately implies the existence of injective finite Gabor frames. Since the window vector $\varphi=(\varphi_1,\ldots,\varphi_s)$ that generates an injective Gabor frame is characterized by the zero sets of a finite collection of nonzero polynomials, we immediately get:

Corollary 6.2 The Gabor frame $\{\pi(k,l)\varphi_r\}_{(k,l)\in\mathbb{Z}^2_N,1\le r\le s}$ is injective for every generic window vector $\varphi=(\varphi_1,\ldots,\varphi_s)$ in $\mathbb{C}^N\oplus\cdots\oplus\mathbb{C}^N$, and hence the set of all such window vectors is dense in $\mathbb{C}^N\oplus\cdots\oplus\mathbb{C}^N$.

Next, we examine the stability of the injective property of Gabor frames under small perturbations. For this purpose, a metric between frames is needed. There are various metrics defined in [19]. For the same reason as in Corollary 6.2, the injective Gabor frames are preserved under small perturbation with respect to the norm $\sum_{i=1}^s\|\varphi_i-\psi_i\|^2$ in $\mathbb{C}^N\oplus\cdots\oplus\mathbb{C}^N$. Here, we will use the following metric and present a quantitative version of the stability result.

Definition 6.5 Let $\mathcal{F}_1=\{\pi(m,n)\varphi_r\}^{1\le r\le s}_{(m,n)\in\mathbb{Z}^2_N}$, $\mathcal{F}_2=\{\pi(m,n)\psi_r\}^{1\le r\le s}_{(m,n)\in\mathbb{Z}^2_N}$ be two full Gabor frames generated by s window vectors for each with $\mathbb{Z}^2_N$. Let P be the set of permutations of $\{1,\ldots,s\}$. The distance between $\mathcal{F}_1,\mathcal{F}_2$ is defined by

$$d^2_G(\mathcal{F}_1,\mathcal{F}_2)=\min_{\sigma\in P}(\sum_{i=1}^s\|\varphi_{\sigma(i)}-\psi_i\|^2).$$

Let $\mathcal{F}_1=\{\pi(m,n)\varphi_r\}_{(m,n)\in\mathbb{Z}^2_N,1\le r\le s}$ be an injective full Gabor frame for $\mathbb{C}^N$. Set $\delta'=\min_{(k,l)\in\mathbb{Z}^2_N}\{\sum_{r=1}^s|\langle\pi(k,l)\varphi_r,\varphi_r\rangle|\}$ and

$$\delta=\sqrt{\sum_{r=1}^s\|\varphi_r\|^2+\delta'}-\sqrt{\sum_{r=1}^s\|\varphi_r\|^2}.$$

Proposition 6.6 ([121]) Let $\mathcal{F}_1=\{\pi(m,n)\varphi_r\}_{(m,n)\in\mathbb{Z}^2_N}$, $1\le r\le s,1$ be an injective full Gabor frame for $\mathbb{C}^N$. Then every full multi-window Gabor frame $\mathcal{F}_2=$

$\{\pi(m,n)\psi_r\}_{(m,n)\in\mathbb{Z}_N^2}$, $1 \leq r \leq s$ *satisfying the condition*

$$d_{\mathcal{G}}(\mathcal{F}_1, \mathcal{F}_2) < \delta$$

is also injective.

Since $\{\pi(m,n)\varphi : (m,n) \in \mathbb{Z}_N^2\}$ is injective for every generic vector φ, it raises a natural question about the existence of injective multi-window Gabor frames such that none of the full Gabor frames generated by proper subsets of the window vectors gives the injectivity. The following is such an example:

Example 6.2.7 Consider two window vectors with their frame matrices of single-window Gabor frames on $\mathbb{Z}_2^2$ and

$$\varphi_1 = \begin{pmatrix} e^{\frac{\pi}{3}i} \\ e^{\frac{\pi}{2}i} \end{pmatrix}, \quad \varphi_2 = \begin{pmatrix} 1 \\ 0 \end{pmatrix}.$$

By Theorem 6.12, neither the Gabor frame generated by φ_1 nor φ_2 is injective, and clearly the bi-window Gabor frame generated by both of them is injective.   □

6.3   Phase Retrieval for Finite Frames

The phase retrieval problem considers recovering a signal of interest from magnitudes of its linear or nonlinear measurements, which arises in various fields of science and engineering applications, such as X-ray crystallography, coherent diffractive imaging, optics, and quantum information, more specific applications like using X-ray diffraction to determine the structure of DNA or crystals, reconstructing astronomical images captured by telescopes, pure state detection and recovery in quantum information theory, etc. Balan et al. are among the pioneers who initiated the investigation of the phase retrieval problem by using linear measurements against a frame (cf. [15,16,21,22,24]) and then frames have been extensively studied in the context of the phase-retrieval problem and many related methods include Gabor transform [105], short-time Fourier transform [8] and wavelet transform [9], etc. are extensively applied into this problem. For linear measurements with a frame $\{f_i\}_{i=1}^N$, it asks to reconstruct f from its intensity measurements $\{|\langle f, f_i\rangle|\}_{i=1}^N$. Clearly, the intensity measurements are the same for both f and λf for every unimodular scalar λ. Therefore, the phase retrieval problem asks to recover f up to an unimodular scalar. In this section, we will be working on this problem in the finite-dimensional Hilbert space $\mathcal{H}$, that is, $\mathbb{R}^n$ or $\mathbb{C}^n$, and with finite frames.

Definition 6.6 A frame $\{f_i\}_{i=1}^N$ for $\mathcal{H}$ is called phase retrievable if the induced quotient map $\mathcal{A} : \mathcal{H}/\mathbb{T} \to \mathbb{F}^N$ defined by

$$\mathcal{A}(f/\mathbb{T}) = \{|\langle f, f_i \rangle|\}_{i=1}^N$$

is injective, where $\mathbb{T} = \{\lambda \in \mathbb{F} : |\lambda| = 1\}$.

One elementary but widely used result on phase retrieval frames is due to Balan, Casazza and Edidin [22] who obtained the first characterization of phase retrievable frames in terms of the complement property of the frame.

Definition 6.7 A frame $\{f_i\}_{i=1}^N$ is said to have the complement property if for every $\Omega \subseteq \{1, \dots, N\}$ we have either $\{f_i\}_{i \in \Omega}$ or $\{f_i\}_{i \in \Omega^c}$ spans $\mathcal{H}$.

Theorem 6.13 *Let $\{f_i\}_{i=1}^N$ be a frame for $\mathcal{H}$. If $\{f_i\}_{i=1}^N$ is phase retrievable, then it satisfies the complement property. The complementary property is also sufficient when $\mathbb{F} = \mathbb{R}$, but not sufficient in the complex case $\mathbb{F} = \mathbb{C}$.*

Proof Suppose that $\{f_i\}_{i=1}^N$ is a phase retrieval frame for $\mathcal{H}$. If it does not satisfy the complement property, then there exists $\Lambda \subset \{1, \dots, N\}$ such that both $\{f_i\}_{i \in \Lambda}$ and $\{f_i\}_{i \in \Lambda^c}$ are incomplete, thus, we can find $f, g \neq 0$ such that

$$\langle f, f_i \rangle = 0, i \in \Lambda, \quad \text{and} \quad \langle g, f_i \rangle = 0, i \in \Lambda^c,$$

then

$$|\langle f + g, f_i \rangle| = |\langle f - g, f_i \rangle|, \quad 1 \leq i \leq N.$$

Since $\{f_i\}_{i=1}^N$ is a phase retrieval frame for $\mathcal{H}$, then there exists a constant c such that $f + g = c(f - g)$. Thus, $f = ((c + 1)/(c - 1))g$. It follows that $\langle f, f_i \rangle = 0$, for all $1 \leq i \leq N$, that is, $f = 0$, which leads to a contradiction.

If $\mathcal{H}$ is a real Hilbert space and $\{f_i\}_{i=1}^N$ is a frame in $\mathcal{H}$ with complementary property. Taking $f, g \in \mathcal{H}$ and suppose that

$$|\langle f, f_i \rangle| = |\langle g, f_i \rangle|, \quad 1 \leq i \leq N.$$

Set $\Lambda = \{i : \langle f, f_i \rangle = \langle g, f_i \rangle\}$. If $\{f_i\}_{i \in \Lambda}$ is complete, then $f = g$. If $\{f_i\}_{i \in \Lambda}$ is incomplete, then $\{f_i\}_{i \in \Lambda^c}$ is complete hence

$$\langle f, f_i \rangle = -\langle g, f_i \rangle, \quad i \in \Lambda^c$$

gives that $f = -g$. Therefore, $\{f_i\}_{i=1}^N$ is a phase retrievable frame.   □

Remark 6.3 By the complement property, if a frame $\mathcal{F} = \{f_i\}_{i=1}^{n}$ is phase retrievable for $\mathbb{R}^d$, then $n \geq 2d - 1$. If $n = 2d - 1$, then $\mathcal{F}$ is phase retrievable if and only if the frame matrix F corresponding to $\mathcal{F}$ has full spark, where the spark of a $d \times n$ matrix M is the size of the smallest linearly dependent subset of columns of M.

The second condition is based on the rank-one operator lifting of the frame $\{f_i\}_{i=1}^{N}$. Let $f, g \in \mathcal{H}$ and $f \otimes g$ be the rank-one operator defined by $(f \otimes g)x = \langle x, g \rangle f$ for every $x \in H$. We use $\langle A, B \rangle = \mathrm{tr}(A^*B)$ to denote the Hilbert-Schmidt inner product on the space of $n \times n$ matrices and let $\mathcal{S}_2$ be the set of all the Hermitian $n \times n$ matrices with rank less than or equal to 2. Let $\Theta_{L(\mathcal{F})}$ be the analysis operator of $L(\mathcal{F}) := \{L(f_i)\}$, where $L(f_i) = f_i \otimes f_i$. From the definition of phase-retrievable frames, it is easy to obtain the following characterization:

Proposition 6.7 *A frame $\{f_j\}_{j \in \mathbb{J}}$ is phase-retrievable if and only if*

$$\ker \Theta_{L(\mathcal{F})} \cap \mathcal{S}_2 = \{0\},$$

where $\mathcal{S}_2$ is the set of all the self-adjoint operators with rank less than or equal to 2.

Balan et al. [21, 22] have shown that every generic frame $\{f_i\}_{i=1}^{N}$ for $\mathbb{F}^n$ is phase retrievable if $N \geq 2n - 1$ in the real case $\mathbb{F} = \mathbb{R}$ or if $N \geq 4n - 2$ in the complex case $\mathbb{F} = \mathbb{C}$.

To examine the robustness of the phase retrieval process for a given frame $\mathcal{F}$, Cahill et al. in [44] introduced the quotient metric on the space $\mathcal{H}/\sim$ induced by the metric on $\mathcal{H}$ given by

$$d(\tilde{f}, \tilde{g}) = \inf_{|\alpha|=1} \|f - \alpha g\|$$

and quantify the robustness in terms of the lower Lipschitz bound of the map $\Theta_{\mathcal{F}}$ (the analysis operator of $\mathcal{F}$) under this metric and obtained the following results.

Proposition 6.8 ([44]) *If $\mathcal{H}$ is a finite-dimensional Hilbert space and $\mathcal{F} = \{f_n\}_{n=1}^{M}$ does phase retrieval for $\mathcal{H}$, then there is some positive constant C (depending only on $\mathcal{F}$) such that $\Theta_{\mathcal{F}}$ has a positive lower Lipschitz bound C, i.e.,*

$$\inf_{|c|=1} \|x - cy\| \leq C \|\Theta_{\mathcal{F}}(x) - \Theta_{\mathcal{F}}(y)\|.$$

While let $\mathcal{H}$ be an infinite-dimensional separable Hilbert space and let $\mathcal{F} = \{f_n\}_{n=1}^{\infty}$ be a frame for $\mathcal{H}$ with frame bounds $0 < A \leq B < \infty$; further suppose that $\|\varphi_n\| \geq c > 0$ for every $n \in \mathbb{N}$. Then, for every $\delta > 0$, there exist $f, g \in \mathcal{H}$ so that $\inf_{|\alpha|=1} \|f - \alpha g\| \geq 1$ but

$$\|\Theta_{\mathcal{F}}(f) - \Theta_{\mathcal{F}}(g)\| < \delta.$$

That is, the phase retrieval frames are stable in finite-dimensional spaces but unstable in infinite-dimensional spaces.

6.3.1 Phase-retrievable Frames on Finite Abelian Groups

In this subsection, we discuss some recent results on the phase retrievability for group representation frames [154]. As shown in Proposition 6.7, phase retrieval can be formulated as a rank-one matrix recovery (phase-lifting) problem if a phase retrievable frame has the maximal span property. In this case, $\{f_i \otimes f_i\}_{i=1}^N$ is a frame for the space of all $n \times n$ self-adjoint matrices equipped with the inner product $\langle A, B \rangle = \text{tr}(A^*B)$. Let $\{A_i\}_{i=1}^N$ be a dual frame for $\{f_i \otimes f_i\}_{i=1}^N$. Then for every $x \in \mathcal{H}$, we have

$$x \otimes x = \sum_{i=1}^N \langle x \otimes x, f_i \otimes f_i \rangle A_i = \sum_{i=1}^N |\langle x, f_i \rangle|^2 A_i$$

and so x can be reconstructed (up to a multiple of an unimodular scalar) by factorizing the above right-hand side rank-one matrix. A key element for this reconstruction requires the existence and construction of frames with the maximal span property. Here, we investigate a special class of frames, namely, projective representation frames, that have the maximal span property.

A projective unitary representation π for a finite group G is a mapping $g \mapsto \pi(g)$ from G into the group $U(\mathcal{H})$ of all the unitary operators on a Hilbert space $\mathcal{H}$ such that $\pi(g)\pi(h) = \mu(g, h)\pi(gh)$ for all $g, h \in G$, where $\mu(g, h)$ is a scalar-valued function on $G \times G$ taking values in the circle group $\mathbb{T}$. This function $\mu(g, h)$ is then called a multiplier of π. In this case, we also say that π is a μ-projective unitary representation. It is clear from the definition that we have

1. $\mu(g_1, g_2 g_3)\mu(g_2, g_3) = \mu(g_1 g_2, g_3)\mu(g_1, g_2)$ for all $g_1, g_2, g_3 \in G$,
2. $\mu(g, e) = \mu(e, g) = 1$ for all $g \in G$, where e denotes the group unit of G.

Any function $\mu : G \times G \mapsto \mathbb{T}$ satisfying (1)–(2) above will be called a multiplier or 2-cocycle of G. It follows from (1) and (2) that we also have,

3. $\mu(g, g^{-1}) = \mu(g^{-1}, g)$ holds for all $g \in G$.

A projective unitary representation π of G on $\mathcal{H}$ is called irreducible if $\text{span}\{\pi(g) : g \in G\} = B(\mathcal{H})$, the space of all the linear operators on $\mathcal{H}$. Let π be a projective group representation. A vector ξ is called a π-maximal spanning frame vector if $\{\pi(g)\xi\}_{g \in G}$

has the maximal spanning property. We will use $\mathcal{M}_\pi$ to denote the set of all π-maximal spanning frame vectors.

Problem 6.1 Let π be a projective unitary representation of a finite group G on $\mathbb{C}^n$. Under what condition does π admit a frame vector with the maximal span property? In particular, is it true that every irreducible μ-projective representation π admits a maximal spanning frame vector?

We will show that for any finite abelian group, every irreducible μ-projective representation π admits a maximal spanning frame vector.

Let μ be a 2-cocycle (or multiplier) of a finite abelian group G, and let C_μ be its associated symmetric multiplier matrix defined by $C_\mu = [c_{g,h}]$ with $c_{g,h} = \mu(g,h)\overline{\mu(h,g)}$. Let π be a projective unitary representation of G on an n-dimensional complex Hilbert space H. For each $\xi \in H$, consider the matrix

$$A(\xi) = [a_{g,h}(\xi)]_{G \times G}$$

with $a_{g,h}(\xi) = \langle \pi(h)\pi(g)\xi, \pi(g)\xi \rangle$.

We first establish the following sufficient conditions.

Lemma 6.1 If there exists $\xi \in H$ such that $A(\xi)$ has rank n^2 (where $n = \dim \mathcal{H}$), then π is irreducible and $\{\pi(g)\xi\}_{g \in G}$ has the maximal span property.

Proof Let $X = \{\pi(g)\}_{g \in G}$ and $Y = \{\pi(g)\xi \otimes \pi(g)\xi\}_{g \in G}$ be two sequences in $B(\mathcal{H})$ equipped with the trace inner product. Note that the mixed Gramian matrix $\Theta_Y \Theta_X^*$ is exactly the matrix $A(\xi)$ which is assumed to have rank n^2. Thus $\text{rank}(\Theta_Y) \geq n^2$ and $\text{rank}(\Theta_X) \geq n^2$. Since we also have $\text{rank}(\Theta_Y) \leq n^2$ and $\text{rank}(\Theta_X) \leq n^2$, we get that $\text{rank}(\Theta_Y) = n^2 = \text{rank}(\Theta_X)$, which implies that π is irreducible and $\{\pi(g)\xi\}_{g \in G}$ has the maximal span property.   □

Lemma 6.2 Suppose that π is a μ-projective unitary representation for a finite group G on an n-dimensional complex Hilbert space $\mathcal{H}$. Then there exists $\xi \in \mathcal{H}$ such $\langle \pi(g)\xi, \xi \rangle \neq 0$ for all $g \in G$. Moreover, the set of all such vectors ξ is open and dense in $\mathcal{H}$.

Proof We can assume that $\mathcal{H} = \mathbb{C}^n$. By the Baire-Category Theorem, it suffices to prove that for each $g \in G$, the set $\{\xi \in \mathbb{C}^n : \langle \pi(g)\xi, \xi \rangle \neq 0\}$ is open and dense in $\mathbb{C}^n$. Since $\langle \pi(g)\xi, \xi \rangle$ is a quadratic polynomial of ξ, we only need to point out that this is a nonzero polynomial. Indeed, if $\langle \pi(g)\xi, \xi \rangle = 0$ for all $\xi \in \mathbb{C}^n$, then we have $\pi(g) = 0$, which is a contradiction.   □

Lemma 6.3 *Suppose that π is a μ-projective unitary representation for an abelian group G. If there exists $\xi \in \mathcal{H}$ such that $\{\pi(g)\xi\}_{g \in G}$ has the maximal spanning property, then $\langle \pi(g)\xi, \xi \rangle \neq 0$ for every $g \in G$.*

Proof Since $\{\pi(g)\xi\}_{g \in G}$ has the maximal spanning property, then $\mathrm{span}\{\pi(g)\xi \otimes \pi(g)\xi : g \in G\} = B(\mathcal{H})$. So if $\langle \pi(h)\xi, \xi \rangle = 0$ for some $h \in G$, then for every $g \in G$ we get

$$|\langle \pi(h)\pi(g)\xi, \pi(g)\xi \rangle| = |\langle \pi(g^{-1})\pi(h)\pi(g)\xi, \xi \rangle| = |\langle \pi(g^{-1}hg)\xi, \xi \rangle|$$

$$= |\langle \pi(h)\xi, \xi \rangle| = 0.$$

So $\mathrm{tr}(\pi(h)(\pi(g)\xi \otimes \pi(g)\xi)) = 0$ and $\pi(h) = 0$, which leads to a contradiction. $\square$

Let μ be a multiplier for an abelian group G. Recall that the *symmetric multiplier matrix* is defined by $C_\mu = [c_{g,h}]$ with $c_{g,h} = \mu(g,h)\overline{\mu(h,g)}$.

Theorem 6.14 ([154]) *Suppose that π is a μ-projective unitary representation for an abelian group G on an n-dimensional Hilbert space $\mathcal{H} = \mathbb{C}^n$. Then $\mathrm{rank}(C_\mu) \leq n^2$. Moreover, π is an irreducible μ-representation if and only if $\mathrm{rank}(C_\mu) = n^2$.*

Proof By Lemma 6.2, there exists $\eta \in \mathbb{C}^n$ such that $\langle \pi(g)\eta, \eta \rangle \neq 0$ for any $g \in G$. Let $\Theta_1 : M_{n \times n}(\mathbb{C}) \mapsto \ell^2(G)$ be the analysis operator for $\{\pi(g)\}_{g \in G}$, and $\Theta_2 : M_{n \times n}(\mathbb{C}) \mapsto \ell^2(G)$ be the analysis operator for $\{\pi(g)\eta \otimes \pi(g)\eta\}_{g \in G}$. We have

$$\Theta_2 \Theta_1^* = [\langle \pi(g)\pi(h)\eta, \pi(h)\eta \rangle]_{G \times G}.$$

Note that

$$\langle \pi(g)\pi(h)\eta, \pi(h)\eta \rangle = c_{g,h}\langle \pi(g)\eta, \eta \rangle$$

and $\langle \pi(g)\eta, \eta \rangle \neq 0$ for every $g \in G$. So we get that

$$\mathrm{rank}(C_\mu) = \mathrm{rank}(\Theta_2 \Theta_1^*) \leq \mathrm{rank}(\Theta_1) = \dim(\mathrm{span}\{\pi(g) : g \in G\}) \leq n^2.$$

Now assume that $\mathrm{rank}(C_\mu) = n^2$. Then the above inequality implies that $\dim(\mathrm{span}\{\pi(g) : g \in G\}) = n^2$, and thus π is irreducible.

Conversely, let us assume that π is irreducible. let $\hat{G}$ be the dual group of G and let $\bar{\pi} : g \mapsto \overline{\pi(g)}$ be the complex conjugation of $\pi(g)$. Let $T_\mu = \{\chi \in \hat{G} : \chi \subset \pi \otimes \bar{\pi}\}$. Then T_μ is a subgroup of $\hat{G}$. Define

$$G_\mu = T_\mu^\perp = \{g \in G : \chi(g) = 1, \forall \chi \in T_\mu\}.$$

Note that $|T_\mu| = \dim \mathcal{H} \times \dim \mathcal{H} = n^2$. Thus $|G_\mu| = [G : G_\mu] = |T_\mu| = n^2$. Since G is abelian, it is easy to verify that $c : G \times G \mapsto \mathbb{T}$ defined by $c(g, h) = c_{gh} = \mu(g, h)\overline{\mu(h, g)}$ is a bi-homomorphism, i.e., $c(gg', h) = c(g, h)c(g'h)$ and $c(g, hh') = c(g, h)c(g, h')$ for all $g, g', h, h' \in G$. This induces a homomorphism $\lambda_\mu : G \mapsto \hat{G}$. By [46, Proposition 2.4] we know that

$$G_\mu = \ker(\lambda_\mu) = \{g \in G : \lambda_\mu(g) = 1\}.$$

Therefore we get

$$|\lambda_\mu(G)| = [G : \ker(\lambda_\mu)] = n^2.$$

Recall that the characters of G are linearly independent. Since each row of the symmetric multiplier matrix C_μ defines a character of G by $h \mapsto c(g, h)$, then the rank of C_μ is equal to the number of different characters that appear in the rows of C_μ. By the definition of λ_μ, we know that this number is exactly the cardinality of the image of λ_μ. This implies that $\mathrm{rank}(C_\mu) = |\lambda_\mu(G)| = n^2$ as claimed.   □

Corollary 6.3 *Let μ be a multiplier of an abelian group G and $n^2 = \mathrm{rank}(C_\mu)$. Then every n-dimensional μ-projective representation π of G is irreducible.*

Proof Let σ be an irreducible subrepresentation of π on a d-dimensional π-invariant subspace. By Theorem 6.14, the representation dimension of σ is equal to $\mathrm{rank}(C_\mu) = d^2$. This implies that $d = n$ and thus $\sigma = \pi$. Therefore, π is irreducible.   □

Theorem 6.15 ([154]) *Suppose that π is a μ-projective unitary representation for a finite abelian group G on an n-dimensional complex Hilbert space H. If π is irreducible, then π admits a frame vector with the maximal span property. Moreover, $\{\pi(g)\xi\}_{g \in G}$ has the maximal span property if and only if*

$$\langle \pi(g)\xi, \xi \rangle \neq 0, \ \forall \, g \in G.$$

Proof Assume that π is an irreducible μ-projective representation of G on $\mathcal{H} = \mathbb{C}^n$. By Lemma 6.3 we know that if $\{\pi(g)\xi\}_{g \in G}$ has the maximal span property, then $\langle \pi(g)\xi, \xi \rangle \neq 0$ for every $g \in G$. Therefore, it suffices to show that $\{\pi(g)\xi\}_{g \in G}$ has the maximal span property when $\langle \pi(g)\xi, \xi \rangle \neq 0$ for every $g \in G$.

Let Θ_1 and $\Theta_2 : M_{n \times n}(\mathbb{C}) \mapsto \ell^2(G)$ be the analysis operators defined in the proof of Theorem 6.14. Then we know that $\mathrm{rank}(\Theta_2\Theta_1^*) = \mathrm{rank}(C_\mu)$. Since π is irreducible, we get that $\mathrm{rank}(\Theta_1^*) = n^2$ and by Theorem 6.14 again that $\mathrm{rank}(C_\mu) = n^2$. This implies that $\mathrm{rank}(\Theta_2\Theta_1^*) = n^2$ which implies that $\mathrm{rank}(\Theta_2) = n^2$ since we also have $\mathrm{rank}(\Theta_2) \leq n^2$.

Therefore, $\{\pi(g)\xi \otimes \pi(g)\xi : g \in G\}$ spans $M_{n \times n}(\mathbb{C})$, i.e., $\{\pi(g)\xi\}_{g \in G}$ has the maximal span property. $\square$

Example 6.3.1 Let G_1 and G_2 be two finite groups. Let $\alpha_i \in Z^2(G_i, \mathbb{T})$ be a 2-cocycle of $G_i (i = 1, 2)$. Define a map $\alpha_1 \times \alpha_2 : (G_1 \times G_2) \times (G_1 \times G_2) \rightarrow \mathbb{T}$ by

$$(\alpha_1 \times \alpha_2)((g_1, g_2), (h_1, h_2)) = \alpha_1(g_1, h_1)\alpha_2(g_2, h_2), \forall g_1, h_1 \in G_1, g_2, h_2 \in G_2.$$

It is easy to check that $\alpha_1 \times \alpha_2 \in Z^2(G_1 \times G_2, \mathbb{T})$. Let (π_1, V_1, α_1) and (π_2, V_2, α_2) be projective unitary representations of G_1 and G_2, respectively, on finite dimensional Hilbert spaces V_1 and V_2. Define a map $\pi_1 \times \pi_2 : G_1 \times G_2 \rightarrow B(V_1 \otimes V_2)$ by

$$\pi_1 \times \pi_2((g_1, g_2)) = \pi_1(g_1) \otimes \pi_2(g_2), \quad \forall g_1 \in G_1, g_2 \in G_2.$$

Then $(\pi_1 \times \pi_2, V_1 \otimes V_2, \alpha_1 \times \alpha_2)$ is a projective representation of $G_1 \times G_2$. If, moreover, π_1 and π_2 are unitary projective representations, then so is $\pi_1 \times \pi_2$. In this situation, we have

1. if π_i is irreducible for $(i = 1, 2)$, then $\pi_1 \times \pi_2$ is an irreducible projective representation of $G_1 \times G_2$;
2. each irreducible projective representation of $G_1 \times G_2$ with multiplier $\alpha = \alpha_1 \times \alpha_2$ is isomorphic to a representation $\pi_1 \times \pi_2$, where π_i is an irreducible projective representation of G_i with multiplier $\alpha_i (i = 1, 2)$.

It is easy to prove:

Proposition 6.9 *If $\xi_i \in V_i$ is a π_i-maximal spanning frame vector $(i = 1, 2)$, then $\xi_1 \otimes \xi_2$ is a $(\pi_1 \times \pi_2)$-maximal spanning frame vector.*

For an abelian group, all the irreducible μ-projective unitary representations have the same representation dimension n, and we have proved that $\mathrm{rank}(C_\mu) = n^2$ which is a key ingredient in the proof of our main theorem. However, this is not the case for non-abelian groups since irreducible representations with respect to the same multiplier could have different representation dimensions. This leads to the following question:

Problem 6.2 Let μ be a multiplier for a finite non-abelian group G such that all the irreducible μ-projective unitary representations have the same representation dimension n. Is it still true that $\mathrm{rank}(C_\mu) = n^2$?

The following is an example which tells us that this indeed is true for some special groups.

Example 6.3.2 ([154]) Consider the metacyclic groups of type $G = \mathbb{Z}_m \ltimes \mathbb{Z}_p$ with p a prime. Fix a presentation of G

$$G = \langle a, b \mid a^m = 1, b^p = 1, bab^{-1} = a^r \rangle,$$

where $r \in \mathbb{Z}_{\geq 0}$ and $r^p \equiv 1 (\text{mod} m)$. By [148, Theorem 2.11.3]

$$H^2(G, \mathbb{C}^\times) = \begin{cases} 0 & \text{if } p \nmid (m, r - 1), \\ \mathbb{Z}_p & \text{if } p \mid (m, r - 1). \end{cases}$$

In the following, we assume that $p \mid (m, r - 1)$. Fix ζ a primitive l-th root of unity, where $l = (m, 1 + r + \cdots + r^{p-1})$. Define $\alpha : G \times G \to \mathbb{T}$ by

$$\alpha(a^i b^j, a^{i'} b^{j'}) = \begin{cases} 1, & \text{if } j = 0, \\ \zeta^{i'(1+r+\cdots+r^{j-1})}, & \text{otherwise.} \end{cases}$$

By [148, Lemmas 2.11.1, 2.11.3 and Theorem 2.11.3], this α is a well-defined element in $Z^2(G, \mathbb{T})$ and it represents a generator of $H^2(G, \mathbb{T})$. By the calculation in [154], the rank of C_α is p^2. On the other hand, by [50, Corollary 3.11], every irreducible α-projective representation of G has dimension p. Hence, the Problem 6.1 has an affirmative answer in this case.

So far, we only considered irreducible projective unitary representations with maximal spanning frame vectors. We ask the following question:

Problem 6.3 Let π be a projective unitary representation of a finite group G on an n-dimensional Hilbert space $\mathcal{H}$.

1. Under what condition does π admit a phase-retrievable frame vector ξ? (i.e., $\{\pi(g)\xi\}$ is a phase-retrievable frame).
2. If π admits a maximal spanning frame vector, must π be irreducible?
3. If π is irreducible, is it possible that there exists a phase-retrievable frame $\{\pi(g)\xi\}$ which does not have the maximal span property?

6.3.2   Exact Phase-retrievable Frames and Subspaces

Recall that an exact frame is a frame with the property that it fails to be a frame when removing any element from the frame. In the finite dimensional case, exact frames are bases. Similarly, we say that a phase-retrievable frame $\{f_i\}_i^N$ for an n-dimensional Hilbert space is exact if it fails to be phase-retrievable when removing any element from the frame sequence. Unlike exact frames, exact phase-retrievable frames could have different lengths. In this subsection, we will show that for the real Hilbert space case, exact phase-retrievable frame of length N exists for every $2n - 1 \leq N \leq n(n + 1)/2$. For a general phase-retrievable frame, we introduce the concept of redundancy with respect to its phase-retrievability and the exact phase-retrievable (PR)-redundancy. For an arbitrary frame which is not necessarily phase-retrievable, we can still investigate the phase-retrievability by studying its maximal phase-retrievable (PR)-subspaces with respect to their dimensions.

Proposition 6.7 indicates that $\ker \Theta_{L(\mathcal{F})} \cap \mathcal{S}_2$ seems to be a good candidate to measure the phase-retrievability for a frame $\mathcal{F}$. This motivates to introduce the following concept of redundancy with respect to phase-retrievability (or PR-redundancy) and the exact PR-redundancy property. Let $\mathcal{F} = \{f_i\}_{i=1}^{N}$ be a frame for $\mathcal{H}$. For each subset Λ of $\{1, \ldots, N\}$, let $\mathcal{F}_\Lambda = \{f_i\}_{i\in\Lambda}$ and use $|\Lambda|$ to denote the cardinality of Λ.

Definition 6.8 Given a frame $\mathcal{F} = \{f_i\}_{i=1}^{N}$ for $\mathcal{H}$. Let k be the smallest integer such that there exists a subset Λ of $\{1, \ldots, N\}$ with the property that $|\Lambda| = k$ and

$$\ker(\Theta_{L(\mathcal{F}_\Lambda)}) \cap \mathcal{S}_2 = \ker(\Theta_{L(\mathcal{F})}) \cap \mathcal{S}_2.$$

Then we call N/k the PR-redundancy of $\mathcal{F}$. A frame is said to have the exact PR-redundancy property if its PR-redundancy is 1. A phase-retrievable frame with the exact PR-redundancy will be called an exact phase-retrievable frame.

6.3.2.1 Exact Phase-Retrievable Frames

Since the dimension of the space of all the Hermitian matrices in $M_n(\mathbb{R})$ is $n(n+1)/2$, the length of any exact phase retrievable frame for $\mathbb{R}^n$ is at most $n(n + 1)/2$. The following simple example shows that $n(n + 1)/2$ is in fact the maximal length of all the exact phase retrievable frames.

Example 6.3.3 Suppose that $\{e_i : 1 \leq i \leq n\}$ is an orthonormal basis for $\mathcal{H}$, then $\{e_i + e_j : 1 \leq i \leq j \leq n\}$ is an exact phase retrieval frame for $\mathcal{H}$.

Proof We first show that $\{e_i + e_j : 1 \leq i \leq j \leq n\}$ is a phase retrieval frame for $\mathcal{H}$. Choosing $f, g \in \mathcal{H}\{0\}$. Set

$$|\langle f, e_i + e_j\rangle| = |\langle g, e_i + e_j\rangle|, \quad 1 \leq i \leq j \leq n. \tag{6.4}$$

Taking $i = j$, it follows that

$$|\langle f, e_i\rangle| = |\langle g, e_i\rangle|, \quad 1 \leq i \leq n.$$

As $f, g \neq 0$, then there exists some i_0 such that $|\langle f, e_{i_0}\rangle| = |\langle g, e_{i_0}\rangle| \neq 0$. If $\langle f, e_{i_0}\rangle = \langle g, e_{i_0}\rangle$, taking $i - i_0$ in (6.4) and squasing both sides of (6.4), we have

$$\langle f, e_j\rangle = \langle g, e_j\rangle, \quad 1 \leq j \leq n.$$

Thus, $f = g$. If $\langle f, e_{i_0}\rangle = -\langle g, e_{i_0}\rangle$, we can similarly prove that $f = -g$. Therefore, $\{e_i + e_j : 1 \leq i \leq j \leq n\}$ is a phase retrieval frame.

Next is to prove the exactness. For any $1 \leq i_0 \leq j_0 \leq n$, we show that $\{e_i + e_j : 1 \leq i \leq j \leq n, (i, j) \neq (i_0, j_0)\}$ is not phase retrievable. We only need to prove that it does not have the complement property. Let

$$E_1 = \{e_i + e_j : 1 \leq i \leq j \leq n, (i, j) \neq (i_0, j_0), i = i_0 \text{ or } j = i_0\},$$

$$E_2 = \{e_i + e_j : 1 \leq i \leq j \leq n, (i, j) \neq (i_0, j_0), i \neq i_0 \text{ and } j \neq i_0\}.$$

As there are exactly n i_0 in the set $\{(i, j) : 1 \leq i \leq j \leq n\}$, thus, $\#E_1 = n - 1$. It follows that E_1 is incomplete in $\mathcal{H}$.

On the other hand, since $e_{i_0} \notin \text{span}(E_2)$, so E_2 is also incomplete. To conclude, $\{e_i + e_j : 1 \leq i \leq j \leq n, (i, j) \neq (i_0, j_0)\}$ does not have the complement property hence it is not phase retrievable.   □

By the complement property, the smallest length of any phase retrievable frame for $\mathbb{R}^n$ is $2n - 1$. Therefore, any phase retrievable frame of length $2n - 1$ is exact. The next example shows that the set of all the exact phase retrievable frames of length $2n - 1$ is open and dense in the space of $\mathbb{R}^n \oplus \cdots \oplus \mathbb{R}^n$ ($2n - 1$-copies).

Example 6.3.4 Let A be an $n \times (2n - 1)$ real matrix, where any n columns are linearly independent, then the column vectors of A form an exact phase retrieval frame.

Now we prove that for any integer N between $2n - 1$ and $n(n + 1)/2$, there exists an exact phase retrieval frame of length N.

Theorem 6.16 *For every integer N with $2n - 1 \leq N \leq n(n + 1)/2$, there exists an exact phase-retrievable frame of length N.*

Proof We only give a sketch of the proof. Since every full-spark frame of length $2n - 1$ is an exact PR-frame, we only need to prove the theorem for $2n \leq N \leq n(n + 1)/2$. First, we show that for $2n \leq N \leq n(n + 1)/2$, there exist $n \times N$ matrices A such that

(P1) A contains the $n \times n$ identity matrix as a submatrix;
(P2) the rest $N - n$ columns of A consisting of independent continuous random variables or zeros and each column contains at least one 0 and two nonzero entries;

(P3) there are exactly n nonzero entries in every row of A;

(P4) for each $1 \leq i \leq n$, there exist mutually different indices $j_1, \ldots, j_n$ such that $a_{i,j_l}, a_{l,j_l} \neq 0$;

(P5) columns of A form an exact PR frame with probability 1.

We first point out that a phase-retrievable frame which satisfies (P3) must be exact. (P4) says that every row contains a nonzero entry in such columns and different rows correspond to different columns. Consider the following example,

$$A = \begin{pmatrix} 1 & 0 & 0 & a_{1,4} & a_{1,5} & 0 \\ 0 & 1 & 0 & a_{2,4} & 0 & a_{2,3} \\ 0 & 0 & 1 & 0 & a_{3,5} & a_{3,3} \end{pmatrix},$$

where $a_{i,j}$ are independent continuous random variables. For $i = 1$, set $\{j_1, j_2, j_3\} = \{1, 4, 5\}$. Then we have $a_{1,j_l}, a_{l,j_l} \neq 0$ for $1 \leq l \leq 3$. It is easy to see that A satisfies (P1) $\sim$ (P5). In other words, such a matrix exists for $n = 3$.

Now we assume that such a matrix A exists for some n and show that for the case of $n + 1$ by induction with $n \geq 3$. Specifically, Han et al. [123] showed step by step that for $(n + 1) \times (N + n + 1)$ matrices, $(n + 1) \times (N + n)$ matrices, $(n + 1) \times (N + 2)$ matrices and for $2n \leq N \leq n(n + 1)/2$, there exist $n \times N$ matrices satisfying (P1) $\sim$ (P5). Finally, since columns of a randomly generated $n \times (2n - 1)$ matrix form an exact PR frame almost surely, we get the conclusion as desired.   □

The following are some explicit examples for $n = 5$ and $10 \leq N \leq 15$. In each case, column vectors of A form an exact PR frame. Moreover, such matrices correspond to exact PR frames almost surely if the nonzero entries are replaced with independent continuous random variables.

$(n, N) = (5, 10)$

$$A = \begin{pmatrix} 1 & 0 & 0 & 0 & 0 & 6 & 4 & 2 & 11 & 0 \\ 0 & 1 & 0 & 0 & 0 & 13 & 10 & 8 & 0 & 3 \\ 0 & 0 & 1 & 0 & 0 & 7 & 7 & 0 & 9 & 8 \\ 0 & 0 & 0 & 1 & 0 & 16 & 0 & 8 & 30 & 13 \\ 0 & 0 & 0 & 0 & 1 & 0 & 4 & 12 & 14 & 18 \end{pmatrix}.$$

$(n, N) = (5, 12)$

$$A = \begin{pmatrix} 1 & 0 & 0 & 0 & 0 & 6 & 4 & 2 & 11 & 0 & 0 \\ 0 & 1 & 0 & 0 & 0 & 13 & 10 & 8 & 0 & 3 & 0 \\ 0 & 0 & 1 & 0 & 0 & 7 & 7 & 0 & 9 & 8 & 0 \\ 0 & 0 & 0 & 1 & 0 & 16 & 0 & 8 & 30 & 13 & 0 \\ 0 & 0 & 0 & 0 & 1 & 0 & 4 & 12 & 14 & 18 & 0 \end{pmatrix}.$$

$(n, N) = (5, 12)$:

$$A = \begin{pmatrix} 1 & 0 & 0 & 0 & 0 & 7 & 0 & 10 & 10 & 11 & 0 & 0 \\ 0 & 1 & 0 & 0 & 0 & 4 & 0 & 7 & 16 & 0 & 15 & 0 \\ 0 & 0 & 1 & 0 & 0 & 0 & 16 & 2 & 0 & 2 & 3 & 0 \\ 0 & 0 & 0 & 1 & 0 & 0 & 1 & 0 & 23 & 3 & 0 & 9 \\ 0 & 0 & 0 & 0 & 1 & 0 & 0 & 12 & 2 & 11 & 0 & 2 \end{pmatrix}.$$

$(n, N) = (5, 13)$:

$$A = \begin{pmatrix} 1 & 0 & 0 & 0 & 0 & 6 & 0 & 4 & 12 & 16 & 0 & 0 & 0 \\ 0 & 1 & 0 & 0 & 0 & 6 & 0 & 8 & 5 & 0 & 0 & 15 & 0 \\ 0 & 0 & 1 & 0 & 0 & 0 & 9 & 5 & 0 & 0 & 11 & 12 & 0 \\ 0 & 0 & 0 & 1 & 0 & 0 & 160 & 6 & 1 & 0 & 0 & 5 \\ 0 & 0 & 0 & 0 & 1 & 0 & 0 & 7 & 6 & 0 & 10 & 0 & 9 \end{pmatrix}.$$

$(n, N) = (5, 14)$:

$$A = \begin{pmatrix} 1 & 0 & 0 & 0 & 0 & 11 & 0 & 20 & 0 & 16 & 4 & 0 & 0 & 0 \\ 0 & 1 & 0 & 0 & 0 & 5 & 0 & 0 & 1 & 16 & 0 & 0 & 4 & 0 \\ 0 & 0 & 1 & 0 & 0 & 0 & 3 & 6 & 0 & 0 & 0 & 13 & 8 & 0 \\ 0 & 0 & 0 & 1 & 0 & 0 & 17 & 0 & 0 & 8 & 8 & 0 & 0 & 4 \\ 0 & 0 & 0 & 0 & 1 & 0 & 0 & 0 & 1 & 2 & 0 & 1 & 0 & 3 \end{pmatrix}.$$

$(n, N) = (5, 15)$:

$$A = \begin{pmatrix} 1 & 0 & 0 & 0 & 0 & 12 & 0 & 4 & 0 & 7 & 0 & 13 & 0 & 0 & 0 \\ 0 & 1 & 0 & 0 & 0 & 17 & 0 & 0 & 3 & 0 & 10 & 0 & 0 & 2 & 0 \\ 0 & 0 & 1 & 0 & 0 & 0 & 18 & 0 & 0 & 0 & 0 & 12 & 17 & 0 \\ 0 & 0 & 0 & 1 & 0 & 0 & 3 & 0 & 0 & 0 & 1 & 15 & 0 & 0 & 2 \\ 0 & 0 & 0 & 0 & 1 & 0 & 0 & 0 & 3 & 1 & 0 & 0 & 13 & 0 & 18 \end{pmatrix}.$$

6.3.2.2 Phase-retrievable Subspaces

For a non-phase-retrievable frame $\mathcal{F}$, naturally we would like to examine its phase-retrievability on the subsets instead of the entire signal space. A typical example is the subset of sparse signals (e.g. [76, 212]). To better understand frame phase-retrievability, Han et al. [123] considered the problem of identifying the largest subspaces M such that $\mathcal{F}$ does the phase-retrieval for all the signals in M. For this purpose, they introduced the following definition:

Definition 6.9 Let $\mathcal{F} = \{f_i\}_{i=1}^N$ be a frame for $\mathcal{H}$ and M be a subspace of $\mathcal{H}$. We say that M is a *phase-retrievable subspace* with respect to $\mathcal{F}$ if $\{P_M f_i\}_{i=1}^N$ is a phase retrievable frame for M, where P_M is the orthogonal projection from $\mathcal{H}$ onto M. A phase-retrievable subspace M is called *maximal* if it is not a proper subspace of any other phase-retrievable subspaces with respect to $\mathcal{F}$.

We use "$\mathcal{F}$-PR subspace" to denote a phase-retrievable subspace with respect to $\mathcal{F}$. For a given frame $\mathcal{F}$, it naturally arises the following questions: What are possible dimensions k such that there exists a k-dimensional maximal $\mathcal{F}$-PR subspace? Can we characterize all the maximal phase-retrievable subspaces? As the dimensions of the maximal PR-subspaces could be different, it is interesting to find out the information about the largest dimension for the maximal $\mathcal{F}$-PR subspaces. We begin with the following special case.

Proposition 6.10 *Let $\mathcal{F} = \{f_i\}_{i=1}^n$ be a basis for $\mathcal{H}$. Then there exists a k-dimensional maximal $\mathcal{F}$-PR subspace if and only if $1 \le k \le [(n+1)/2]$.*

Now let us consider the general frame case: Let $\mathcal{F}$ be a frame for $\mathcal{H}$. For each subset Λ of $\{1, \ldots, N\}$, let

$$d_\Lambda = \max\left\{\dim \operatorname{span}(\mathcal{F}_\Lambda), \dim \operatorname{span}(\mathcal{F}_{\Lambda^c})\right\}.$$

Define

$$d(\mathcal{F}) = \min\left\{d_\Lambda : \Lambda \subset \{1, \ldots, N\}\right\}.$$

Theorem 6.17 *Let $\mathcal{F}$ be a frame for $\mathcal{H}$. Then k is the largest integer such that there exists a k-dimensional maximal $\mathcal{F}$-PR subspace if and only if $k = d(\mathcal{F})$.*

Clearly, $d(\mathcal{F}) = n$ if and only if $\mathcal{F}$ has the complement property. Thus, the above theorem is a natural generalization of Theorem 6.13. Now we would like to know what are the possible values of $d(\mathcal{F})$. Since every frame contains a basis, we get by Proposition 6.10 that $d(\mathcal{F}) \ge [n+1/2]$. The following theorem tells us that for every k between $[(n+1)/2]$ and n, there is a frame $\mathcal{F}$ with the exact PR-redundancy property such that $k = d(\mathcal{F})$.

Theorem 6.18 *Let $\mathcal{H} = \mathbb{R}^n$ and k be an integer such that $n \ge k \ge [n+1/2]$. Then for each N between $2k - 1$ and $k(k + 1)/2 + (n - k)(n - k + 1)/2$, there exists a frame $\mathcal{F}$ of length N such that it has the exact PR-redundancy property and $d(\mathcal{F}) = k$, i.e., k is the largest integer such that there exists a k-dimensional maximal $\mathcal{F}$-PR subspace.*

The above theorem characterized the possible value for the dimension of a maximal PR-subspace for a general frame. We would like to examine more about the maximal phase-retrievable subspaces with respect to a basis $\mathcal{F} = \{f_i\}_{i=1}^n$.

We first focus on orthonormal bases and then use the similarity to pass to general bases. Now we assume that $\mathcal{E} = \{e_1, e_2, \cdots, e_n\}$ is an orthonormal basis for $\mathbb{R}^n$. For a vector $x = \sum_{i=1}^n \alpha_i e_i \in \mathbb{R}^n$, the support of x is defined by $\operatorname{supp}_{\mathcal{E}}(x) := \{i \mid \alpha_i \neq 0\}$ ($\operatorname{supp}_{\mathcal{F}}(x)$ is defined similarly for any basis $\mathcal{F}$). We will also use $\operatorname{supp}(x)$ to denote $\operatorname{supp}_{\mathcal{E}}(x)$ if $\mathcal{E}$ is well understood in the statements,

Proposition 6.11 *Suppose that M is a k-dimensional $\mathcal{E}$-PR subspace. Then for any nonzero vector $x \in M$, we have $|\operatorname{supp}(x)| \geq k$. Consequently, if there exists $x \in M$ such that $|\operatorname{supp}(x)| = k$, then M is maximal.*

Let $\mathcal{F} = \{f_1, \ldots, f_n\}$ be a basis for $\mathbb{R}^n$, and $\mathcal{F}^* = \{f_1^*, \ldots, f_n^*\}$ be its dual basis. Let T be the invertible matrix such that $f_i = Te_i$ for all i. Next are some simple observations:

1. M is a PR-subspace with respect to $\mathcal{F}$ if and only if $T^t M$ is a PR-subspace with respect to $\mathcal{E}$.
2. The dual basis of $\mathcal{F}$ is $\mathcal{F}^* = \{(T^{-1})^t T^{-1} e_i\}_{i=1}^n$.
3. The coordinate vector of x with respect to the basis $\mathcal{F}^*$ is the same as the coordinate vector of $T^t x$ with respect to the basis $\mathcal{E}$.

Combining the above proposition and observations we get the following theorem:

Theorem 6.19 *Let $\mathcal{F} = \{f_1, \ldots, f_n\}$ be a basis for $\mathbb{R}^n$, and $\mathcal{F}^* = \{f_1^*, \ldots, f_n^*\}$ be its dual basis. Then we have*

1. *If M is a k-dimensional PR-subspace with respect to $\mathcal{F}$, then $|\operatorname{supp}_{\mathcal{F}^*}(x)| \geq k$ for any nonzero vector $x \in M$. Consequently, M is maximal if there is a vector $x \in M$ such that $|\operatorname{supp}_{\mathcal{F}^*}(x)| = k$.*
2. *For any vector $x \in \mathbb{R}^n$ such that $|\operatorname{supp}_{\mathcal{F}^*}(x)| = k$, there is a k-dimensional maximal PR-subspace M with respect to $\mathcal{J}$ such that $x \in M$.*
3. *If $k < [(n+1)/2]$ and M is a k-dimensional PR-subspace with respect to $\mathcal{F}$, then M is maximal if and only if there is a vector $x \in M$ such that $|\operatorname{supp}_{\mathcal{F}^*}(x)| = k$.*

Samplings in Mixed Lebesgue Spaces

7

This chapter investigates various aspects of signals in mixed Lebesgue spaces $L^{p,q}(\mathbb{R}^{d+1})$. Firstly, the definition of mixed Lebesgue spaces is given, then the structures of shift-invariant spaces in $L^{p,q}(\mathbb{R}^{d+1})$ are characterized in terms of semi-discrete convolutions of their generators with sequences in $\ell^{p,q}(\mathbb{Z}^{d+1})$, and it is shown that if the generators are compactly supported, these spaces are the intersection of $L^{p,q}(\mathbb{R}^{d+1})$ and the linear space of functions formed with these convolutions. Secondly, the $L^{p,q}$-stability of shifts of finite functions is considered, with necessary and sufficient conditions given for stability. Thirdly, it studies the nonuniform sampling and reconstruction problem in shift-invariant subspaces of these spaces, proving that the sampling problem is well-posed. A fast reconstruction algorithm is provided, allowing exact reconstruction of functions from sufficiently dense samples. Finally, the sampling theory for non-decaying signals is explored, with Riesz-type bounds connecting the weighted discrete norms of samples to the weighted continuous norms of original and interpolated signals, and stability of sampling and reconstruction is proven for kernels in appropriate mixed Wiener amalgam spaces.

7.1   Mixed Lebesgue Spaces

Mixed Lebesgue spaces were proposed by Benedek and Panzone in [36], and then Robio de Francia et al. also studied them in [87,203]. Mixed Lebesgue spaces are a generalization of Lebesgue spaces that arise naturally when considering functions that depend on independent quantities with different properties [23, 36, 80, 87]. For example, a function which depends on spacial and time variables may belong to mixed Lebesgue spaces. If a function comes from mixed Lebesgue spaces, we can consider the integrability of each variable separately. This is different from Lebesgue spaces which mainly require the same

level of control over all the variables of a function. The flexibility allows these mixed
Lebesgue spaces to play an important role in the study of time-based partial differential
equations and dynamical samplings.

The shift-invariant spaces play an important role in analysis theory, most notably in the
areas of spline approximation, wavelets, Gabor (Weyl-Heisenberg) systems, subdivision
schemes and sampling theory [193, 200, 214, 217]. Therefore, it is interesting to determine
the structure of shift-invariant spaces in mixed Lebesgue spaces $L^{p,q}(\mathbb{R}^{d+1})$.

Usually, there are two ways to describe the structure of a finitely generated shift-
invariant space. One way is to use the Fourier transforms of the generators. Another way
is to use the semi-discrete convolutions of its generators with sequences in $\ell^p(\mathbb{Z}^{d+1})$. De
Boor, DeVore, Ron and Shen gave some characterizations of a finitely generated shift-
invariant subspace of $L^2(\mathbb{R}^{d+1})$ in terms of the Fourier transforms of the generators
[25, 183, 185]. For other characterizations of the shift-invariant subspaces of $L^2(\mathbb{R}^{d+1})$ in
terms of the Fourier transforms of generators see [32, 35, 37, 150, 193, 216]. Jia described
the structure of a finitely generated shift-invariant space in $L^p(\mathbb{R}^{d+1})$ employing the semi-
discrete convolutions of its generators with sequences in $\ell^p(\mathbb{Z}^{d+1})$.

It is generally known that the stability is an expected property for functions in sampling
problems [191, 194, 199, 207, 208, 214, 215, 218]. In [143], Jia and Micchelli gave a
characterization for L^2-stability, and they used the Hölder inequality to extend the result
to the case $1 < p \leq \infty$. Then they studied the characterization of L^p-stability ($1 \leq
p \leq \infty$) of the shifts of a finite number of compactly supported functions in [142, 143].
Furthermore, Jia [141] proved that the shifts of $\phi_1, \cdots, \phi_n$ are L^p-stable if and only if for
any $\xi \in \mathbb{Z}^d$, the sequences $\{\phi_k(\xi + 2\beta\pi)\}_{\beta \in \mathbb{Z}^d}$, ($k = 1, \cdots, n$) are linearly independent.
This is a generalization of the previous results of Jia and Micchelli in [142, 143].

The sampling and reconstruction theory plays an important role in signal processing
since it bridges the modern digital world and the analogue world of continuous functions.
Sampling refers to the process of taking values of time-continuous signals at certain time
intervals during the conversion of signals. The standard problem in sampling is to recover
a function $f \in V \subset L^2(\mathbb{R})$ from a set of samples $\{f(x_i) : i \in \Lambda\}$, where V is a suitable
set of functions and Λ is a countable indexing set. In other words, sampling converts
the continuous signal $f(x)$ into discrete signal $c(k)$. The well-known classical Shannon
sampling theorem [139, 191, 206] states that for any $f \in L^2(\mathbb{R})$ with $\text{supp} \hat{f} \subseteq [-T, T]$,

$$f(x) = \sum_{n \in \mathbb{Z}} f\left(\frac{n}{2T}\right) \frac{\sin \pi(2Tx - n)}{\pi(2Tx - n)} = \sum_{n \in \mathbb{Z}} f\left(\frac{n}{2T}\right) \text{sinc}(2Tx - n),$$

where the series converges in $L^2(\mathbb{R})$ and uniformly on compact sets. Here for any $f \in
L^2(\mathbb{R})$,

$$\hat{f}(\xi) = \int_{\mathbb{R}} f(x)e^{-2\pi i x \xi} dx, \quad \xi \in \mathbb{R}.$$

This theorem tells us that we can recover the entirety of a function from a well-chosen sample. The sampling set $\{\frac{n}{2T} : n \in \mathbb{Z}\}$ is uniform. However, in many realistic situations, the data are known only on a nonuniform sampling set. For example, the loss of data packets during transmission through the internet from satellites can be viewed as a nonuniform sampling problem. The nonuniform sampling problem has been studied in recent years [27, 114, 162, 194, 199, 207, 214, 218]. The problems of uniform and nonuniform sampling and reconstruction also have been extended to more general shift-invariant spaces of the form

$$V(\phi) = \left\{ \sum_{k \in \mathbb{Z}} c(k)\phi(x - k) : \{c(k) : k \in \mathbb{Z}\} \in \ell^2(\mathbb{Z}) \right\}.$$

Recently, many researchers studied the sampling theory on general shift-invariant spaces in $L^p(\mathbb{R}^d)$, $1 \le p < +\infty$, the details can be seen in [2–4, 10, 208, 215].

The uniform sampling problem for band-limited functions in mixed Lebesgue spaces has been studied by Rodolfo H. Torres and Erika L. Ward [203, 209].

If a function comes from a mixed Lebesgue space, then we can consider the integrability of each variable separately [23]. This property guarantees us to study the multidimensional non-decaying signals in weighted $L^{p,q}(\mathbb{R}^{d+1})$ spaces. In fact, the classical Shannon sampling theorem was extended from the spectral finite function spaces to the more general shift-invariant subspaces [204, 205], and many mathematicians studied the sampling in shift-invariant subspaces [3, 4, 6, 41, 167]. But the sampling theory requires that the input signal is square-integrable, so it is impossible to apply it to signals that do not decay or even grow infinitely.

In order to simulate the natural decay conditions of real signals and images, Aldroubi and Gröchenig proposed the sampling theory in weighted L^p spaces with moderate weights in [4]. Then, Nguyen and Unser applied this property and studied the sampling theory of the non-decaying signal in [168]. They modeled the non-decaying signals in weighted $L^p(\mathbb{R}^d)$ spaces, where the signals growth was controlled by the decaying weighting function.

The definition of mixed Lebesgue spaces $L^{p,q}(\mathbb{R}^{d+1})$ is as follows.

Definition 7.1 Let $1 < p, q < +\infty$. Then $L^{p,q}(\mathbb{R}^{d+1})$ consists of all measurable functions f on $\mathbb{R}^{d+1}$ such that

$$\|f\|_{L^{p,q}} = \left[\int_{\mathbb{R}} \left(\int_{\mathbb{R}^d} |f(x, y)|^q dy \right)^{\frac{p}{q}} dx \right]^{\frac{1}{p}} < +\infty.$$

The corresponding sequence spaces of mixed Lebesgue spaces $L^{p,q}(\mathbb{R}^{d+1})$ are

$$\ell^{p,q}(\mathbb{Z}^{d+1}) = \left\{ c : \|c\|_{\ell^{p,q}} = \left[\sum_{k_1 \in \mathbb{Z}} \left(\sum_{k_2 \in \mathbb{Z}^d} |c(k_1, k_2)|^q \right)^{\frac{p}{q}} \right]^{\frac{1}{p}} < +\infty \right\}.$$

It is easy to see $\| \cdot \|_{L^{p,q}}$ is a norm, then for $f \in L^{p,q}(\mathbb{R}^{d+1})$ and $g \in L^{p,q}(\mathbb{R}^{d+1})$,

$$\|f + g\|_{L^{p,q}} \leq \|f\|_{L^{p,q}} + \|g\|_{L^{p,q}}. \tag{7.1}$$

In order to control the local behaviour of a function, Wiener amalgam spaces [113] and mixed Wiener amalgam spaces are introduced. The following is the definition of Wiener amalgam spaces $W(L^p)(\mathbb{R}^{d+1})$.

Definition 7.2 For $1 \leq p < \infty$, a function f belongs to Wiener amalgam spaces $W(L^p) = W(L^p)(\mathbb{R}^{d+1})$, if it satisfies

$$\|f\|_{W(L^p)}^p := \sum_{k \in \mathbb{Z}^{d+1}} \text{esssup}_{x \in [0,1]^{d+1}} |f(x + k)|^p < \infty.$$

For $p = \infty$, a measurable function f belongs to $W(L^\infty) = W(L^\infty)(\mathbb{R}^{d+1})$ if it satisfies

$$\|f\|_{W(L^\infty)} := \sup_{k \in \mathbb{Z}^{d+1}} \text{esssup}_{x \in [0,1]^{d+1}} |f(x + k)| < \infty.$$

Let $W_0(L^p)$ ($1 \leq p \leq \infty$) be the space of all continuous functions in $W(L^p)$.
The following is the definition of mixed Wiener amalgam spaces $W(L^{p,q})(\mathbb{R}^{d+1})$.

Definition 7.3 A measurable function f belongs to mixed Wiener amalgam spaces $W(L^{p,q}) = W(L^{p,q})(\mathbb{R}^{d+1})$, $1 \leq p, q < \infty$, if it satisfies

$$\|f\|_{W(L^{p,q})(\mathbb{R}^{d+1})}^p := \sum_{n \in \mathbb{Z}} \sup_{x \in [0,1]} \left[\sum_{l \in \mathbb{Z}^d} \sup_{y \in [0,1]^d} |f(x + n, y + l)|^q \right]^{p/q} < \infty. \tag{7.2}$$

Let $W_0(L^{p,q})$ ($1 \leq p, q < \infty$) be the space of all continuous functions in $W(L^{p,q})$. It is easy to see that $W(L^p) \subset W(L^{p,p})$.
Here and after, let $\hat{f}(\xi)$ denote the Fourier transform of $f \in L^1(\mathbb{R}^{d+1})$:

$$\hat{f}(\xi) = \int_{\mathbb{R}^{d+1}} f(x)e^{-2\pi i \xi x} dx.$$

For any $c \in \ell^2(\mathbb{Z}^{d+1})$(or $c \in \ell^1(\mathbb{Z}^{d+1})$), define the discrete Fourier transform on torus $\mathbb{T}^{d+1} \simeq [0, 1)^{d+1}$ by

$$\hat{c}(\xi) = \sum_{k \in \mathbb{Z}^{d+1}} c(k)e^{-2\pi i k\xi}.$$

For any $f, g \in L^2(\mathbb{R}^{d+1})$, define their convolution

$$(f * g)(x) = \int_{\mathbb{R}^{d+1}} f(x - y)g(y)dy = \int_{\mathbb{R}^{d+1}} f(y)g(x - y)dy.$$

Proposition 7.1 ([86, Proposition 8.9]) *Suppose that* $f \in L^p, g \in L^1 (1 \le p \le \infty)$, *then*

$$\|f * g\|_{L^p} \le \|f\|_{L^p}\|g\|_{L^1}.$$

The following lemma provides a mixed spaces version of Young's inequality.

Lemma 7.1 *If* $f \in L^{p,q}(\mathbb{R}^{d+1})$, $g \in L^1(\mathbb{R}^{d+1})$ $(1 < p, q < \infty)$, *then we have*

$$\|f * g\|_{L^{p,q}} \le \|f\|_{L^{p,q}}\|g\|_{L^1}.$$

Proof Let $f_{x_1} = f(x_1, \cdot)$ and $g_{x_1} = f(x_1, \cdot)$, then

$$\|f * g\|_{L^{p,q}}^p = \int_{\mathbb{R}} \left[\int_{\mathbb{R}^d} \left| \int_{\mathbb{R}} \int_{\mathbb{R}^d} f(y_1, y_2)g(x_1 - y_1, x_2 - y_2)dy_2 dy_1 \right|^q dx_2 \right]^{p/q} dx_1$$

$$= \int_{\mathbb{R}} \left[\int_{\mathbb{R}^d} \left| \int_{\mathbb{R}} f_{y_1} * g_{x_1-y_1}(x_2)dy_1 \right|^q dx_2 \right]^{p/q} dx_1$$

$$= \int_{\mathbb{R}} \left\| \int_{\mathbb{R}} f_{y_1} * g_{x_1-y_1}(\cdot)dy_1 \right\|_{L^q}^p dx_1.$$

Minkowski inequality and Proposition 7.1 lead to

$$\|f * g\|_{L^{p,q}}^p \le \int_{\mathbb{R}} \left(\int_{\mathbb{R}} \|f_{y_1} * g_{x_1-y_1}(\cdot)\|_{L^q} dy_1 \right)^p dx_1$$

$$\le \int_{\mathbb{R}} \left(\int_{\mathbb{R}} \|f_{y_1}\|_{L^q} \|g_{x_1-y_1}\|_{L^1} dy_1 \right)^p dx_1.$$

Let $h(x_1) = \|f_{x_1}\|_{L^q}$ and $b(x_1) = \|g_{x_1}\|_{L^1}$, then by Proposition 7.1 again

$$\int_{\mathbb{R}} \left(\int_{\mathbb{R}} \|f_{y_1}\|_{L^q} \|g_{x_1-y_1}\|_{L^1} dy_1 \right)^p dx_1 = \int_{\mathbb{R}} |(h*b)(x_1)|^p dx_1 = \|h*b\|_{L^p}^p \le \|h\|_{L^p}^p \|b\|_{L^1}^p.$$

where

$$\|h\|_{L^p}^p = \int_{\mathbb{R}} \|f_{x_1}\|_{L^q}^p dx_1 = \int_{\mathbb{R}} \left(\int_{\mathbb{R}^d} |f(x_1, x_2)|^q dx_2 \right)^{p/q} dx_1 = \|f\|_{L^{p,q}}^p$$

and

$$\|b\|_{L^1}^p = \left(\int_{\mathbb{R}} \|g_{x_1}\|_{L^1} dx_1 \right)^p = \left(\int_{\mathbb{R}} \int_{\mathbb{R}^d} |g(x_1, x_2)| dx_2 dx_1 \right)^p = \|g\|_{L^1}^p.$$

Finally, one has

$$\|f * g\|_{L^{p,q}}^p \leq \|f\|_{L^{p,q}}^p \|g\|_{L^1}^p.$$

$\square$

For any $c \in \ell^p, d \in \ell^1 \ (1 \leq p \leq +\infty)$, define the discrete convolution

$$(c *_d d)(l) = \sum_{k \in \mathbb{Z}^{d+1}} c(k) d(l - k).$$

The following is the discrete version of Young's inequality.

Proposition 7.2 ([86, Theorem 6.18]) *If $c \in \ell^p, d \in \ell^1 \ (1 \leq p \leq \infty)$, then*

$$\|c * d\|_{\ell^p} \leq \|c\|_{\ell^p} \|d\|_{\ell^1}.$$

The semi-discrete convolution $a *_{sd} \phi$ is the sum

$$(a *_{sd} \phi)(\cdot) = \sum_{k \in \mathbb{Z}^{d+1}} a(k) \phi(\cdot - k).$$

for a sequence a on $\mathbb{Z}^{d+1}$ and a compactly supported function ϕ on $\mathbb{R}^{d+1}$.

Given a function $\phi \in W(L^1)$, the following lemma shows that $\phi *_{sd}$ maps ℓ^p to L^p and the semi-discrete version of Young's inequality is provided.

Lemma 7.2 *Let $\phi \in W(L^1)$ and $c \in \ell^p \ (1 \leq p \leq \infty)$. Then*

$$\|c *_{sd} \phi\|_{L^p} \leq \|c\|_{\ell^p} \|\phi\|_{W(L^1)}. \tag{7.3}$$

Proof For $p = \infty$, (7.3) holds obviously.

In the case $1 \leq p < \infty$, by Fubini's theorem,

$$\|c *_{sd} \phi\|_{L^p}^p = \int_{\mathbb{R}^{d+1}} \left| \sum_{k \in \mathbb{Z}^{d+1}} c(k)\phi(x-k) \right|^p dx$$

$$= \sum_{l \in \mathbb{Z}^{d+1}} \int_{[0,1]^{d+1}} \left| \sum_{k \in \mathbb{Z}^{d+1}} c(k)\phi(x+l-k) \right|^p dx$$

$$= \int_{[0,1]^{d+1}} \sum_{l \in \mathbb{Z}^{d+1}} \left| \sum_{k \in \mathbb{Z}^{d+1}} c(k)\phi(x+l-k) \right|^p dx.$$

Fix x for the moment, let d_x be the sequence $\{d_x(k) = \phi(x+k) : k \in \mathbb{Z}^{d+1}\}$. Then

$$\sum_{k \in \mathbb{Z}^{d+1}} c(k)\phi(x+l-k) = \sum_{k \in \mathbb{Z}^{d+1}} c(k)d_x(l-k) = (c *_d d_x)(l).$$

Proposition 7.2 leads to

$$\sum_{l \in \mathbb{Z}^{d+1}} \left| \sum_{k \in \mathbb{Z}^{d+1}} c(k)\phi(x+l-k) \right|^p = \|c *_d d_x\|_{\ell^p}^p$$

$$\leq \|c\|_{\ell^p}^p \|d_x\|_{\ell^1}^p = \|c\|_{\ell^p}^p \left(\sum_{k \in \mathbb{Z}^{d+1}} |\phi(x+k)| \right)^p$$

$$\leq \|c\|_{\ell^p}^p \|\phi\|_{W(L^1)}^p.$$

Therefore, one has

$$\|c *_{sd} \phi\|_{L^p}^p \leq \int_{[0,1]^{d+1}} \|c\|_{\ell^p}^p \|\phi\|_{W(L^1)}^p dx = \|c\|_{\ell^p}^p \|\phi\|_{W(L^1)}^p.$$

One completes the proof. □

The following lemma provides a discrete version of Young's inequality in discrete mixed spaces.

Lemma 7.3 *If $c \in \ell^{p,q}, d \in \ell^1$, $(1 \leq p, q < \infty)$, then*

$$\|c *_d d\|_{\ell^{p,q}} \leq \|c\|_{\ell^{p,q}} \|d\|_{\ell^1}.$$

Proof Fix $k_1 \in \mathbb{Z}$ for the moment, denote $c_{k_1} = \{c_{k_1}(l) = c(k_1, l) : l \in \mathbb{Z}^d\}$ and $d_{k_1} = \{d_{k_1}(l) = d(k_1, l) : l \in \mathbb{Z}^d\}$, then

$$\|c *_d d\|_{\ell^{p,q}}^p = \sum_{n_1 \in \mathbb{Z}} \left[\sum_{n_2 \in \mathbb{Z}^d} \left| \sum_{k_1 \in \mathbb{Z}} \sum_{k_2 \in \mathbb{Z}^d} c(k_1, k_2) d(n_1 - k_1, n_2 - k_2) \right|^q \right]^{p/q}$$

$$= \sum_{n_1 \in \mathbb{Z}} \left\| \sum_{k_1 \in \mathbb{Z}} (c_{k_1} *_d d_{n_1 - k_1})(\cdot) \right\|_{\ell^q}^p$$

$$\leq \sum_{n_1 \in \mathbb{Z}} \left(\sum_{k_1 \in \mathbb{Z}} \|(c_{k_1} *_d d_{n_1 - k_1})(\cdot)\|_{\ell^q} \right)^p.$$

The discrete version of Young's inequality (Proposition 7.2) leads to

$$\|c *_d d\|_{\ell^{p,q}}^p \leq \sum_{n_1 \in \mathbb{Z}} \left(\sum_{k_1 \in \mathbb{Z}} \|c_{k_1}\|_{\ell^q} \|d_{n_1 - k_1}\|_{\ell^1} \right)^p.$$

Let $a = \{a(n) = \|c_n\|_{\ell^q} : n \in \mathbb{Z}\}$ and $b = \{b(n) = \|d_n\|_{\ell^1} : n \in \mathbb{Z}\}$. Using Proposition 7.2 again, one has

$$\sum_{n_1 \in \mathbb{Z}} \left(\sum_{k_1 \in \mathbb{Z}} \|c_{k_1}\|_{\ell^q} \|d_{n_1 - k_1}\|_{\ell^1} \right)^p = \|a *_d b\|_{\ell^p}^p \leq \|a\|_{\ell^p}^p \|b\|_{\ell^1}^p$$

$$= \sum_{k_1 \in \mathbb{Z}} \|c_{k_1}\|_{\ell^q}^p \left(\sum_{k_1 \in \mathbb{Z}} \|d_{k_1}\|_{\ell^1} \right)^p.$$

Therefore,

$$\|c *_d d\|_{\ell^{p,q}}^p \leq \sum_{k_1 \in \mathbb{Z}} \|c_{k_1}\|_{\ell^q}^p \left(\sum_{k_1 \in \mathbb{Z}} \|d_{k_1}\|_{\ell^1} \right)^p$$

$$= \sum_{k_1 \in \mathbb{Z}} \left(\sum_{k_2 \in \mathbb{Z}^d} |c(k_1, k_2)|^q \right)^{p/q} \left(\sum_{k_1 \in \mathbb{Z}} \sum_{k_2 \in \mathbb{Z}^d} |d(k_1, k_2)| \right)^p$$

$$= \|c\|_{\ell^{p,q}}^p \|d\|_{\ell^1}^p.$$

$\square$

The following is Hölder's inequality for mixed Lebesgue spaces.

Proposition 7.3 ([209, Theorem 1.1.3]) *Let* $1 \leq p, q \leq \infty$, *with* $\frac{1}{p} + \frac{1}{p'} = 1$ *and* $\frac{1}{q} + \frac{1}{q'} = 1$. *Then*

$$\|fg\|_{L^{1,1}} \leq \|f\|_{L^{p,q}} \|g\|_{L^{p',q'}}.$$

Throughout this section, p' and q' are respectively the conjugate exponents of p and q (i.e., $1/p + 1/p' = 1$, $1/q + 1/q' = 1$).

7.2 Shift-Invariant Subspaces

In this section, we mainly research the shift-invariant spaces in mixed Lebesgue spaces $L^{p,q}(\mathbb{R}^{d+1})$. In the first part, the principal generated shift-invariant spaces in mixed Lebesgue spaces $L^{p,q}(\mathbb{R}^{d+1})$ are discussed. In the second part, the finitely generated shift-invariant spaces in mixed Lebesgue spaces $L^{p,q}(\mathbb{R}^{d+1})$ are researched. Then the shift-invariant spaces with Compactly Supported Generators and the shift-invariant spaces with general generators are discussed.

The stability of the function system can ensure that the signal can be recovered when the disturbance occurs. The following is the definition of $L^{p,q}$-stable which is useful for the shift-invariant spaces. For the research about $L^{p,q}$-stable, please refer to the next section on research for $L^{p,q}$-stable.

Definition 7.4 Let $\phi_1, \cdots, \phi_n \in L^{p,q}(\mathbb{R}^{d+1})$ $(1 < p, q < \infty)$. The shifts $\phi_j(\cdot - k)$ $(1 \leq j \leq n, k \in \mathbb{Z}^{d+1})$ are said to be $L^{p,q}$-stable if there are two positive constants C_1 and C_2 such that

$$C_1 \sum_{j=1}^{n} \|c_j\|_{\ell^{p,q}} \leq \left\| \sum_{j=1}^{n} c_j *_{sd} \phi_j \right\|_{L^{p,q}} \leq C_2 \sum_{j=1}^{n} \|c_j\|_{\ell^{p,q}}$$

for all $c_j \in \ell^{p,q}(\mathbb{Z}^{d+1})$, $1 \leq j \leq n$.

In particular, if $p = q$, then we call the shifts $\phi_j(\cdot - k)$ $(1 \leq j \leq n, k \in \mathbb{Z}^{d+1})$ L^p-stable.

For $\phi \in W(L^{1,1})$, the principal generated shift-invariant subspaces in mixed Lebesgue spaces $L^{p,q}$ are defined as follows

$$S_{p,q}(\phi) = \left\{ \sum_{k_1 \in \mathbb{Z}} \sum_{k_2 \in \mathbb{Z}^d} c(k_1, k_2) \phi(\cdot - k_1, \cdot - k_2) : \left\{ c(k_1, k_2) : k_1 \in \mathbb{Z}, k_2 \in \mathbb{Z}^d \right\} \in \ell^{p,q} \right\}.$$

It is obvious that the double sum pointwisely converges almost everywhere. In fact, $c(k_1, k_2) \in \ell^{p,q}$ tells that $c(k_1, k_2) \in \ell^\infty$. This with $\phi \in W(L^{1,1})$ leads to

$$\sum_{k_1 \in \mathbb{Z}} \sum_{k_2 \in \mathbb{Z}^d} |c(k_1, k_2)\phi(x - k_1, y - k_2)| \leq \|c\|_\infty \sum_{k_1 \in \mathbb{Z}} \sum_{k_2 \in \mathbb{Z}^d} |\phi(x - k_1, y - k_2)|$$

$$\leq \|c\|_\infty \|\phi\|_{W(L^{1,1})} \quad (a.e.).$$

The following theorem shows that the principal generated shift-invariant subspaces are well-defined in $L^{p,q}(\mathbb{R}^{d+1})$.

Theorem 7.1 *Let $1 \leq p, q < \infty$ and $\phi \in W(L^{1,1})$. Then for any $c \in \ell^{p,q}$, the function $f = \sum_{k_1 \in \mathbb{Z}} \sum_{k_2 \in \mathbb{Z}^d} c(k_1, k_2)\phi(\cdot - k_1, \cdot - k_2)$ belongs to $L^{p,q}$. Furthermore,*

$$\|f\|_{L^{p,q}} \leq \|c\|_{\ell^{p,q}} \|\phi\|_{W(L^{1,1})}.$$

Proof For fixed $x \in \mathbb{R}$ and $k_1 \in \mathbb{Z}$, let $c_{k_1}(k_2) = \{c(k_1, k_2) : k_2 \in \mathbb{Z}^d\}$, $\phi_{x-k_1}(y) = \phi(x - k_1, y)$. Then

$$\sum_{k_2 \in \mathbb{Z}^d} c(k_1, k_2)\phi(x - k_1, y - k_2) = \sum_{k_2 \in \mathbb{Z}^d} c_{k_1}(k_2)\phi_{x-k_1}(y - k_2) = (c_{k_1} *_{sd} \phi_{x-k_1})(y).$$

This and Lemma 7.2 lead to

$$\|f\|_{L^{p,q}}^p = \int_{\mathbb{R}} \left[\int_{\mathbb{R}^d} \left| \sum_{k_1 \in \mathbb{Z}} \sum_{k_2 \in \mathbb{Z}^d} c(k_1, k_2)\phi(x - k_1, y - k_2) \right|^q dy \right]^{\frac{p}{q}} dx$$

$$= \int_{\mathbb{R}} \left\| \sum_{k_1 \in \mathbb{Z}} (c_{k_1} *_{sd} \phi_{x-k_1})(\cdot) \right\|_{L^q}^p dx$$

$$\leq \int_{\mathbb{R}} \left[\sum_{k_1 \in \mathbb{Z}} \|(c_{k_1} *_{sd} \phi_{x-k_1})(\cdot)\|_{L^q} \right]^p dx$$

$$\leq \int_{\mathbb{R}} \left[\sum_{k_1 \in \mathbb{Z}} \|c_{k_1}\|_{\ell^q} \|\phi_{x-k_1}\|_{W(L^1)} \right]^p dx$$

$$= \int_{\mathbb{R}} \left[\sum_{k_1 \in \mathbb{Z}} \|c(k_1, \cdot)\|_{\ell^q} \|\phi(x - k_1, \cdot)\|_{W(L^1)} \right]^p dx. \tag{7.4}$$

Let $d = \{d(k_1) = \|c(k_1, \cdot)\|_{\ell^q} : k_1 \in \mathbb{Z}\}$ and $h(x) = \|\phi(x, \cdot)\|_{W(L^1)}$. Then by (7.4) and Lemma 7.2 again, one has

$$\int_{\mathbb{R}} \left| \sum_{k_1 \in \mathbb{Z}} \|c(k_1, \cdot)\|_{\ell^q} \|\phi(x - k_1, \cdot)\|_{W(L^1)} \right|^p dx$$

$$= \int_{\mathbb{R}} \left| \sum_{k_1 \in \mathbb{Z}} d(k_1) h(x - k_1) \right|^p dx = \|d *_{sd} h\|_{L^p}^p$$

$$\leq \|d\|_{\ell^p}^p \|h\|_{W(L^1)}^p = \|c\|_{\ell^{p,q}}^p \|\phi\|_{W(L^{1,1})}^p.$$

Hence, $\|f\|_{L^{p,q}} \leq \|c\|_{\ell^{p,q}} \|\phi\|_{W(L^{1,1})}$. □

In order to obtain the main theorem, we introduce the shift-invariant subspaces of L^p spaces generated by finitely generators. A measurable function $f : \mathbb{R}^{d+1} \to \mathbb{C}$ is called locally integrable, if

$$\int_K |f(x)| dx < \infty$$

for every compact subset K of $\mathbb{R}^{d+1}$. We denote by $L_{loc}(\mathbb{R}^{d+1})$ the linear space of all locally integrable functions on $\mathbb{R}^{d+1}$. Let $\phi_1, \cdots, \phi_n$ be functions on $\mathbb{R}^{d+1}$. Denote

$$\Phi = \{\phi_1, \cdots, \phi_n\}.$$

Definition 7.5 Define

$$S_0(\Phi) = span\{\phi_j(\cdot - k), 1 \leq j \leq n, k \in \mathbb{Z}^{d+1}\}.$$

Clearly, $S_0(\Phi)$ is the smallest shift-invariant space containing Φ.

Let $\Phi \subset L^p(\mathbb{R}^{d+1})$ $(1 < p < +\infty)$. Define

$$S_p(\Phi) = \overline{span\{\phi_j(\cdot - k), 1 \leq j \leq n, k \in \mathbb{Z}^{d+1}\}}.$$

That is, $S_p(\Phi)$ is the closed subspace in $L^p(\mathbb{R}^{d+1})$ spanned by $\{\phi_j(\cdot - k), 1 \leq j \leq n, k \in \mathbb{Z}^{d+1}\}$. Moreover, if the functions in Φ are compactly supported, Jia R. Q. proved that $S_p(\Phi) = S(\Phi) \cap L^p(\mathbb{R}^{d+1})$ in [140, 142].

Definition 7.6 Let Φ be a finite collection of compactly supported functions on $\mathbb{R}^{d+1}$. Define

$$S(\Phi) = \left\{ f \mid f(x) = \sum_{j=1}^{n} (a_j *_{sd} \phi_j)(x) \text{ for any sequence } a_j \text{ on } \mathbb{Z}^{d+1} \right\}.$$

It was shown by de Boor, DeVore and Ron in [25] that $f \in S_2(\Phi)$ if and only if the Fourier transform of f can be written as

$$\hat{f}(\omega) = \sum_{j=1}^{n} c_j(\omega) \hat{\phi}_j(\omega),$$

where $c_j(\omega)$ is a 1-periodic function.

Definition 7.7 Let $\Phi \subset L^{p,q}\left(\mathbb{R}^{d+1}\right) (1 < p, q < \infty)$. Define

$$S_{p,q}(\Phi) = \overline{span\{\phi_j(\cdot - k), 1 \le j \le n, k \in \mathbb{Z}^{d+1}\}},$$

for the closure of $S_0(\Phi)$ in $L^{p,q}\left(\mathbb{R}^{d+1}\right)$.

Then, $S_{p,q}(\Phi)$ is the smallest closed shift-invariant subspace of $L^{p,q}\left(\mathbb{R}^{d+1}\right)$ that contains Φ.

We characterize $S_{p,q}(\Phi)$ in terms of the semi-discrete convolutions of their generators with sequences in $\ell^{p,q}\left(\mathbb{Z}^{d+1}\right)$.

Proposition 7.4 ([189, p.66]) *Let X, Y be Banach spaces, and A be a one-to-one bounded linear operator from X to Y. Then a necessary and sufficient condition that the range of A be closed in Y is that $\|x\| \le C\|Ax\|$ holds.*

The following theorem is the generalization of Theorem 7.1 in [140].

Theorem 7.2 *Let $\Phi \subset L^{p,q}\left(\mathbb{R}^{d+1}\right) (1 < p, q < \infty)$. Suppose that the shifts $\phi_j(\cdot - k) (1 \le j \le n, k \in \mathbb{Z}^{d+1})$ are $L^{p,q}$-stable. Then a function f lies in $S_{p,q}(\Phi)$ if and only if*

$$f = \sum_{j=1}^{n} a_j *_{sd} \phi_j$$

for some sequences a_j in $\ell^{p,q}\left(\mathbb{Z}^{d+1}\right) (1 \le j \le n)$.

Proof We first prove the necessity. Let T_Φ be the mapping from $\left(\ell^0\left(\mathbb{Z}^{d+1}\right)\right)^n$ to $L^{p,q}\left(\mathbb{R}^{d+1}\right)$ given by

$$T_\Phi(a_1,\cdots,a_n) = \sum_{j=1}^n a_j *_{sd} \phi_j, \quad a_1,\cdots,a_n \in \ell^0\left(\mathbb{Z}^{d+1}\right).$$

Since the shifts $\phi_j(\cdot - k)$ $\left(1 \le j \le n,\ k \in \mathbb{Z}^{d+1}\right)$ are $L^{p,q}$-stable and $\ell^0\left(\mathbb{Z}^{d+1}\right) \subset \ell^{p,q}\left(\mathbb{Z}^{d+1}\right)$, the domain of T_Φ can be extended to $\left(\ell^{p,q}\left(\mathbb{Z}^{d+1}\right)\right)^n$ by Hahn-Banach theorem.

Because $\phi_j(\cdot - k)$ $(1 \le j \le n,\ k \in \mathbb{Z}^{d+1})$ are $L^{p,q}$-stable, T_Φ is a one to one continuous linear operator from $\left(\ell^{p,q}\left(\mathbb{Z}^{d+1}\right)\right)^n$ to $L^{p,q}\left(\mathbb{R}^{d+1}\right)$ and there exists a positive constant C such that

$$\sum_{j=1}^n \|a_j\|_{\ell^{p,q}} \le C\,\|T_\Phi(a_1,\cdots,a_n)\|_{L^{p,q}}$$

for all $(a_1,\cdots,a_n) \in \left(\ell^{p,q}\left(\mathbb{Z}^{d+1}\right)\right)^n$. We consider $\left(\ell^{p,q}\left(\mathbb{Z}^{d+1}\right)\right)^n$ to be endowed with the ℓ^1 norm, then the range of T_Φ is closed by Proposition 7.4. Therefore, $\Phi \subset T_\Phi\left(\left(\ell^{p,q}\left(\mathbb{Z}^{d+1}\right)\right)^n\right)$ leads to $S_0(\Phi) \subseteq T_\Phi\left(\left(\ell^{p,q}\left(\mathbb{Z}^{d+1}\right)\right)^n\right)$ and

$$S_{p,q}(\Phi) \subseteq T_\Phi\left(\left(\ell^{p,q}\left(\mathbb{Z}^{d+1}\right)\right)^n\right). \tag{7.5}$$

Now we prove the other side. Let $f \in T_\Phi\left(\left(\ell^{p,q}\left(\mathbb{Z}^{d+1}\right)\right)^n\right)$, then there exist $c_j \in \ell^{p,q}\left(\mathbb{Z}^{d+1}\right)(1 \le j \le n)$ such that

$$f = \sum_{j=1}^n c_j *_{sd} \phi_j.$$

Since $\ell^0\left(\mathbb{Z}^{d+1}\right)$ is dense in $\ell^{p,q}\left(\mathbb{Z}^{d+1}\right)$, for any $c_j(1 \le j \le n)$, there exists a sequence $\left\{c_j^m : c_j^m \in \ell^0\left(\mathbb{Z}^{d+1}\right),\ m \in \mathbb{Z}\right\}$ such that

$$c_j^m \xrightarrow{\ell^{p,q}} c_j, \quad m \to \infty. \tag{7.6}$$

Let $f_m = \sum_{j=1}^n c_j^m *_{sd} \phi_j$, then $f_m \in S_0(\Phi) \subset S_{p,q}(\Phi)$. We want to prove

$$f_m \xrightarrow{L^{p,q}} f, \quad m \to \infty.$$

Since the shifts $\phi_j(\cdot - k)$ $\left(1 \leq j \leq n, \, k \in \mathbb{Z}^{d+1}\right)$ are $L^{p,q}$-stable, with (7.6) we have,

$$\|f - f_m\|_{L^{p,q}} \leq C_2 \sum_{j=1}^{n} \left\|c_j - c_j^m\right\|_{\ell^{p,q}} \to 0, \quad m \to \infty.$$

By the definition of $S_{p,q}(\Phi)$, we know that $S_{p,q}(\Phi)$ is closed. Therefore

$$f \in S_{p,q}(\Phi).$$

It follows that

$$T_\Phi\left(\left(\ell^{p,q}\left(\mathbb{Z}^{d+1}\right)\right)^n\right) \subseteq S_{p,q}(\Phi).$$

This with (7.5) completes the proof of the theorem. $\square$

When the generators in Φ of $S_{p,q}(\Phi)$ are compactly supported, we will prove that $S_{p,q}(\Phi) = S(\Phi) \cap L^{p,q}\left(\mathbb{R}^{d+1}\right)$. Firstly, we show that $S(\Phi) \cap L^{p,q}\left(\mathbb{R}^{d+1}\right)$ is closed in $L^{p,q}\left(\mathbb{R}^{d+1}\right)$. The following two propositions are needed.

Proposition 7.5 ([140, Theorem 3.1]) *Let Φ be the set with compactly supported integrable functions on $\mathbb{R}^{d+1}$. Then $S(\Phi)$ is a closed subspace of $L_{loc}\left(\mathbb{R}^{d+1}\right)$.*

Theorem 7.3 *Let $\Phi \subset L^{p,q}\left(\mathbb{R}^{d+1}\right)$ $(1 < p, q < \infty)$ be the set with compactly supported functions. Then $S(\Phi) \cap L^{p,q}\left(\mathbb{R}^{d+1}\right)$ is closed in $L^{p,q}\left(\mathbb{R}^{d+1}\right)$.*

Proof Suppose that $\Phi \subset L^{p,q}\left(\mathbb{R}^{d+1}\right)$ $(1 < p, q < \infty)$ are compactly support on $K = K_1 \times K_2$, where $K_1 \subset \mathbb{R}$, and $K_2 \subset \mathbb{R}^d$. Let $\{f_n\}_{n=1,2,\ldots}$ be a sequence in $S(\Phi) \cap L^{p,q}\left(\mathbb{R}^{d+1}\right)$ converging to f in $L^{p,q}\left(\mathbb{R}^{d+1}\right)$. By Proposition 7.3,

$$\|(f - f_n)I_K\|_{L^1} \leq \|f - f_n\|_{L^{p,q}}\|I_K\|_{L^{p',q'}} = |K_1|^{1/p'}|K_2|^{1/q'}\|f - f_n\|_{L^{p,q}}.$$

Here, I_K stands for the indicator function of a set K and $|K|$ means the measure of K. Therefore, f_n converges to f in the topology of $L_{loc}\left(\mathbb{R}^{d+1}\right)$. By Proposition 7.5, f lies in $S(\Phi)$. This shows that $S(\Phi) \cap L^{p,q}\left(\mathbb{R}^{d+1}\right)$ is closed in $L^{p,q}\left(\mathbb{R}^{d+1}\right)$. $\square$

The following lemma can be proved similarly to [149, Ch. 1, Theorem 4.4]. And we leave the details to interested readers.

Lemma 7.4 *If f is $m = (m_1, \cdots, m_d)$ times differentiable almost everywhere and $(\frac{\partial}{\partial x})^m f = \frac{\partial^{|m|}}{\partial x_1^{m_1} \cdots \partial x_d^{m_d}} f \in L^1(\mathbb{T}^d)$, then*

$$|\hat{f}(k)| \le \frac{\|(\frac{\partial}{\partial x})^m f\|_{L^1}}{|k^m|},$$

where $k^m = k_1^{m_1} \cdots k_d^{m_d}$ and $k_i \neq 0$ for $i = 1, \cdots, d$.

Using Lemmas 7.3 and 7.4, we get the following Lemma.

Lemma 7.5 Let $a \in \ell^0(\mathbb{Z}^{d+1})$ such that $\hat{a}(\xi) \neq 0$ for all $\xi \in \mathbb{T}^{d+1}$. For given $v \in \ell^{p,q}(\mathbb{Z}^{d+1})$, the discrete convolution equation $a *_d u = v$ is uniquely solvable in $\ell^{p,q}(\mathbb{Z}^{d+1})$.

Proof Since $a \in \ell^0(\mathbb{Z}^{d+1})$, $\hat{a}$ has arbitrary order continuous partial derivative. Considering $\hat{a}(\xi) \neq 0$, one has $|\hat{a}(\xi)| \ge C_1$ for some constant $C_1 > 0$. Then $\left|\left[\frac{1}{\hat{a}(\xi)}\right]^{(2,\cdots,2)}\right|$ has an upper bound. Let

$$c(k) = \frac{1}{(2\pi)^{d+1}} \int_{[0,2\pi)^{d+1}} \frac{1}{\hat{a}(\xi)} e^{ik\xi} d\xi, \quad k \in \mathbb{Z}^{d+1}.$$

Then by Lemma 7.4, the sequence $c \in \ell^1(\mathbb{Z}^{d+1})$. From the Fourier inversion theorem, one has $\hat{c}(\xi)\hat{a}(\xi) = 1$ for all $\xi \in \mathbb{T}^{d+1}$. Hence $c *_d a = \delta$, where δ is the sequence given by $\delta(k) = 1$ for $k = 0$ and $\delta(k) = 0$ for $k \in \{\mathbb{Z}^{d+1}\} \setminus \{0\}$. If $a *_d u = v$, then we have

$$u = \delta *_d u = (c *_d a) *_d u = c *_d (a *_d u) = c *_d v.$$

From $c \in \ell^1(\mathbb{Z}^{d+1})$ and $v \in \ell^{p,q}(\mathbb{Z}^{d+1})$, then $u \in \ell^{p,q}(\mathbb{Z}^{d+1})$ by Lemma 7.3. This shows that the discrete convolution equation $a *_d u = v$ has a unique solution in $\ell^{p,q}(\mathbb{Z}^{d+1})$. $\square$

Consider the system of discrete convolution equations

$$\sum_{k=1}^n a_{j,k} *_d u_k = v_j, \quad j = 1, \cdots, m. \tag{7.7}$$

where $a_{j,k} \in \ell^0(\mathbb{Z}^{d+1})$ $(j = 1, \cdots, m; k = 1, \cdots, n)$ and $v_j \in \ell(\mathbb{Z}^{d+1})$ $(j = 1, \cdots, m)$. We say that this system of equations is compatible if for any $c_1, \cdots, c_m \in \ell^0(\mathbb{Z}^{d+1})$ with $\sum_{j=1}^m c_j *_d a_{j,k} = 0$, $k = 1, \cdots, n$, one must have $\sum_{j=1}^m c_j *_d v_j = 0$. When $n = 1$ in (7.7), we have the following Lemma

Lemma 7.6 Let $a_{j,1}$ and v_j be as system of equations in (7.7), $c_j \in \ell^0(\mathbb{Z}^{d+1})$ $(j = 1, \cdots, m)$, $a_{0,1} = \sum_{j=1}^m c_j *_d a_{j,1}$ and $v_0 = \sum_{j=1}^m c_j *_d v_j$. Suppose that the system of equations in (7.7) is compatible, then we have $a_{0,1} *_d v_j = a_{j,1} *_d v_0$, $j = 1, \cdots, m$.

Proof For fixed $1 \leq j' \leq m$, let $d_j^{j'} = \begin{cases} \sum_{k=1}^{m} c_k *_d a_{k,1}, & j = j'; \\ 0, & j \neq j', \end{cases}$ for $1 \leq j \leq m$. Since

$$\sum_{j=1}^{m} (d_j^{j'} - a_{j',1} *_d c_j) *_d a_{j,1} = \left(\sum_{k=1}^{m} c_k *_d a_{k,1} \right) *_d a_{j',1} - \sum_{j=1}^{m} a_{j',1} *_d c_j *_d a_{j,1}$$

$$= \sum_{j=1}^{m} a_{j',1} *_d c_j *_d a_{j,1} - \sum_{j=1}^{m} a_{j',1} *_d c_j *_d a_{j,1} = 0,$$

by the assumption that the system of equations in (7.7) is compatible, we have

$$\sum_{j=1}^{m} (d_j^{j'} - a_{j',1} *_d c_j) *_d v_j = 0.$$

On the other hand, by the definition of $d_j^{j'}$,

$$\sum_{j=1}^{m} (d_j^{j'} - a_{j',1} *_d c_j) *_d v_j = \sum_{k=1}^{m} c_k *_d a_{k,1} *_d v_{j'} - \sum_{j=1}^{m} a_{j',1} *_d c_j *_d v_j$$

$$= a_{0,1} *_d v_{j'} - a_{j',1} *_d v_0.$$

Therefore, we obtain $a_{0,1} *_d v_{j'} = a_{j',1} *_d v_0,\ j' = 1, \cdots, m$.   □

The following lemma extends [140, Theorem 6.1] to the case of mixed Lebesgue spaces.

Lemma 7.7 *Let* $v_1, \cdots, v_m \in \ell^{p,q} (\mathbb{Z}^{d+1})$. *Suppose that the system of discrete convolution equations in (7.7) is compatible. If the matrix*

$$A(\xi) = (\hat{a}_{j,k}(\xi))_{1 \leq j \leq m, 1 \leq k \leq n}$$

has rank n for every $\xi \in \mathbb{T}^{d+1}$, *then (7.7) is uniquely solvable in* $\ell^{p,q} (\mathbb{Z}^{d+1})$.

Proof For $j = 1, \cdots, m$, let c_j be the sequence given by

$$c_j(k) = \overline{a_{j,1}(-k)}, \quad k \in \mathbb{Z}^{d+1}.$$

Then for every $\xi \in \mathbb{T}^{d+1}$, $\hat{c}_j(\xi) = \sum_{k \in \mathbb{Z}^{d+1}} \overline{a_{j,1}(-k)} e^{-ik\xi} = \overline{\sum_{k \in \mathbb{Z}^{d+1}} a_{j,1}(k) e^{-ik\xi}} = \overline{\hat{a}_{j,1}(\xi)}$. Suppose that

$$a_{0,k} = \sum_{j=1}^{m} c_j *_d a_{j,k}, \quad k = 1, \cdots, n,$$

then $a_{0,k}$ belongs to $\ell^0(\mathbb{Z}^{d+1})$. By the hypothesis that matrix $A(\xi)$ has rank n for every $\xi \in \mathbb{T}^{d+1}$, we have

$$\hat{a}_{0,1}(\xi) = \sum_{j=1}^{m} \hat{c}_j(\xi)\hat{a}_{j,1}(\xi) = \sum_{j=1}^{m} |\hat{a}_{j,1}(\xi)|^2 > 0, \quad \xi \in \mathbb{T}^{d+1}. \tag{7.8}$$

Let us consider the case $n = 1$ first. In this case, (7.7) implies

$$a_{0,1} *_d u_1 = \sum_{j=1}^{m} c_j *_d a_{j,1} *_d u_1 = \sum_{j=1}^{m} c_j *_d v_j =: v_0.$$

It is easy to see $v_0 \in \ell^{p,q}\left(\mathbb{Z}^{d+1}\right)$ by Lemma 7.3. Since $a_{0,1} \in \ell^0(\mathbb{Z}^{d+1})$ and $\hat{a}_{0,1}(\xi) > 0$ for all $\xi \in \mathbb{T}^{d+1}$, the equation $a_{0,1} *_d u_1 = v_0$ is uniquely solvable in $\ell^{p,q}\left(\mathbb{Z}^{d+1}\right)$ by Lemma 7.5. That is, u_1 is the unique solution. Using the assumption that the system of equations in (7.7) is compatible and Lemma 7.6,

$$a_{0,1} *_d v_j = a_{j,1} *_d v_0.$$

It follows that

$$a_{0,1} *_d a_{j,1} *_d u_1 = a_{j,1} *_d v_0 = a_{0,1} *_d v_j.$$

Hence, by (7.8), we have $a_{j,1} *_d u_1 = v_j$ for $j = 1, \cdots, m$. This proves that u_1 is the unique solution to the system of equations in (7.7). In fact, if there exists $\omega \neq u_1$ such that $a_{j,1} * \omega = v_j$, $j = 1, \cdots, m$. Then $\sum_{j=1}^{m}(a_{j,1} * \omega - v_j) * c_j = 0$, that is, $a_{0,1} * \omega = v_0$. This conflicts with u_1 being the unique solution of $a_{0,1} * u_1 = v_0$.

The proof proceeds with induction on n. Suppose $n > 1$ and the desired result is valid for $n - 1$. Let $w_0 = v_0 = \sum_{j=1}^{m} c_j *_d v_j$, $w_j = a_{0,1} *_d v_j - a_{j,1} *_d w_0$ $(j = 1, \cdots, m)$ and

$$b_{j,k} = a_{0,1} *_d a_{j,k} - a_{j,1} *_d a_{0,k}, \quad j = 1, \cdots, m; \ k = 2, \cdots, n.$$

Then $w_0, w_1, \cdots, w_m \in \ell^{p,q}\left(\mathbb{Z}^{d+1}\right)$. Consequently, (7.7) is equivalent to the following system of equations:

$$\sum_{k=1}^{n} a_{0,k} *_d u_k = w_0 \tag{7.9}$$

and

$$\sum_{k=2}^{n} b_{j,k} *_d u_k = w_j, \quad j = 1, \cdots, m. \tag{7.10}$$

We prove it in the following paragraph.

Taking the convolution of both sides of (7.9) with $a_{j,1}$,

$$\sum_{k=1}^{n} a_{0,k} *_d u_k *_d a_{j,1} = w_0 *_d a_{j,1}. \tag{7.11}$$

Take the expression of $b_{j,k}$, w_j into (7.10), we have

$$\sum_{k=2}^{n} (a_{0,1} *_d a_{j,k} - a_{j,1} *_d a_{0,k}) *_d u_k = a_{0,1} *_d v_j - a_{j,1} *_d w_0.$$

This with (7.11) leads to

$$\sum_{k=1}^{n} a_{0,1} *_d a_{j,k} *_d u_k = a_{0,1} *_d v_j.$$

That is

$$a_{0,1} *_d \left(\sum_{k=1}^{n} a_{j,k} *_d u_k - v_j \right) = 0.$$

Since $a_{0,1} \in \ell^0 \left(\mathbb{Z}^{d+1} \right)$ and $\hat{a}_{0,1}(\xi) \neq 0$ for all $\xi \in \mathbb{T}^{d+1}$, by Lemma 7.5, we have (7.7)

$$\sum_{k=1}^{n} a_{j,k} *_d u_k = v_j.$$

On the other hand, (7.7) leads to $c_j *_d \left(\sum_{k=1}^{n} a_{j,k} *_d u_k \right) = c_j *_d v_j$ for $j = 1, \cdots, m$. Hence

$$\sum_{k=1}^{n} \sum_{j=1}^{m} c_j *_d a_{j,k} *_d u_k = \sum_{j=1}^{m} c_j *_d v_j.$$

By the definitions of $a_{0,k}$, v_0, (7.9) holds. Then we can get (7.10) by a similar process.

If (7.10) is compatible and the matrix $B(\xi) = (\hat{b}_{j,k}(\xi))_{1 \leq j \leq m, 2 \leq k \leq n}$ has rank $n - 1$ for every $\xi \in \mathbb{T}^{d+1}$. Thus, by the induction hypothesis, (7.10) is uniquely solvable in $\ell^{p,q} \left(\mathbb{Z}^{d+1} \right)$. Once $u_2, \cdots, u_n$ are obtained, u_1 is uniquely determined from (7.9). This completes the induction procedure.

Now, one proves (7.10) is compatible and $B(\xi) = (\hat{b}_{j,k}(\xi))_{1\le j\le m, 2\le k\le n}$ has rank $n-1$ for every $\xi \in \mathbb{T}^{d+1}$. For any $d_j \in \ell^0(\mathbb{Z}^{d+1})$ $(j = 1, \cdots, m)$ satisfying $\sum_{j=1}^{m} d_j *_d b_{j,k} = 0$,

$$\sum_{j=1}^{m} d_j *_d b_{j,k} = \sum_{j=1}^{m} d_j *_d (a_{0,1} *_d a_{j,k} - a_{j,1} *_d \sum_{l=1}^{m} c_l *_d a_{l,k})$$

$$= \sum_{j=1}^{m} (d_j *_d a_{0,1} - c_j *_d \sum_{l=1}^{m} d_l *_d a_{l,1}) *_d a_{j,k} = 0.$$

Since (7.7) is compatible, we have

$$\sum_{j=1}^{m} (d_j *_d a_{0,1} - c_j *_d \sum_{l=1}^{m} d_l *_d a_{l,1}) *_d v_j = 0.$$

That is, $\sum_{j=1}^{m} d_j * \omega_j = 0$. Let

$$C(\xi) = \begin{pmatrix} 1 & -\hat{a}_{0,2}(\xi) & -\hat{a}_{0,3}(\xi) & \cdots & -\hat{a}_{0,n-1}(\xi) & -\hat{a}_{0,n}(\xi) \\ 0 & \hat{a}_{0,1}(\xi) & 0 & \cdots & 0 & 0 \\ \vdots & \vdots & \vdots & \vdots & \vdots & \vdots \\ 0 & 0 & 0 & \cdots & \hat{a}_{0,1}(\xi) & 0 \\ 0 & 0 & 0 & \cdots & 0 & \hat{a}_{0,1}(\xi) \end{pmatrix}.$$

Then $A(\xi)C(\xi) = (A_1(\xi)\ B(\xi))$, where $A_1(\xi) = (\hat{a}_{1,1}(\xi), \cdots, \hat{a}_{m,1}(\xi))^T$. Hence $B(\xi)$ has rank $n-1$ for any $\xi \in \mathbb{T}^{d+1}$, which completes the proof. □

By Theorem 7.7, we have the following lemma.

Lemma 7.8 *Let $\Phi \subset L^{p,q}(\mathbb{R}^{d+1})$ be the set with functions vanishing outside unit cube $[0, 1)^{d+1}$ and linearly independent. Then the shifts $\phi_j(\cdot - k)$ $(1 \le j \le n,\ k \in \mathbb{Z}^{d+1})$ are $L^{p,q}$-stable.*

Proof Since $\phi_j \in L^{p,q}(\mathbb{R}^{d+1})$ $(1 \le j \le n)$ are vanish outside the unit cube $[0, 1)^{d+1}$, $\Phi \subset L^{p,q}(\mathbb{R}^{d+1})$ $(1 < p, q < \infty)$. Assume that the shifts $\phi_j(\cdot - k)$ $(1 \le j \le n,\ k \in \mathbb{Z}^{d+1})$ are $L^{p,q}$-unstable. By Theorem 7.7, there exists some $\xi_0 \in \mathbb{R}^{d+1}$ such that the sequences $(\hat{\phi}_j(\xi_0 + 2\pi k))_{k \in \mathbb{Z}^{d+1}}$ $(1 \le j \le n)$ are linearly dependent. In other words, there exist constants r_j $(1 \le j \le n)$, not all zero, such that

$$\sum_{j=1}^{n} r_j \hat{\phi}_j(\xi_0 + 2\pi k) = 0, \quad k \in \mathbb{Z}^{d+1}.$$

By the definition of the Fourier transform, we have

$$\sum_{j=1}^{n} r_j \int_{[0,1)^{d+1}} \phi_j(x) e^{-ix(\xi_0 + 2\pi k)} dx = 0, \quad k \in \mathbb{Z}^{d+1},$$

namely

$$\int_{[0,1)^{d+1}} \sum_{j=1}^{n} r_j \phi_j(x) e^{-ix\xi_0} e^{-ix2\pi k} dx = 0, \quad k \in \mathbb{Z}^{d+1}.$$

Since $\Phi \subset L^{p,q}\left(\mathbb{R}^{d+1}\right)$ is the set with functions vanishing outside the unit cube $[0, 1)^{d+1}$, by Proposition 7.3, $\Phi \subset L^1\left([0, 1)^{d+1}\right)$, that is $\sum_{j=1}^{n} r_j \phi_j(x) e^{-ix\xi_0} \in L^1\left([0, 1)^{d+1}\right)$. Therefore,

$$\sum_{j=1}^{n} r_j \phi_j(x) e^{-ix\xi_0} = 0$$

by the uniqueness of the Fourier series. Then

$$\sum_{j=1}^{n} r_j \phi_j(x) = 0, \ \forall \, x \in \mathbb{R}^{d+1}.$$

This contradicts the fact that $\phi_1, \cdots, \phi_n$ are linearly independent. $\square$

Theorem 7.4 *Let* $\Phi \subset L^{p,q}\left(\mathbb{R}^{d+1}\right) (1 < p, q < \infty)$ *be the set with compactly supported functions. Suppose that the shifts* $\phi_j(\cdot - k) \left(1 \leq j \leq n, \, k \in \mathbb{Z}^{d+1}\right)$ *are* $L^{p,q}$*-stable. Then* $S_{p,q}(\Phi) = S(\Phi) \cap L^{p,q}\left(\mathbb{R}^{d+1}\right).$

Proof By Theorem 7.3, $S(\Phi) \cap L^{p,q}\left(\mathbb{R}^{d+1}\right)$ is closed in $L^{p,q}\left(\mathbb{R}^{d+1}\right) (1 < p, q < \infty)$. Hence $S_{p,q}(\Phi)$ is contained in $S(\Phi) \cap L^{p,q}\left(\mathbb{R}^{d+1}\right)$.

For $\phi_j \in L^{p,q}\left(\mathbb{R}^{d+1}\right) (1 \leq j \leq n)$ be compactly supported. We can find functions $\psi_1, \cdots, \psi_m \in L^{p,q}\left(\mathbb{R}^{d+1}\right)$ such that they vanish outside the unit cube $[0, 1)^{d+1}$ and $\{\psi_j|_{[0,1)^{d+1}} : j = 1, \cdots, m\}$ forms a basis for $S(\Phi)|_{[0,1)^{d+1}}$ (see Section 4 of [143]). Thus each ϕ_k $(k = 1, \cdots, n)$ can be represented as

$$\phi_k = \sum_{j=1}^{m} a_{j,k} *_{sd} \psi_j, \tag{7.12}$$

where $a_{j,k}$ $(j = 1, \cdots, m; \, k = 1, \cdots, n)$ are finitely supported sequences on $\mathbb{Z}^{d+1}$.

Now, we assume that $f \in S(\Phi) \cap L^{p,q}(\mathbb{R}^{d+1})$. Then there exist sequence $u_1, \cdots, u_n$ on $\mathbb{Z}^{d+1}$ such that

$$f = \sum_{k=1}^{n} u_k *_{sd} \phi_k. \tag{7.13}$$

Since $S(\Phi) \subset S(\psi_1, \cdots, \psi_m)$,

$$f = \sum_{j=1}^{m} v_j *_{sd} \psi_j, \tag{7.14}$$

where $v_1, \cdots, v_m$ are sequences on $\mathbb{Z}^{d+1}$. By Lemma 7.8 and $f \in L^{p,q}(\mathbb{R}^{d+1})$ $(1 < p, q < \infty)$, the sequences $v_1, \cdots, v_m \in \ell^{p,q}(\mathbb{Z}^{d+1})$. By (7.12), (7.13) and (7.14), $u_1, \cdots, u_n$ satisfy the following system of discrete convolution equations:

$$\sum_{k=1}^{n} a_{j,k} *_d u_k = v_j, \quad j = 1, \cdots, m. \tag{7.15}$$

Clearly, above system of equations is compatible. We shall show that the matrix

$$A(\xi) = (\hat{a}_{j,k}(\xi))_{1 \leq j \leq m, 1 \leq k \leq n}$$

has rank n for any $\xi \in \mathbb{T}$, provided that the shifts $\phi_j(\cdot - k)$ $(1 \leq j \leq n, k \in \mathbb{Z}^{d+1})$ are $L^{p,q}$-stable. In fact, it follows from (7.12) that for $k = 1, \cdots, n$,

$$\hat{\phi}_k(\xi + 2\pi\alpha) = \sum_{j=1}^{m} \hat{a}_{j,k}(\xi)\hat{\psi}_j(\xi + 2\pi\alpha), \quad \xi \in \mathbb{R}^{d+1}, \alpha \in \mathbb{Z}^{d+1}. \tag{7.16}$$

If $A(\xi)$ had rank less than n for some $\xi_0 \in \mathbb{R}^{d+1}$, then there exists a nonzero vector $r = (r_1, r_2, \cdots, r_n)^T$ such that $A(\xi_0)r = 0$, that is

$$\sum_{k=1}^{n} r_k \hat{a}_{1,k}(\xi_0) = 0, \quad \sum_{k=1}^{n} r_k \hat{a}_{2,k}(\xi_0) = 0 \cdots \quad \sum_{k=1}^{n} r_k \hat{a}_{m,k}(\xi_0) = 0.$$

Therefore, for any $\alpha \in \mathbb{Z}^{d+1}$,

$$\sum_{k=1}^{n} r_k \sum_{j=1}^{m} \hat{a}_{j,k}(\xi_0)\hat{\psi}_j(\xi_0 + 2\pi\alpha) = \sum_{j=1}^{m} \sum_{k=1}^{n} r_k \hat{a}_{j,k}(\xi_0)\hat{\psi}_j(\xi_0 + 2\pi\alpha) = 0.$$

By (7.16), we have

$$\sum_{k=1}^{n} r_k \hat{\phi}_k(\xi_0 + 2\pi\alpha) = 0, \quad \alpha \in \mathbb{Z}^{d+1}.$$

Namely, the sequences $\{\hat{\phi}_k(\xi_0 + 2\pi\alpha) : \alpha \in \mathbb{Z}^{d+1}\}$ $(k = 1, \cdots, n)$ are linearly dependent. By Theorem 7.7, this contradicts the assumption of $L^{p,q}$-stability. Therefore, $A(\xi)$ has rank n for every $\xi \in \mathbb{T}^{d+1}$. This and the fact that the system of equations in (7.15) is compatible show that (7.15) is uniquely solvable in $\ell^{p,q}(\mathbb{Z}^{d+1})$ and $u_1, \cdots, u_n$ is the solution by Lemma 7.7. Then

$$f = \sum_{k=1}^{n} \phi_k *_{sd} u_k$$

lies in $S_{p,q}(\Phi)$ by Theorem 7.2. This proves that $S_{p,q}(\Phi) = S(\Phi) \cap L^{p,q}(\mathbb{R}^{d+1})$.   □

Now we give the main result in this section which extends [140, Theorem 7.2] to the case of mixed Lebesgue spaces.

7.3   The $L^{p,q}$-Stability of the Shift-invariant Subspaces

In this section, we study the $L^{p,q}$-stability of the shift-invariant spaces in Lebesgue spaces $L^{p,q}(\mathbb{R}^{d+1})$. This section is divided into two parts. In the first part, we mainly investigate the stability of the principal generated shift-invariant spaces in mixed Lebesgue spaces $L^{p,q}(\mathbb{R}^{d+1})$. In the second part, we mainly investigate the stability of the finitely generated shift-invariant spaces in mixed Lebesgue spaces $L^{p,q}(\mathbb{R}^{d+1})$.

Now, we give the definition of the spaces $\mathcal{L}^p(\mathbb{R}^{d+1})$.

Definition 7.8 Given a function $f : \mathbb{R}^{d+1} \to \mathbb{R}$, define $f \in \mathcal{L}^p(\mathbb{R}^{d+1})$ means that

$$\|f\|_{\mathcal{L}^p} = \left[\int_{[0,1]^{d+1}} \left(\sum_{k \in \mathbb{Z}^{d+1}} |f(x+k)| \right)^p dx \right]^{1/p} < \infty.$$

For any $1 \le p \le \infty$, let $\mathcal{L}^p = \mathcal{L}^p(\mathbb{R}^{d+1})$ be the linear space of all functions f for which $\|f\|_{\mathcal{L}^p} < \infty$. The norms are defined above and with usual modification in the case of $p = \infty$.

Now, we give the definition of the spaces $\mathcal{L}^{p,q}(\mathbb{R}^{d+1})$, which are useful for the $L^{p,q}$-stable.

Definition 7.9 Given a function f, define

$$
\|f\|_{\mathcal{L}^{p,q}} := \left\| \left[\sum_{k_1 \in \mathbb{Z}} \left[\int_{[0,1]^d} \left(\sum_{k_2 \in \mathbb{Z}^d} |f(\cdot + k_1, x_2 + k_2)| \right)^q dx_2 \right]^{1/q} \right] \right\|_{L^p([0,1])}.
$$

For $1 \le p, q \le \infty$, let $\mathcal{L}^{p,q} = \mathcal{L}^{p,q}(\mathbb{R}^{d+1})$ be the linear space of all functions f for which $\|f\|_{\mathcal{L}^{p,q}} < \infty$. The norms are defined above and with usual modification in the case of $q = \infty$.

Considering triangle inequality, for $f \in \mathcal{L}^{p,q}$ and $g \in \mathcal{L}^{p,q}$,

$$
\|f + g\|_{\mathcal{L}^{p,q}} \le \|f\|_{\mathcal{L}^{p,q}} + \|g\|_{\mathcal{L}^{p,q}}. \tag{7.17}
$$

Clearly, for $1 \le p \le \infty$, $1 \le q \le \infty$,

$$
\mathcal{L}^{\infty,\infty} \subset \mathcal{L}^{p,q} \subset \mathcal{L}^{1,1}. \tag{7.18}
$$

7.3.1   Principal Generated Case

We first give some basic notations and definitions which are useful in this subsection. Meanwhile, some useful lemmas and propositions are presented in this subsection.

Here, $\mathbb{T}$ denotes the additive group of the reals modulo 1 (that is $\mathbb{R}^{d+1}/\mathbb{Z}^{d+1}$). For $g \in L^1(\mathbb{T})$, define

$$
\hat{g}(k) = \int_{\mathbb{T}} g(x)e^{-i2\pi kx} dx.
$$

Let $\mathcal{B} = \{\hat{a}(\xi) : a \text{ is a sequence, and } a \in \ell^1(\mathbb{Z}^{d+1})\}$.

Lemma 7.9 *If $f \in \mathcal{B}$ and $f(\xi) \ne 0$ for every $\xi \in \mathbb{T}$, then by Wiener's lemma (see [187, p.266]) $1/f$ is also in $\mathcal{B}$.*

Proposition 7.6 ([142, Theorem 2.1]) *Let $\phi \in \mathcal{L}^p$ and $c \in \ell^p(\mathbb{Z}^{d+1})$ $(1 \le p \le \infty)$. Then*

$$
\|c *_{sd} \phi\|_{\mathcal{L}^p} \le \|c\|_{\ell^p} \|\phi\|_{\mathcal{L}^p}.
$$

The following lemma provides a mixed spaces version of the above proposition.

Lemma 7.10 Let $\phi \in \mathcal{L}^{p,q}(\mathbb{R}^{d+1})$, where $1 < p, q < \infty$. Then for any $c \in \ell^{p,q}(\mathbb{Z}^{d+1})$,

$$\|c *_{sd} \phi\|_{L^{p,q}} \leq \|c\|_{\ell^{p,q}} \|\phi\|_{\mathcal{L}^{p,q}}.$$

Proof Let $c = \{c(k_1, k_2) : k_1 \in \mathbb{Z}, k_2 \in \mathbb{Z}^d\} \in \ell^{p,q}$. Then

$$\|c *_{sd} \phi\|_{L^{p,q}(\mathbb{R}\times\mathbb{R}^d)}^p = \int_{\mathbb{R}} \left[\int_{\mathbb{R}^d} \left| \sum_{k_1 \in \mathbb{Z}} \sum_{k_2 \in \mathbb{Z}^d} c(k_1, k_2)\phi(x_1 - k_1, x_2 - k_2) \right|^q dx_2 \right]^{\frac{p}{q}} dx_1.$$

For fixed k_1 and x_1, let $c_{k_1} = \{c_{k_1}(k_2) = c(k_1, k_2) : k_2 \in \mathbb{Z}^d\}$ and $\phi_{x_1 - k_1}(x_2) = \phi(x_1 - k_1, x_2)$. Then

$$\sum_{k_2 \in \mathbb{Z}^d} c(k_1, k_2)\phi(x_1 - k_1, x_2 - k_2) = \sum_{k_2 \in \mathbb{Z}^d} c_{k_1}(k_2)\phi_{x_1 - k_1}(x_2 - k_2) = (c_{k_1} *_{sd} \phi_{x_1 - k_1})(x_2).$$

Using Proposition 7.6, one has

$$\|c *_{sd} \phi\|_{L^{p,q}}^p = \int_{\mathbb{R}} \left\| \sum_{k_1 \in \mathbb{Z}} (c_{k_1} *_{sd} \phi_{x_1 - k_1})(\cdot) \right\|_{L^q}^p dx_1$$

$$\leq \int_{\mathbb{R}} \left[\sum_{k_1 \in \mathbb{Z}} \|(c_{k_1} *_{sd} \phi_{x_1 - k_1})(\cdot)\|_{L^q} \right]^p dx_1$$

$$\leq \int_{\mathbb{R}} \left| \sum_{k_1 \in \mathbb{Z}} \|c_{k_1}\|_{\ell^q} \|\phi_{x_1 - k_1}\|_{L^q} \right|^p dx_1$$

$$= \int_{\mathbb{R}} \left| \sum_{k_1 \in \mathbb{Z}} \|c(k_1, \cdot)\|_{\ell^q} \|\phi(x_1 - k_1, \cdot)\|_{L^q} \right|^p dx_1. \qquad (7.19)$$

Denote $d = \{d(k_1) = \|c(k_1, \cdot)\|_{\ell^q} : k_1 \in \mathbb{Z}\}$ and $h(x) = \|\phi(x, \cdot)\|_{L^q}$. Then by Proposition 7.6,

$$\int_{\mathbb{R}} \left| \sum_{k_1 \in \mathbb{Z}} \|c(k_1, \cdot)\|_{\ell^q} \|\phi(x_1 - k_1, \cdot)\|_{L^q} \right|^p dx_1 = \int_{\mathbb{R}} |(d *_{sd} h)(x_1)|^p dx_1$$

$$\leq \|d\|_{\ell^p}^p \|h\|_{L^p}^p$$

$$= \|c\|_{\ell^{p,q}}^p \|\phi\|_{\mathcal{L}^{p,q}}^p.$$

This with (7.19) leads to $\|c *_{sd} \phi\|_{L^{p,q}} \leq \|\phi\|_{\mathcal{L}^{p,q}} \|c\|_{\ell^{p,q}}$. $\square$

The following proposition gives two necessary and sufficient conditions of L^2-stability for the shifts of the function $\phi \in \mathcal{L}^{\infty,\infty}$. For $\phi \in \mathcal{L}^{\infty,\infty}$ and $c \in \ell^1(\mathbb{Z}^{d+1})$, define

$$\tilde{S}_1(\phi) = \{g(x) = (c *_{sd} \phi)(x) = \sum_{k \in \mathbb{Z}^{d+1}} c(k)\phi(x - k)\}.$$

Proposition 7.7 ([142, Theorem 3.3]) *Let $\phi \in \mathcal{L}^{\infty,\infty}$ be L^2-stable, if and only if one of the following conditions holds:*

1. for every $\xi \in \mathbb{R}^{d+1}$,

$$\sum_{k \in \mathbb{Z}^{d+1}} \left|\widehat{\phi}(\xi + 2\pi k)\right|^2 > 0;$$

2. there exists a function $g \in \tilde{S}_1(\phi)$ such that

$$\langle \phi(\cdot - \alpha), g \rangle = \delta_{0,\alpha}.$$

Lemma 7.11 *The g in proposition 7.7 belongs to $\mathcal{L}^{\infty,\infty}$. That is, for $c = \{c(k), k \in \mathbb{Z}^{d+1}\} \in \ell^1$ and $\phi \in \mathcal{L}^{\infty,\infty}$,*

$$g = \sum_{k \in \mathbb{Z}^{d+1}} c(k)\phi(\cdot - k) \in \mathcal{L}^{\infty,\infty}.$$

Proof It is easy to see

$$\|g\|_{\mathcal{L}^{\infty,\infty}}$$

$$= \text{esssup}_{x_1 \in [0,1]} \sum_{n_1 \in \mathbb{Z}} \text{esssup}_{x_2 \in [0,1]^d} \sum_{n_2 \in \mathbb{Z}^d} |g(x_1 + n_1, x_2 + n_2)|$$

$$\leq \text{esssup}_{x \in [0,1]} \sum_{n_1 \in \mathbb{Z}} \text{esssup}_{x_2 \in [0,1]^d} \sum_{n_2 \in \mathbb{Z}^d} \sum_{k_1 \in \mathbb{Z}} \sum_{k_2 \in \mathbb{Z}^d}$$

$$|c(k_1, k_2)||\phi(x_1 + n_1 - k_1, x_2 + n_2 - k_2)|$$

$$\leq \sum_{k_1 \in \mathbb{Z}} \sum_{k_2 \in \mathbb{Z}^d} |c(k_1, k_2)| \text{esssup}_{x \in [0,1]} \sum_{n_1 \in \mathbb{Z}} \text{esssup}_{x_2 \in [0,1]^d}$$

$$\sum_{n_2 \in \mathbb{Z}^d} |\phi(x_1 + n_1 - k_1, x_2 + n_2 - k_2)|$$

$$= \|c\|_{\ell^1} \|\phi\|_{\mathcal{L}^{\infty,\infty}}.$$

$\square$

The following lemma shows that $\mathcal{L}^{p,q}$ is a subspace of L^1.

Lemma 7.12 *If $f \in \mathcal{L}^{p,q}$, then $f \in L^1$.*

Proof Let $x = (x_1, x_2)$ where $x_1 \in \mathbb{R}$ and $x_2 \in \mathbb{R}^d$, then

$$\|f\|_{L^1} = \int_{\mathbb{R}} \int_{\mathbb{R}^d} |f(x_1, x_2)| dx_2 dx_1$$

$$= \int_{\mathbb{R}} \sum_{k_2 \in \mathbb{Z}^d} \int_{[0,1]^d} |f(x_1, x_2 + k_2)| dx_2 dx_1$$

$$= \int_{\mathbb{R}} \int_{[0,1]^d} \sum_{k_2 \in \mathbb{Z}^d} |f(x_1, x_2 + k_2)| dx_2 dx_1.$$

Applying Hölder inequality twice, one sees that

$$\|f\|_{L^1} \leq \int_{\mathbb{R}} \left[\int_{[0,1]^d} \left| \sum_{k_2 \in \mathbb{Z}^d} |f(x_1, x_2 + k_2)| \right|^q dx_2 \right]^{1/q} dx_1$$

$$= \int_{\mathbb{R}} \|f(x_1, \cdot)\|_{\mathcal{L}^q} dx_1$$

$$= \int_{[0,1]} \sum_{k_1 \in \mathbb{Z}} \|f(x_1 + k_1, \cdot)\|_{\mathcal{L}^q} dx_1$$

$$\leq \left[\int_{[0,1]} \left| \sum_{k_1 \in \mathbb{Z}} \|f(x_1 + k_1, \cdot)\|_{\mathcal{L}^q} \right|^p dx_1 \right]^{1/p}$$

$$= \|f\|_{\mathcal{L}^{p,q}}.$$

$\square$

Based on Hölder's inequality and Fubini theorem, we have the following lemma.

Lemma 7.13 *Let $1 < p, q < \infty$, with $\frac{1}{p} + \frac{1}{p'} = 1$ and $\frac{1}{q} + \frac{1}{q'} = 1$. If $f \in \mathcal{L}^{p,q}$ and $g \in \mathcal{L}^{p',q'}$, then*

$$\|f * g\|_{\mathcal{L}^{\infty,\infty}} \leq \|f\|_{\mathcal{L}^{p,q}} \|g\|_{\mathcal{L}^{p',q'}}.$$

Proof Let $x = (x_1, x_2)$, where $x_1 \in \mathbb{R}$ and $x_2 \in \mathbb{R}^d$. From the definition of convolution, one has

$$(f * g)(x + k) = (f * g)(x_1 + k_1, x_2 + k_2)$$

$$= \int_{\mathbb{R}} \int_{\mathbb{R}^d} f(y_1, y_2) g(x_1 + k_1 - y_1, x_2 + k_2 - y_2) dy_2 dy_1.$$

It is easy to see

$$|(f * g)(x_1 + k_1, x_2 + k_2)|$$

$$\leq \int_{\mathbb{R}} \sum_{n_2 \in \mathbb{Z}^d} \int_{[0,1]^d} |f(y_1, y_2 + n_2)| |g(x_1 + k_1 - y_1, x_2 + k_2 - y_2 - n_2)| dy_2 dy_1.$$

Repeated use Fubini theorem, one obtains

$$\sum_{k_2 \in \mathbb{Z}^d} |(f * g)(x_1 + k_1, x_2 + k_2)|$$

$$= \int_{\mathbb{R}} \int_{[0,1]^d} \sum_{n_2 \in \mathbb{Z}^d} |f(y_1, y_2 + n_2)| \sum_{k_2 \in \mathbb{Z}^d} |g(x_1 + k_1 - y_1, x_2 + k_2 - y_2 - n_2)| dy_2 dy_1$$

$$= \int_{\mathbb{R}} \int_{[0,1]^d} \sum_{n_2 \in \mathbb{Z}^d} |f(y_1, y_2 + n_2)| \sum_{k_2 \in \mathbb{Z}^d} |g(x_1 + k_1 - y_1, x_2 + k_2 - y_2)| dy_2 dy_1.$$

Hölder inequality leads to

$$\int_{[0,1]^d} \sum_{n_2 \in \mathbb{Z}^d} |f(y_1, y_2 + n_2)| \sum_{k_2 \in \mathbb{Z}^d} |g(x_1 + k_1 - y_1, x_2 + k_2 - y_2)| dy_2$$

$$\leq \left\| \sum_{n_2 \in \mathbb{Z}^d} |f(y_1, \cdot + n_2)| \right\|_{L^q[0,1]^d} \left\| \sum_{k_2 \in \mathbb{Z}^d} |g(x_1 + k_1 - y_1, \cdot + k_2)| \right\|_{L^{q'}[0,1]^d}$$

$$= \left\| f(y_1, \cdot) \right\|_{\mathcal{L}^q} \left\| g(x_1 + k_1 - y_1, \cdot) \right\|_{\mathcal{L}^{q'}}$$

where $\frac{1}{q} + \frac{1}{q'} = 1$. Then,

$$\operatorname{esssup}_{x_2 \in [0,1]^d} \sum_{k_2 \in \mathbb{Z}^d} |(f*g)(x_1 + k_1, x_2 + k_2)| \leq \int_{\mathbb{R}} \left\| f(y_1, \cdot) \right\|_{\mathcal{L}^q} \left\| g(x_1 + k_1 - y_1, \cdot) \right\|_{\mathcal{L}^{q'}} dy_1.$$

Similarly, Fubini theorem and Hölder inequality tell that, for any $x \in [0, 1]^{d+1}$

$$\sum_{k \in \mathbb{Z}} \operatorname{esssup}_{x_2 \in [0,1]^d} \sum_{k_2 \in \mathbb{Z}^d} |(f * g)(x_1 + k_1, x_2 + k_2)|$$

$$\leq \sum_{k_1 \in \mathbb{Z}} \sum_{n_1 \in \mathbb{Z}} \int_{[0,1]} \| f(y_1 + n_1, \cdot) \|_{\mathcal{L}^q} \| g(x_1 + k_1 - y_1 - n_1, \cdot) \|_{\mathcal{L}^{q'}} dy_1$$

$$= \int_{[0,1]} \sum_{n_1 \in \mathbb{Z}} \|f(y_1 + n_1, \cdot)\|_{\mathcal{L}^q} \sum_{k_1 \in \mathbb{Z}} \|g(x_1 + k_1 - y_1, \cdot)\|_{\mathcal{L}^{q'}} \, dy_1$$

$$\leq \left[\int_{[0,1]} \left| \sum_{n_1 \in \mathbb{Z}} \|f(y_1 + n_1, \cdot)\|_{\mathcal{L}^q} \right|^p dy_1 \right]^{1/p} \left[\int_{[0,1]} \left| \sum_{k_1 \in \mathbb{Z}} \|g(y_1 + k_1, \cdot)\|_{\mathcal{L}^{q'}} \right|^{p'} dy_1 \right]^{1/p'}$$

$$= \|f\|_{\mathcal{L}^{p,q}} \|f\|_{\mathcal{L}^{p',q'}}.$$

Therefore, one gets the conclusion

$$\|f * g\|_{\mathcal{L}^{\infty,\infty}} = \operatorname{esssup}_{x_1 \in [0,1]} \sum_{k_1 \in \mathbb{Z}} \operatorname{esssup}_{x_2 \in [0,1]^d} \sum_{k_2 \in \mathbb{Z}^d} |(f * g)(x_1 + k_1, x_2 + k_2)|$$

$$\leq \|f\|_{\mathcal{L}^{p,q}} \|g\|_{\mathcal{L}^{p',q'}}.$$

$\square$

Now we prove some useful lemmas which are needed in the proofs of Theorem 7.5.

Proposition 7.8 ([72, Corollary 1.11]) *If $f \in L^1(\mathbb{T})$ and all of the Fourier coefficients of f are equal to zero, then f is identically zero.*

Lemma 7.14 *If $f \in L^1(\mathbb{R}^{d+1})$ and $\widehat{f}(2\pi n) = 0$ for all $n \in \mathbb{Z}^{d+1}$, then $\sum_{k \in \mathbb{Z}^{d+1}} f(\cdot - k) = 0$.*

Proof Let $g(x) = \sum_{k \in \mathbb{Z}^{d+1}} f(x - k) \in L^1(\mathbb{T})$, its Fourier coefficient is

$$c(n) = \int_{[0,1]^{d+1}} \sum_{k \in \mathbb{Z}^{d+1}} f(x - k) e^{-i2\pi nx} dx = \int_{\mathbb{R}^{d+1}} f(x) e^{-i2\pi nx} dx = \widehat{f}(2\pi n) = 0.$$

From Proposition 7.8, we have $g(x) = \sum_{k \in \mathbb{Z}^{d+1}} f(x - k) = 0$. $\square$

Now we show the main result which give a necessary and sufficient condition for the shifts $\phi(\cdot - k)$ $(k \in \mathbb{Z}^{d+1})$ to be $L^{p,q}$-stable.

Theorem 7.5 *Let $\phi \in \mathcal{L}^{p,q}\left(\mathbb{R}^{d+1}\right) (1 < p, q < \infty)$. Then the shifts $\phi(\cdot - k) (k \in \mathbb{Z}^{d+1})$ are $L^{p,q}$-stable if and only if for any $\xi \in \mathbb{R}^{d+1}$,*

$$\sum_{k \in \mathbb{Z}^{d+1}} |\widehat{\phi}(\xi + 2\pi k)|^2 > 0. \tag{7.20}$$

Proof ($\Leftarrow$). From Lemma 7.10, it is easy to see

$$\|c *_{sd} \phi\|_{L^{p,q}} \leq M_2 \|c\|_{\ell^{p,q}},$$

where $M_2 = \|\phi\|_{\mathcal{L}^{p,q}}$. Now we prove that there exists a positive constant M_1 such that $M_1 \|c\|_{\ell^{p,q}} \leq \|c *_{sd} \phi\|_{L^{p,q}}$.

First, we assume that $\phi \in \mathcal{L}^{\infty,\infty}$ and $f = c *_{sd} \phi$. By Proposition 7.7, there exists a function $g \in \tilde{S}_1(\phi)$ such that

$$\langle \phi(\cdot - \alpha), g \rangle = \delta_{0,\alpha}.$$

Therefore, Lemma 7.12 tells that

$$\int_{\mathbb{R}^{d+1}} f(x) \overline{g(x-k)} dx = \int_{\mathbb{R}^{d+1}} \sum_{l \in \mathbb{Z}^{d+1}} c(l) \phi(x-l) \overline{g(x-k)} dx$$

$$= \sum_{l \in \mathbb{Z}^{d+1}} c(l) \int_{\mathbb{R}^{d+1}} \phi(y-l+k) \overline{g(y)} dy$$

$$= c(k).$$

Let $b = \left\{ b(k) : k \in \mathbb{Z}^{d+1} \right\} \in \ell^{p',q'}$, where $\frac{1}{p} + \frac{1}{p'} = 1$ and $\frac{1}{q} + \frac{1}{q'} = 1$. It is easy to see

$$|\langle c, b \rangle| = \left| \sum_{k \in \mathbb{Z}^{d+1}} c(k) \overline{b(k)} \right|$$

$$= \left| \sum_{k \in \mathbb{Z}^{d+1}} \overline{b(k)} \int_{\mathbb{R}^{d+1}} f(x) \overline{g(x-k)} dx \right|$$

$$= \left| \int_{\mathbb{R}^{d+1}} f(x) \sum_{k \in \mathbb{Z}^{d+1}} \overline{b(k)} g(x-k) dx \right|.$$

Applying Proposition 7.3, one has

$$|\langle c, b \rangle| \leq \|f\|_{L^{p,q}} \left\| \sum_{k \in \mathbb{Z}^{d+1}} b(k) g(x-k) \right\|_{L^{p',q'}}$$

$$\leq \|f\|_{L^{p,q}} \|b\|_{\ell^{p',q'}} \|g\|_{\mathcal{L}^{p',q'}}.$$

The last inequality followed by Lemmas 7.10 and 7.11 and (7.18). Thus

$$\|c\|_{\ell^{p,q}} \leq \|f\|_{L^{p,q}} \|g\|_{\mathcal{L}^{p',q'}}. \tag{7.21}$$

Let $M_1 = 1/\|g\|_{\mathcal{L}^{p',q'}}$. Then one obtains $M_1\|c\|_{\ell^{p,q}} \leq \|f\|_{L^{p,q}}$, that is

$$M_1\|c\|_{\ell^{p,q}} \leq \|c *_{sd} \phi\|_{L^{p,q}}.$$

To deal with the case $\phi \in \mathcal{L}^{p,q}\left(\mathbb{R}^{d+1}\right)(1 < p, q < \infty)$, one smooths ϕ by convolving it with the function φ. Here $\varphi(x) := e^{-\pi|x|^2}$ and $|x|$ is the Euclidean norm of x. Let $\rho = \phi * \varphi$. Then by Lemma 7.13, $\|\rho\|_{\mathcal{L}^{\infty,\infty}} \leq \|\phi\|_{\mathcal{L}^{p,q}}\|\varphi\|_{\mathcal{L}^{p',q'}}$, where $\frac{1}{p} + \frac{1}{p'} = 1$ and $\frac{1}{q} + \frac{1}{q'} = 1$. Moreover, since $\widehat{\varphi}(\xi) = e^{-|\xi|^2/(4\pi)}$ never vanishes, ρ also satisfies (7.20). These with what has been proved, $\|\varphi\|_{L^1} = 1$ and Lemma 7.1, one has

$$M_1\|c\|_{\ell^{p,q}} \leq \|c *_{sd} \rho\|_{L^{p,q}} = \|c *_{sd} \phi * \varphi\|_{L^{p,q}} \leq \|c *_{sd} \phi\|_{L^{p,q}}\|\varphi\|_{L^1} = \|c *_{sd} \phi\|_{L^{p,q}}.$$

Thus, we obtain for any $\phi \in \mathcal{L}^{p,q}\left(\mathbb{R}^{d+1}\right)$,

$$M_1\|c\|_{\ell^{p,q}} \leq \|c *_{sd} \phi\|_{L^{p,q}}.$$

Hence the shifts $\phi(\cdot - k)\,(k \in \mathbb{Z}^{d+1})$ are $\ell^{p,q}$-stable.

($\Rightarrow$). Suppose that for some $\xi_0 \in \mathbb{R}^{d+1}$, $\sum_{k\in\mathbb{Z}^{d+1}} \left|\widehat{\phi}(\xi_0 + 2\pi k)\right|^2 = 0$, that is for all $k \in \mathbb{Z}^{d+1}$, $\widehat{\phi}(\xi_0 + 2\pi k) = 0$. Now we prove that the shifts $\phi(\cdot - k)\,(k \in \mathbb{Z}^{d+1})$ are not $\ell^{p,q}$-stable. Without loss of generality, assume that $\widehat{\phi}(2\pi k) = 0$ for all $k \in \mathbb{Z}^{d+1}$ (by considering the function $e^{-i\xi_0 x}\phi(x)$ if necessary). By Lemma 7.12 and Lemma 7.14, it follows that

$$\sum_{k\in\mathbb{Z}^{d+1}} \phi(\cdot - k) = 0. \tag{7.22}$$

For each integer $n > 0$, let e_n be the sequence on $\mathbb{Z}^{d+1}$ given by

$$e_n(k) = \begin{cases} 1, & \text{if } |k_i| \leq n \text{ for every } i = 1, 2, \cdots, d+1, \\ 0, & \text{otherwise} \end{cases}$$

with $k = (k_1, k_2, \cdots, k_{d+1}) \in \mathbb{Z}^{d+1}$, then $e_n \in \ell^{p,q}$. To prove that the shifts $\phi(\cdot - k)$ are not $\ell^{p,q}$-stable, it suffices to show that

$$\frac{\|e_n *_{sd} \phi\|_{L^{p,q}}}{\|e_n\|_{\ell^{p,q}}} \to 0 \quad \text{as } n \to \infty. \tag{7.23}$$

To this end, one first truncates ϕ as follows. For each integer $N > 0$, let ϕ_N be the function on $\mathbb{R}^{d+1}$ given by

$$\phi_N(x) = \begin{cases} \phi(x), & \text{if } |x_i| \leq N \text{ for every } i = 1, 2, \cdots, d+1, \\ 0, & \text{otherwise.} \end{cases}$$

and ψ_N is the function on $\mathbb{R}^{d+1}$ given by

$$\psi_N(x) = \begin{cases} \sum_{k \in \mathbb{Z}^{d+1}} (\phi - \phi_N)(x - k), & \text{if } x \in [0, 1)^{d+1}, \\ 0, & \text{otherwise.} \end{cases}$$

The construction of ψ_N implies that

$$\|\psi_N\|_{\mathcal{L}^{p,q}} = \left\| \left\| \sum_{k_1 \in \mathbb{Z}} \sum_{k_2 \in \mathbb{Z}^d} (\phi - \phi_N)(\cdot - k_1, \cdot - k_2) \right\|_{L^q[0,1]^d} \right\|_{L^p[0,1]}$$

$$\leq \left\| \left\| \sum_{k_1 \in \mathbb{Z}} \sum_{k_2 \in \mathbb{Z}^d} |(\phi - \phi_N)(\cdot - k_1, \cdot - k_2)| \right\|_{L^q[0,1]^d} \right\|_{L^p[0,1]}$$

$$= \|\phi - \phi_N\|_{\mathcal{L}^{p,q}}. \tag{7.24}$$

Now we define

$$\varphi_N := \phi_N + \psi_N.$$

(7.17) and (7.24) lead to

$$\|\phi - \varphi_N\|_{\mathcal{L}^{p,q}} = \|\phi - \phi_N - \psi_N\|_{\mathcal{L}^{p,q}} \leq \|\phi - \phi_N\|_{\mathcal{L}^{p,q}} + \|\psi_N\|_{\mathcal{L}^{p,q}} \leq 2\|\phi - \phi_N\|_{\mathcal{L}^{p,q}}. \tag{7.25}$$

Lemma 7.10 shows that

$$\|e_n *_{sd} (\phi - \varphi_N)\|_{L^{p,q}} \leq \|\phi - \varphi_N\|_{\mathcal{L}^{p,q}} \|e_n\|_{\ell^{p,q}} \leq 2\|\phi - \phi_N\|_{\mathcal{L}^{p,q}} \|e_n\|_{\ell^{p,q}}.$$

Therefore by (7.1), one has

$$\frac{\|e_n *_{sd} \phi\|_{L^{p,q}}}{\|e_n\|_{\ell^{p,q}}} \leq \frac{\|e_n *_{sd} \varphi_N\|_{L^{p,q}}}{\|e_n\|_{\ell^{p,q}}} + 2\|\phi - \phi_N\|_{\mathcal{L}^{p,q}}.$$

Take N to be the integer part of $\sqrt{n}$. By the dominated convergence theorem, when $N \to +\infty$ (or $n \to +\infty$), $\|\phi - \phi_N\|_{\mathcal{L}^{p,q}} \to 0$. Thus it remains to estimate

$$\frac{\|e_n *_{sd} \varphi_N\|_{L^{p,q}}}{\|e_n\|_{\ell^{p,q}}}.$$

For this purpose, one observes that φ_N is compactly supported, that is

$$\varphi_N(x) = 0 \quad \text{for some } i, \ |x_i| > N. \tag{7.26}$$

Second, using (7.22) and the construction of φ_N,

$$\sum_{k \in \mathbb{Z}^{d+1}} \varphi_N(\cdot - k) = \sum_{k \in \mathbb{Z}^{d+1}} (\phi_N + \psi_N)(\cdot - k) = \sum_{k \in \mathbb{Z}^{d+1}} \phi(\cdot - k) = 0. \tag{7.27}$$

By (7.26), (7.27) and $n > N$, we know that as long as for some i, $|x_i| > n + N$, then

$$(e_n *_{sd} \varphi_N)(\cdot) = \sum_{k \in \mathbb{Z}^{d+1}} e_n(k)\, \varphi_N(\cdot - k) = 0. \tag{7.28}$$

At the same time,

$$(e_n *_{sd} \varphi_N)(\cdot) = \sum_{k \in \mathbb{Z}^{d+1}} e_n(k)\, \varphi_N(\cdot - k) = 0, \tag{7.29}$$

when $|x_i| < n - N$ for every $i = 1, 2, \cdots, d + 1$.

Put $x = (x_1, x_2, \cdots, x_{d+1})$ and $k = (k_1, k_2, \cdots, k_{d+1})$, it follows from (7.28) and (7.29) that

$$\|e_n *_{sd} \varphi_N\|_{L^{p,q}}^p$$

$$= \int_{\mathbb{R}} \left(\int_{\mathbb{R}^d} |(e_n *_{sd} \varphi_N)(x_1, x_2, \cdots, x_{d+1})|^q \, dx_2 \cdots dx_{d+1} \right)^{p/q} dx_1$$

$$\leq \int_{n-N \leq |x_1| \leq n+N}$$

$$\left(\int \cdots \int_{\substack{|x_i| \leq n+N, \\ 2 \leq i \leq d+1}} |(e_n *_{sd} \varphi_N)(x_1, x_2, \cdots, x_{d+1})|^q \, dx_2 \cdots dx_{d+1} \right)^{p/q} dx_1$$

$$+ \int_{|x_1| < n+N} \left(\sum_{l=2}^{d+1} \int \cdots \int_{\substack{|x_i| \leq n+N, \ i \neq l \\ \text{and } i = 2, \cdots, d+1}} \int_{n-N < |x_l| < n+N} \right.$$

$$\left. |(e_n *_{sd} \varphi_N)(x_1, x_2, \cdots, x_{d+1})|^q \, dx_2 \cdots dx_{d+1} \right)^{p/q} dx_1$$

$$\leq \int_{n-N \leq |x_1| \leq n+N} \left(\int \cdots \int_{|x_i| \leq n+N, \ i=2,\cdots,d+1} \right.$$

$$\left(\sum_{k_1 \in \mathbb{Z}} \cdots \sum_{k_{d+1} \in \mathbb{Z}} |\varphi_N(x_1 - k_1, \cdots, x_{d+1} - k_{d+1})| \right)^q dx_2 \cdots dx_{d+1} \right)^{p/q} dx_1$$

$$+ \int_{|x_1| < n+N} \left(\sum_{l=2}^{d+1} \int \cdots \int_{\substack{|x_i| \leq n+N, \, i \neq l \\ \text{and } i = 2, \cdots, d+1}} \int_{n-N < |x_l| < n+N} \right.$$

$$\left. \left(\sum_{k_1 \in \mathbb{Z}} \cdots \sum_{k_{d+1} \in \mathbb{Z}} |\varphi_N(x_1 - k_1, \cdots, x_{d+1} - k_{d+1})| \right)^q dx_2 \cdots dx_{d+1} \right)^{p/q} dx_1.$$

Since $\sum_{k_1 \in \mathbb{Z}} \cdots \sum_{k_{d+1} \in \mathbb{Z}} |\varphi_N(x_1 - k_1, \cdots, x_{d+1} - k_{d+1})|$ is periodic function, then

$$\|e_n *_{sd} \varphi_N\|_{L^{p,q}}^p$$

$$\leq 2N \int_{[0,1]} \left(2^d (n+N)^d \int_{[0,1]} \cdots \int_{[0,1]} \right.$$

$$\left. \left(\sum_{k_1 \in \mathbb{Z}} \cdots \sum_{k_{d+1} \in \mathbb{Z}} |\varphi_N(x_1 - k_1, \cdots, x_{d+1} - k_{d+1})| \right)^q dx_2 \cdots dx_{d+1} \right)^{p/q} dx_1$$

$$+ 2(n+N) \int_{[0,1]} \left(d2^{d-1}(n+N)^{d-1} \int_{[0,1]} \cdots \int_{[0,1]} 2N \int_{[0,1]} \right.$$

$$\left. \left(\sum_{k_1 \in \mathbb{Z}} \cdots \sum_{k_{d+1} \in \mathbb{Z}} |\varphi_N(x_1 - k_1, \cdots, x_{d+1} - k_{d+1})| \right)^q dx_2 \cdots dx_{d+1} \right)^{p/q} dx_1$$

$$\leq [2^{dp/q+1} N (n+N)^{dp/q} + 2^{dp/q+1} d^{p/q} N^{p/q} (n+N)^{(d-1)p/q+1}] \int_{[0,1]} \left(\int_{[0,1]} \cdots \int_{[0,1]} \right.$$

$$\left. \left(\sum_{k_1 \in \mathbb{Z}} \cdots \sum_{k_{d+1} \in \mathbb{Z}} |\varphi_N(x_1 - k_1, \cdots, x_{d+1} - k_{d+1})| \right)^q dx_2 \cdots dx_{d+1} \right)^{p/q} dx_1.$$

Since $n > N$, then there exists constant C_1 such that

$$\|e_n *_{sd} \varphi_N\|_{L^{p,q}}^p \leq C_1 \left(Nn^{dp/q} + n \left(Nn^{d-1} \right)^{p/q} \right) \|\varphi_N\|_{\mathcal{L}^{p,q}}^p.$$

Moreover,

$$\|e_n\|_{\ell^{p,q}}^p = \sum_{k_1 \in \mathbb{Z}} \left(\sum_{k_2 \in \mathbb{Z}} \cdots \sum_{k_{d+1} \in \mathbb{Z}} |e_n(k_1, \cdots, k_{d+1})|^q \right)^{p/q}$$

$$= \sum_{|k_1| \le n} \left(\sum_{|k_2| \le n} \cdots \sum_{|k_{d+1}| \le n} 1 \right)^{p/q}$$

$$\ge C_2 n^{1+dp/q},$$

where C_2 is constant. Thus one has

$$\frac{\|e_n *_{sd} \varphi_N\|_{L^{p,q}}^p}{\|e_n\|_{l^{p,q}}^p} \le \frac{C_1}{C_2} \left(\frac{N}{n} + \left(\frac{N}{n} \right)^{p/q} \right) \|\varphi_N\|_{\mathcal{L}^{p,q}}^p. \tag{7.30}$$

By (7.25),

$$\|\varphi_N\|_{\mathcal{L}^{p,q}} \le \|\phi\|_{\mathcal{L}^{p,q}} + 2\|\phi - \phi_N\|_{\mathcal{L}^{p,q}}. \tag{7.31}$$

Take N to be the integer part of $\sqrt{n}$ in consideration, then one concludes from (7.30) and (7.31) that

$$\frac{\|e_n *_{sd} \varphi_N\|_{L^{p,q}}^p}{\|e_n\|_{l^{p,q}}^p} \to 0 \quad \text{as} \quad n \to \infty.$$

This verifies (7.23), therefore "($\Rightarrow$)" has been proved. $\square$

We first present the following two useful propositions and lemmas.

Proposition 7.9 ([142]) *Let $\phi \in W\left(L^{1,1}\right)$. Then ϕ satisfies*

$$\sum_{k \in \mathbb{Z}^{d+1}} |\hat{\phi}\,(\xi + 2\pi k)|^2 > 0, \quad \xi \in \mathbb{R}^{d+1},$$

if and only if there exists a function

$$g = \sum_{k_1 \in \mathbb{Z}} \sum_{k_2 \in \mathbb{Z}^d} d(k_1, k_2)\phi(\cdot - k_1, \cdot - k_2)$$

such that

$$\langle \phi\,(\cdot - \alpha)\,, g \rangle = \delta_{0,\alpha},$$

where $d = \{d(k_1, k_2) : k_1 \in \mathbb{Z}, k_2 \in \mathbb{Z}^d\} \in \ell^1$.

Lemma 7.15 *The function g in Proposition 7.9 belongs to $W(L^{1,1})$.*

Proof It is easy to see

$$\|g\|_{W(L^{1,1})} = \sum_{n\in\mathbb{Z}} \sup_{x\in[0,1]} \sum_{l\in\mathbb{Z}^d} \sup_{y\in[0,1]^d} |g(x+n, y+l)|$$

$$= \sum_{n\in\mathbb{Z}} \sup_{x\in[0,1]} \sum_{l\in\mathbb{Z}^d} \sup_{y\in[0,1]^d} \left| \sum_{k_1\in\mathbb{Z}} \sum_{k_2\in\mathbb{Z}^d} d(k_1, k_2)\phi(x+n-k_1, y+l-k_2) \right|$$

$$\leq \sum_{n\in\mathbb{Z}} \sup_{x\in[0,1]} \sum_{l\in\mathbb{Z}^d} \sup_{y\in[0,1]^d} \sum_{k_1\in\mathbb{Z}} \sum_{k_2\in\mathbb{Z}^d} |d(k_1, k_2)| |\phi(x+n-k_1, y+l-k_2)|$$

$$\leq \sum_{k_1\in\mathbb{Z}} \sum_{k_2\in\mathbb{Z}^d} |d(k_1, k_2)| \sum_{n\in\mathbb{Z}} \sup_{x\in[0,1]} \sum_{l\in\mathbb{Z}^d} \sup_{y\in[0,1]^d} |\phi(x+n-k_1, y+l-k_2)|$$

$$= \|d\|_{\ell^1} \|\phi\|_{W(L^{1,1})} < \infty.$$

$\square$

Now, we show another main result about the $L^{p,q}$-stability of the principal generated shift-invariant spaces in Lebesgue spaces $L^{p,q}\left(\mathbb{R}^{d+1}\right)$. As usual, for quantities X and Y, $X \approx Y$ and $X \lesssim Y$ mean that there exist constants c_1, c_2, c such that $c_1 X \leq Y \leq c_2 X$ and $X \leq cY$, respectively.

The following theorem gives another main result about the stability of the principal generated shift-invariant spaces in mixed Lebesgue spaces $L^{p,q}\left(\mathbb{R}^{d+1}\right)$.

Theorem 7.6 *Let $1 < p, q < \infty$. Assume that $\phi \in W(L^1)$ satisfies*

$$\sum_{k\in\mathbb{Z}^{d+1}} \left|\widehat{\phi}\left(\xi + 2\pi k\right)\right|^2 > 0, \quad \xi \in \mathbb{R}^{d+1}.$$

Then for any $c \in \ell^{p,q}$, the function $f = \sum_{k_1\in\mathbb{Z}} \sum_{k_2\in\mathbb{Z}^d} c(k_1, k_2)\phi(\cdot - k_1, \cdot - k_2)$ belongs to $L^{p,q}$, and

$$\|c\|_{\ell^{p,q}} \approx \|f\|_{L^{p,q}} \approx \|f\|_{W(L^{p,q})}.$$

Proof We first consider $\|c\|_{\ell^{p,q}} \approx \|f\|_{L^{p,q}}$.

Theorem 7.1 leads to $\|f\|_{L^{p,q}} \lesssim \|c\|_{\ell^{p,q}}$, thus one only needs to prove $\|c\|_{\ell^{p,q}} \lesssim \|f\|_{L^{p,q}}$. By Proposition 7.9, there exists a function g such that

$$\langle \phi\left(\cdot - \alpha\right), g \rangle = \delta_{0,\alpha}.$$

Therefore,

$$c(k_1, k_2) = \int_{\mathbb{R}} \int_{\mathbb{R}^d} f(x, y)\overline{g\left(x - k_1, y - k_2\right)}dxdy.$$

Let $b = \{b(k_1, k_2) : k_1 \in \mathbb{Z}, k_2 \in \mathbb{Z}^d\} \in \ell^{p',q'}$ with $\frac{1}{p} + \frac{1}{p'} = 1$ and $\frac{1}{q} + \frac{1}{q'} = 1$. Then

$$|\langle c, b \rangle| = \left| \sum_{k_1 \in \mathbb{Z}, k_2 \in \mathbb{Z}^d} c(k_1, k_2)\overline{b(k_1, k_2)} \right|$$

$$= \left| \sum_{k_1 \in \mathbb{Z}, k_2 \in \mathbb{Z}^d} \overline{b(k_1, k_2)} \int_{\mathbb{R}} \int_{\mathbb{R}^d} f(x, y)\overline{g(x - k_1, y - k_2)}dxdy \right|$$

$$= \left| \int_{\mathbb{R}} \int_{\mathbb{R}^d} f(x, y) \sum_{k_1 \in \mathbb{Z}, k_2 \in \mathbb{Z}^d} \overline{b(k_1, k_2)g(x - k_1, y - k_2)}dxdy \right|.$$

Using Proposition 7.3, Lemma 7.15 and Theorem 7.1, one has

$$|\langle c, b \rangle| \le \|f\|_{L^{p,q}} \left\| \sum_{k_1 \in \mathbb{Z}, k_2 \in \mathbb{Z}^d} b(k_1, k_2)g(x - k_1, y - k_2) \right\|_{L^{p',q'}}$$

$$\le \|f\|_{L^{p,q}} \|b\|_{\ell^{p',q'}} \|g\|_{W(L^{1,1})}.$$

Thus

$$\|c\|_{\ell^{p,q}} \le \|f\|_{L^{p,q}} \|g\|_{W(L^{1,1})}, \tag{7.32}$$

that is, $\|c\|_{\ell^{p,q}} \lesssim \|f\|_{L^{p,q}}$. Therefore, one obtains $\|c\|_{\ell^{p,q}} \approx \|f\|_{L^{p,q}}$.

Now, we prove $\|f\|_{L^{p,q}} \approx \|f\|_{W(L^{p,q})}$.

It is easy to see that

$$\|f\|_{L^{p,q}}^p = \int_{\mathbb{R}} \left(\int_{\mathbb{R}^d} |f(x, y)|^q dy \right)^{\frac{p}{q}} dx$$

$$= \sum_{n \in \mathbb{Z}} \int_0^1 \left(\sum_{l \in \mathbb{Z}^d} \int_{[0,1]^d} |f(x + n, y + l)|^q dy \right)^{p/q} dx$$

$$\le \sum_{n \in \mathbb{Z}} \sup_{x \in [0,1]} \left[\sum_{l \in \mathbb{Z}^d} \sup_{y \in [0,1]^d} |f(x + n, y + l)|^q \right]^{p/q}$$

$$= \|f\|_{W(L^{p,q})}^p. \tag{7.33}$$

On the other hand,

$$\|f\|_{W(L^{p,q})}^p = \left\|\sum_{k_1\in\mathbb{Z}}\sum_{k_2\in\mathbb{Z}^d} c(k_1,k_2)\phi(\cdot-k_1,\cdot-k_2)\right\|_{W(L^{p,q})}^p$$

$$= \sum_{n\in\mathbb{Z}} \sup_{x\in[0,1]}\left[\sum_{l\in\mathbb{Z}^d} \sup_{y\in[0,1]^d}\left|\sum_{k_1\in\mathbb{Z}}\sum_{k_2\in\mathbb{Z}^d} c(k_1,k_2)\phi(x+n-k_1,y+l-k_2)\right|^q\right]^{p/q}$$

$$\le \sum_{n\in\mathbb{Z}} \sup_{x\in[0,1]}\left[\sum_{l\in\mathbb{Z}^d} \sup_{y\in[0,1]^d}\left(\sum_{k_1\in\mathbb{Z}}\sum_{k_2\in\mathbb{Z}^d} |c(k_1,k_2)||\phi(x+n-k_1,y+l-k_2)|\right)^q\right]^{p/q}$$

$$\le \sum_{n\in\mathbb{Z}}\left[\sum_{l\in\mathbb{Z}^d}\left(\sum_{k_1\in\mathbb{Z}}\sum_{k_2\in\mathbb{Z}^d} |c(k_1,k_2)| \sup_{(x,y)\in[0,1]\times[0,1]^d}|\phi(x+n-k_1,y+l-k_2)|\right)^q\right]^{p/q}.$$

Denote $d(n,l) = \sup_{(x,y)\in[0,1]\times[0,1]^d}|\phi(x+n,y+l)|$, then by Lemma 7.3, one obtains

$$\|f\|_{W(L^{p,q})}^p = \sum_{n\in\mathbb{Z}}\left[\sum_{l\in\mathbb{Z}^d}\left(\sum_{k_1\in\mathbb{Z}}\sum_{k_2\in\mathbb{Z}^d} |c(k_1,k_2)|d(n-k_1,l-k_2)\right)^q\right]^{p/q} = \||c|*_d d\|_{\ell^{p,q}}^p$$

$$\le \||c|\|_{\ell^{p,q}}^p\|d\|_{\ell^1}^p = \|c\|_{\ell^{p,q}}^p\|\phi\|_{W(L^1)}^p \le \|f\|_{L^{p,q}}^p\|g\|_{W(L^{1,1})}^p\|\phi\|_{W(L^1)}^p, \quad (7.34)$$

where the last inequality follows by (7.32) and $|c| = \{|c(k_1,k_2)| : k_1 \in \mathbb{Z}, k_2 \in \mathbb{Z}^d\}$. Combining (7.33) and (7.34), one has $\|f\|_{L^{p,q}} \approx \|f\|_{W(L^{p,q})}$. Therefore, one completes the proof.   □

7.3.2   Finitely Generated Case

In this subsection, we mainly investigate the stability of the finitely generated shift-invariant spaces in mixed Lebesgue spaces $L^{p,q}(\mathbb{R}^{d+1})$.

For the $L^{p,q}$-stability of the shifts of finite functions in mixed Lebesgue spaces $L^{p,q}(\mathbb{R}^{d+1})$, we have the following main result.

Theorem 7.7 Let $\phi_1,\cdots,\phi_n \in \mathcal{L}^{p,q}(\mathbb{R}^{d+1})(1 < p,q < \infty)$. Then the shifts $\phi_j(\cdot - k)(1 \le j \le n, k \in \mathbb{Z}^{d+1})$ are $L^{p,q}$-stable if and only if for any $\xi \in \mathbb{R}^{d+1}$, the sequences $\{\hat{\phi}_j(\xi + 2\pi k)\}_{k\in\mathbb{Z}^{d+1}} (1 \le j \le n)$ are linearly independent.

Proof ($\Rightarrow$). One proves this by contradiction. If for some $\xi_0 \in \mathbb{R}^{d+1}$, the sequences $\{\hat{\phi}_j(\xi_0 + 2\pi k)\}_{k\in\mathbb{Z}^{d+1}} (1 \le j \le n)$ are linearly dependent, then there exist constants

$r_j, (j = 1, \cdots, n)$, not all zero, such that

$$\sum_{j=1}^{n} r_j \widehat{\phi}_j (\xi_0 + 2\pi k) = 0 \quad \text{for all } k \in \mathbb{Z}^{d+1}.$$

Let $\phi := \sum_{j=1}^{n} r_j \phi_j$. Then by Theorem 7.5, ϕ is not $\ell^{p,q}$-stable, hence the shifts $\phi_j(\cdot - k)$ $(1 \leq n, k \in \mathbb{Z}^{d+1})$ are not $\ell^{p,q}$-stable. This proves "($\Rightarrow$)".

($\Leftarrow$). Given $a_1, \cdots, a_n \in \ell^{p,q}$, let $f = \sum_{j=1}^{n} a_j *_{sd} \phi_j$. Then by (7.1) and Lemma 7.10,

$$\|f\|_{L^{p,q}} = \left\| \sum_{j=1}^{n} a_j *_{sd} \phi_j \right\|_{L^{p,q}} \leq \sum_{j=1}^{n} \|a_j *_{sd} \phi_j\|_{L^{p,q}} \leq \sum_{j=1}^{n} \|\phi_j\|_{\mathcal{L}^{p,q}} \|a_j\|_{\ell^{p,q}}$$

$$\leq C \sum_{j=1}^{n} \|a_j\|_{\ell^{p,q}},$$

where $C = \max_{1 \leq j \leq n} \{\|\phi_1\|_{\mathcal{L}^{p,q}}, \cdots, \|\phi_n\|_{\mathcal{L}^{p,q}}\}$.

Next, we prove the lower bound. Assume that $\phi_j \in \mathcal{L}^{\infty,\infty}$ $(1 \leq j \leq n)$, then $\phi_j \in \mathcal{L}^{2,2}$. Therefore,

$$\int_{[0,2\pi]^{d+1}} \sum_{k \in \mathbb{Z}^{d+1}} |\widehat{\phi}_j(\xi + 2\pi k)|^2 d\xi = \int_{\mathbb{R}^{d+1}} |\widehat{\phi}_j(\xi)|^2 d\xi$$

$$= (2\pi)^{\frac{d+1}{2}} \int_{\mathbb{R}^{d+1}} |\phi_j(x)|^2 dx = (2\pi)^{\frac{d+1}{2}} \int_{[0,1]^{d+1}} \sum_{k \in \mathbb{Z}^{d+1}} |\phi_j(x+k)|^2 dx$$

$$\leq (2\pi)^{\frac{d+1}{2}} \|\phi_j\|_{\mathcal{L}^2} < \infty.$$

Then in this case, one has $\{\widehat{\phi}_j(\xi + 2\pi k)\}_{k \in \mathbb{Z}^{d+1}} \in \ell^2(\mathbb{Z}^{d+1})$ $(1 \leq j \leq n)$. Since the sequences $\{\widehat{\phi}_j(\xi + 2\pi k)\}_{k \in \mathbb{Z}^{d+1}}$ $(1 \leq j \leq n)$ are linearly independent, its Gram matrix $([\phi_j, \phi_k](\xi))_{1 \leq j,k \leq n}$ is nonsingular for $\xi \in \mathbb{T}$, where

$$[\phi_j, \phi_k](\xi) = \sum_{l \in \mathbb{Z}^{d+1}} \widehat{\phi}_j(\xi + 2\pi l)\overline{\widehat{\phi}_k(\xi + 2\pi l)}.$$

By Lemma 7.9, there exist $b_{j,k} \in \ell^1(\mathbb{Z}^{d+1})$ $(j, k = 1, \cdots, n)$ such that the matrix $(\widehat{b}_{j,k}(\xi))_{1 \leq j,k \leq n}$ is the inverse of $([\phi_j, \phi_k](\xi))_{1 \leq j,k \leq n}$. Let

$$g_j := \sum_{k=1}^{n} b_{j,k} *_{sd} \phi_k.$$

Lemma 7.11 leads to $g_j \in \mathcal{L}^{\infty,\infty}(\mathbb{R}^{d+1})$. Therefore, for $1 \leq j, k \leq n$, inverse matrix leads to

$$[g_j, \phi_k](\xi) = \sum_{m=1}^{n} \widehat{b}_{j,m}(\xi)[\phi_m, \phi_k](\xi) = \delta_{j,k}.$$

Hence

$$\langle g_j, \phi_k(\cdot - \alpha) \rangle = (2\pi)^{-\frac{d+1}{2}} \int_{\mathbb{R}^{d+1}} \widehat{g}_j(\xi) \overline{\widehat{\phi}_k(\xi)} e^{i\alpha\xi} d\xi$$

$$= (2\pi)^{-\frac{d+1}{2}} \int_{[0,2\pi]^{d+1}} \sum_{l \in \mathbb{Z}^{d+1}} \widehat{g}_j(\xi + 2\pi l) \overline{\widehat{\phi}_k(\xi + 2\pi l)} e^{i\alpha\xi} d\xi$$

$$= (2\pi)^{-\frac{d+1}{2}} \delta_{j,k} \delta_{0,\alpha}$$

and for all $1 \leq j \leq n$, $k \in \mathbb{Z}^{d+1}$,

$$\langle f, g_j(\cdot - k) \rangle = \left\langle \sum_{l=1}^{n} a_l *_{sd} \phi_l, g_j(\cdot - k) \right\rangle$$

$$= \left\langle \sum_{l=1}^{n} \sum_{m \in \mathbb{Z}^{d+1}} a_l(m) \phi_l(x - m), g_j(\cdot - k) \right\rangle$$

$$= (2\pi)^{-\frac{d+1}{2}} a_j(k).$$

Similar to the prove of (7.21), one has $\|a_j\|_{\ell^{p,q}} \leq \|f\|_{L^{p,q}} \|g_j\|_{\mathcal{L}^{p',q'}}$. Let $M_j = 1/\|g_j\|_{\mathcal{L}^{p',q'}}$. Then we obtain $M_j \|a_j\|_{\ell^{p,q}} \leq \|f\|_{L^{p,q}}$. Put $M = \min_{1 \leq j \leq n}\{M_1, \cdots, M_n\}$, then

$$M \sum_{j=1}^{n} \|a_j\|_{\ell^{p,q}} \leq \|f\|_{L^{p,q}}.$$

Similar to the the argument of ($\Leftarrow$) in Theorem 7.5, one extends the result for $\phi_j \in \mathcal{L}^{\infty}$ to the case $\phi_j \in \mathcal{L}^{p,q}(\mathbb{R}^{d+1})$ $(1 \leq j \leq n)$ by convolving ϕ_j with the function φ, where $\varphi(x) := e^{-\pi|x|^2}$ and $|x|$ is the Euclidean norm of x. Thus, one obtains $\widetilde{M} \|c\|_{\ell^{p,q}} \leq \left\| \sum_{j=1}^{n} c *_{sd} \phi_j \right\|_{L^{p,q}}$. Hence the shifts $\phi(\cdot - k)$ $(k \in \mathbb{Z}^{d+1})$ are $L^{p,q}$-stable. $\square$

7.4   Nonuniform Samplings

In this section, we mainly discuss nonuniform sampling in shift-invariant subspaces.

7.4.1   The $L^{p,q}$-Stability of Nonuniform Samplings

Definition 7.10 A set $X = \{(x_k, y_j) : k, j \in \mathbb{J}\} \subset \mathbb{R}^{d+1}$ is said to be strongly separated if $\inf_{(k,j)\neq(k',j')} |(x_k, y_j) - (x_{k'}, y_{j'})| = \delta_1 > 0$ and $\inf_{k\neq k'} |x_k - x_{k'}| = \delta_2 > 0$, where $|(x_k, y_j) - (x_{k'}, y_{j'})| = \sqrt{(x_k - x_{k'})^2 + |y_j - y_{j'}|^2}$, and $\mathbb{J}$ is a countable index set.

The following theorem provides a sufficient condition under which the problem of nonuniform sampling in shift-invariant subspaces is well-defined. It tells us the stability of nonuniform sampling in shift-invariant subspaces.

Theorem 7.8 If $X = \{(x_k, y_j) : k, j \in \mathbb{J}, x_k \in \mathbb{R}, y_j \in \mathbb{R}^d\}$ is strongly separated and $f \in W_0(L^{p,q})$. Then

$$\left[\sum_{k\in\mathbb{J}} \left(\sum_{j\in\mathbb{J}} |f(x_k, y_j)|^q \right)^{p/q} \right]^{1/p} \leq \left(\frac{1}{\delta_1} + 1 \right)^{d/q} \left(\frac{1}{\delta_2} + 1 \right)^{1/p} \| f \|_{W(L^{p,q})}.$$

Proof For fixed x_k,

$$\sum_{j\in\mathbb{J}} |f(x_k, y_j)|^q = \sum_{l\in\mathbb{Z}^d} \sum_{y_j\in[l,l+1]^d} |f(x_k, y_j)|^q$$

$$\leq \sum_{l\in\mathbb{Z}^d} \sum_{y_j\in[l,l+1]^d} \sup_{y\in[l,l+1]^d} |f(x_k, y)|^q$$

$$\leq \sum_{l\in\mathbb{Z}^d} \left(\frac{1}{\delta_1} + 1 \right)^d \sup_{y\in[l,l+1]^d} |f(x_k, y)|^q$$

$$= \left(\frac{1}{\delta_1} + 1 \right)^d \sum_{l\in\mathbb{Z}^d} \sup_{y\in[l,l+1]^d} |f(x_k, y)|^q.$$

Therefore,

$$\sum_{k\in\mathbb{J}} \left(\sum_{j\in\mathbb{J}} |f(x_k, y_j)|^q \right)^{p/q}$$

$$\leq \left(\frac{1}{\delta_1}+1\right)^{pd/q} \sum_{k\in\mathbb{J}} \left[\sum_{l\in\mathbb{Z}^d} \sup_{y\in[l,l+1]^d} |f(x_k,y)|^q\right]^{p/q}$$

$$= \left(\frac{1}{\delta_1}+1\right)^{pd/q} \sum_{n\in\mathbb{Z}} \sum_{x_k\in[n,n+1]} \left[\sum_{l\in\mathbb{Z}^d} \sup_{y\in[l,l+1]^d} |f(x_k,y)|^q\right]^{p/q}$$

$$\leq \left(\frac{1}{\delta_1}+1\right)^{pd/q} \sum_{n\in\mathbb{Z}} \left(\frac{1}{\delta_2}+1\right) \sup_{x\in[n,n+1]} \left[\sum_{l\in\mathbb{Z}^d} \sup_{y\in[l,l+1]^d} |f(x,y)|^q\right]^{p/q}$$

$$= \left(\frac{1}{\delta_1}+1\right)^{pd/q} \left(\frac{1}{\delta_2}+1\right) \sum_{n\in\mathbb{Z}} \sup_{x\in[0,1]} \left[\sum_{l\in\mathbb{Z}^d} \sup_{y\in[0,1]^d} |f(x+n,y+l)|^q\right]^{p/q}$$

$$= \left(\frac{1}{\delta_1}+1\right)^{pd/q} \left(\frac{1}{\delta_2}+1\right) \|f\|^p_{W(L^{p,q})}.$$

Then one has

$$\left[\sum_{k\in\mathbb{J}} \left(\sum_{j\in\mathbb{J}} |f(x_k,y_j)|^q\right)^{p/q}\right]^{1/p} \leq \left(\frac{1}{\delta_1}+1\right)^{d/q} \left(\frac{1}{\delta_2}+1\right)^{1/p} \|f\|_{W(L^{p,q})}.$$

$\square$

7.4.2 The Reconstruction Algorithm

Let f be a continuous function. We define the oscillation (or modulus of continuity) of f by $osc_\delta(f)(x) = \sup_{|y|\leq\delta} |f(x+y)-f(x)|$.

Lemma 7.16 *Let $\phi \in W(L^1)$, then $osc_\delta(\phi) \in W(L^1)$.*

Proof Without loss of generality, assume that $\delta \leq 1$. Then for $j \in \mathbb{Z}^{d+1}$,

$$\sup_{x\in[0,1]^{d+1}} |osc_\delta(\phi)(x+j)| = \sup_{x\in[0,1]^{d+1}} \sup_{|y|\leq\delta} |\phi(x+j+y)-\phi(x+j)|$$

$$\leq \sup_{x\in[0,1]^{d+1}} \sup_{|y|\leq\delta} |\phi(x+y+j)| + \sup_{x\in[0,1]^{d+1}} |\phi(x+j)|$$

$$\leq 2 \sup_{x\in[-1,2]^{d+1}} |\phi(x+j)|$$

$$\leq 2 \sum_{k \in [-1,2]^{d+1} \cap \mathbb{Z}^{d+1}} \sup_{x \in [0,1]^{d+1}} |\phi(x + j + k)|.$$

Summing over $j \in \mathbb{Z}^{d+1}$, one obtains

$$\sum_{j \in \mathbb{Z}^{d+1}} \sup_{x \in [0,1]^{d+1}} |osc_\delta(\phi)(x + j)| \leq 2 \sum_{j \in \mathbb{Z}^{d+1}} \sum_{k \in [-1,2]^{d+1} \cap \mathbb{Z}^{d+1}} \sup_{x \in [0,1]^{d+1}} |\phi(x + j + k)|$$

$$= 2 \sum_{k \in [-1,2]^{d+1} \cap \mathbb{Z}^{d+1}} \sum_{j \in \mathbb{Z}^{d+1}} \sup_{x \in [0,1]^{d+1}} |\phi(x + j + k)|$$

$$\leq 2^{2d+3} \|\phi\|_{W(L^1)}$$

$$< +\infty.$$

Therefore, $osc_\delta(\phi) \in W(L^1)$. $\square$

Lemma 7.17 *Let $\phi \in W_0(L^1)$. Then $\lim_{\delta \to 0} \|osc_\delta(\phi)\|_{W(L^1)} = 0$.*

Proof By Lemma 7.16, one has $osc_\delta(\phi) \in W(L^1)$. Given $\epsilon > 0$, there exists an integer $L_0 > 0$ such that

$$\sum_{|j| \geq L_0} \sup_{x \in [0,1]^{d+1}} |osc_\delta(\phi)(x + j)| < \frac{\epsilon}{2}. \tag{7.35}$$

Moreover, since ϕ is continuous, there exists $\delta_0 > 0$ such that

$$\sup_{x \in [0,1]^{d+1}} \sup_{|y| \leq \delta} |\phi(x + j + y) - \phi(x + j)| \leq \frac{\epsilon}{2(2L_0)^{d+1}}$$

for all $|j| < L_0$ and all $\delta < \delta_0$. Thus

$$\sum_{|j| < L_0} \sup_{x \in [0,1]^{d+1}} |osc_\delta(\phi)(x + j)| = \sum_{|j| < L_0} \sup_{x \in [0,1]^{d+1}} \sup_{|y| \leq \delta} |\phi(x + j + y) - \phi(x + j)|$$

$$\leq \sum_{|j| < L_0} \frac{\epsilon}{2(2L_0)^{d+1}}$$

$$\leq \frac{\epsilon}{2}. \tag{7.36}$$

Combining (7.35) and (7.36), one obtains

$$\|osc_\delta(\phi)\|_{W(L^1)} \leq \sum_{j \in \mathbb{Z}^{d+1}} \sup_{x \in [0,1]^{d+1}} |osc_\delta(\phi)(x + j)| < \epsilon, \quad \forall \, 0 < \delta \leq \delta_0.$$

Thus $\lim_{\delta \to 0} \|osc_\delta(\phi)\|_{W(L^1)} = 0.$ □

Lemma 7.18 *Let $\phi \in W_0(L^1)$. If $f \in S_{p,q}(\phi)$, then the oscillation (or modulus of continuity) $osc_\delta(f)$ belongs to $L^{p,q}$. Moreover for all $\epsilon > 0$, there exists $\delta_0 > 0$ such that*

$$\|osc_\delta(f)\|_{L^{p,q}} \le \epsilon \|f\|_{L^{p,q}}$$

uniformly for all $f \in S_{p,q}(\phi)$ and $0 < \delta < \delta_0$.

Proof If $f = \sum_{k \in \mathbb{Z}^{d+1}} c(k)\phi(\cdot - k) \in S_{p,q}(\phi)$, then

$$\begin{aligned}
osc_\delta(f)(x) &= \sup_{|y| \le \delta} |f(x+y) - f(x)| \\
&\le \sup_{|y| \le \delta} \sum_{k \in \mathbb{Z}^{d+1}} |c(k)||\phi(x+y-k) - \phi(x-k)| \\
&\le \sum_{k \in \mathbb{Z}^{d+1}} |c(k)| \sup_{|y| \le \delta} |\phi(x-k+y) - \phi(x-k)| \\
&= \sum_{k \in \mathbb{Z}^{d+1}} |c(k)| osc_\delta(\phi)(x-k) \\
&= [|c| *_{sd} (osc_\delta)(\phi)](x),
\end{aligned}$$

where $|c| = \{|c(k)| : k \in \mathbb{Z}^{d+1}\}$. By Theorem 7.1, Theorem 7.6 and Lemma 7.16, there exists $M > 0$ such that

$$\begin{aligned}
\|osc_\delta(f)\|_{L^{p,q}} &\le \||c| *_{sd} (osc_\delta)(\phi)\|_{L^{p,q}} \\
&\le \||c|\|_{\ell^{p,q}} \|osc_\delta(\phi)\|_{W(L^{1,1})} \\
&= \|c\|_{\ell^{p,q}} \|osc_\delta(\phi)\|_{W(L^{1,1})} \\
&\le M \|f\|_{L^{p,q}} \|osc_\delta(\phi)\|_{W(L^{1,1})} \\
&\le M \|f\|_{L^{p,q}} \|osc_\delta(\phi)\|_{W(L^1)}.
\end{aligned} \qquad (7.37)$$

Using Lemma 7.17, for any $\epsilon > 0$, there exists $\delta_0 > 0$ such that

$$\|osc_\delta(\phi)\|_{W(L^1)} < \frac{\epsilon}{M}, \qquad \forall\, 0 < \delta \le \delta_0.$$

This with (7.37) leads to $\|osc_\delta(f)\|_{L^{p,q}} \le \epsilon \|f\|_{L^{p,q}}.$ □

Definition 7.11 A set $X = \{(x_j, y_k) : k, j \in \mathbb{J}, x_k \in \mathbb{R}, y_j \in \mathbb{R}^d\}$ is γ_0-dense in $\mathbb{R}^{d+1}$ if

$$\mathbb{R}^{d+1} = \cup_{j,k} B_\gamma(x_j, y_k) \quad \text{for every } \gamma > \gamma_0,$$

where $B_\gamma(x_j, y_k)$ is the open ball with center (x_j, y_k) and radius γ, and $\mathbb{J}$ is a countable index set.

A bounded uniform partition of unity $\{\beta_{j,k}\}_{j,k\in\mathbb{J}}$ associated to a strongly separated sampling set $X = \{(x_j, y_k) : j, k \in \mathbb{J}\}$ is a set of functions such that

1. $0 \le \beta_{j,k} \le 1, \forall\, j, k \in \mathbb{J}$,
2. $\text{supp}\beta_{j,k} \subset B_\gamma(x_j, y_k)$,
3. $\sum_{j\in\mathbb{J}} \sum_{k\in\mathbb{J}} \beta_{j,k} = 1$.

If $f \in W_0(L^{p,q})$, we write

$$Q_X f = \sum_{j\in\mathbb{J}} \sum_{k\in\mathbb{J}} f(x_j, y_k)\beta_{j,k}$$

for the quasi-interpolant of the sequence $c_{j,k} = f(x_j, y_k)$.

In order to obtain the reconstruction algorithm, we need the following lemma.

Lemma 7.19 *Let P be any bounded projection from $L^{p,q}$ onto $S_{p,q}(\phi)$ with $\phi \in W_0(L^1)$. Then there exists $\gamma_0 = \gamma_0(p, q, P)$ such that the operator $I - PQ_X$ is a contraction on $S_{p,q}(\phi)$ for every strongly-separated γ-dense set X with $\gamma \le \gamma_0$.*

Proof For $f \in S_{p,q}(\phi)$, one has

$$\begin{aligned}
\|f - PQ_X f\|_{L^{p,q}} &= \|Pf - PQ_X f\|_{L^{p,q}} \\
&\le \|P\|_{op}\|f - Q_X f\|_{L^{p,q}} \\
&\le \|P\|_{op}\|\text{osc}_\gamma f\|_{L^{p,q}} \\
&\le \epsilon\|P\|_{op}\|f\|_{L^{p,q}}.
\end{aligned}$$

Here, the last inequality follows by Lemma 7.18. One can choose γ_0 such that for any $\gamma < \gamma_0$, $\epsilon\|P\|_{op} < 1$. Hence, it leads to a contraction. $\square$

Now we give a fast iterative algorithm to reconstruct $f \in S_{p,q}(\phi)$ from its samples $\{f(x_j, y_k) : j, k \in \mathbb{J}\}$.

Theorem 7.9 *Let $\phi \in W_0(L^1)$. Assume that P is a bounded projection from $L^{p,q}$ onto $S_{p,q}(\phi)$. Then there exists a density $\gamma > 0$ ($\gamma = \gamma(p, q, P)$) such that any $f \in S_{p,q}(\phi)$ can be recovered from its samples $\{f(x_j, y_k) : (x_j, y_k) \in X\}$ on any γ-dense set $X = \{(x_j, y_k) : j, k \in \mathbb{J}\}$ by the following iterative algorithm:*

$$\begin{cases} f_1 = PQ_X f \\ f_{n+1} = PQ_X(f - f_n) + f_n. \end{cases} \tag{7.38}$$

The iterates f_n converges to f uniformly and in $L^{p,q}$ norms. Furthermore, the convergence is geometric, that is,

$$\|f - f_n\|_{L^{p,q}} \leq M\alpha^n$$

for some $\alpha = \alpha(\gamma) < 1$ and $M < \infty$.

Proof Let $e_n = f - f_n$ be the error after n iterations. By (7.38),

$$\begin{aligned} e_{n+1} &= f - f_{n+1} \\ &= f - f_n - PQ_X(f - f_n) \\ &= (I - PQ_X)e_n. \end{aligned}$$

By Lemma 7.19, one may choose γ so small that $\|I - PQ_X\|_{op} = \alpha < 1$. Then one obtains

$$\|e_{n+1}\|_{L^{p,q}} \leq \alpha\|e_n\|_{L^{p,q}} \leq \alpha^n\|e_0\|_{L^{p,q}}.$$

Thus $\|e_n\|_{L^{p,q}} \to 0$, when $n \to \infty$. The proof is completed. □

7.5 Samplings for Non-decaying Signals

First, we give some Riesz-type bounds for the shift-invariant subspace of non-decaying signals, which link the weighted discrete norm of the samples to the weighted continuous norms of the original signals and the interpolated signals. Then, we prove that the sampling and reconstruction through Riesz-type bounds are stable, as long as the generated kernel ϕ is the element in the appropriate mixed Wiener amalgam spaces.

In order to present the main results we extended in [168], we will introduce some notations and definitions. Throughout this paper, the non-decaying signals are in weighted mixed Lebesgue spaces $L^{p,q}(\mathbb{R}^{d+1})$, where their growth is controlled by some weighting functions. Now, we give the definition of the weighting function.

Now, we give the definition of the weighting function.

Definition 7.12 (Submultiplicative Weights) A weighting function $\omega : \mathbb{R}^{d+1} \to \mathbb{R}$ is called (*weakly*) submultiplicative if it is positive, symmetric, continuous, and there exists a constant C_ω such that for any $x_1, y_1 \in \mathbb{R}, x_2, y_2 \in \mathbb{R}^d$.

$$\omega(x_1 + y_1, x_2 + y_2) \leq C_\omega \, \omega(x_1, x_2)\omega(y_1, y_2). \tag{7.39}$$

According to the definition, we can draw the following conclusions. Because of the submultiplicativity of ω, we can get

$$\omega(x_1, x_2) \leq C_\omega \, \omega(x_1, x_2)\omega(0, 0) \ (i.e., \omega(0, 0) \geq 1/C_\omega).$$

On the other hand, the symmetry of ω indicates

$$\omega(0, 0) = \omega(x_1 - x_1, x_2 - x_2) \leq C_\omega \, \omega(x_1, x_2)\omega(-x_1, -x_2) = C_\omega(\omega(x_1, x_2))^2,$$

hence, $\omega(x_1, x_2) \geq \sqrt{\omega(0, 0)/C_\omega} \geq 1/C_\omega$, for all x_1, x_2. This means that every submultiplicative weighting function has a lower bound.

The following is an example for the weighting function.

Example 7.5.1 Let $\omega(x) = \left(1 + \|x\|^2\right)^{1/2}, \forall x \in \mathbb{R}^{d+1}$, then $\omega(x)$ is a weighting function.   □

Now we give the definitions of weighted mixed Lebesgue spaces $L^{p,q}(\mathbb{R}^{d+1})$ and $\ell^{p,q}(\mathbb{Z}^{d+1})$.

Definition 7.13 (Weighted $L^{p,q}(\mathbb{R}^{d+1})$ and $\ell^{p,q}(\mathbb{Z}^{d+1})$ spaces). For $1 \leq p, q < +\infty$ and a weighting function ω, a function f is in $L^{p,q,\omega}(\mathbb{R}^{d+1})$ if $(f\omega)$ is in $L^{p,q}(\mathbb{R}^{d+1})$; a sequence $\{c(k_1, k_2)\}_{k_1 \in \mathbb{Z}, k_2 \in \mathbb{Z}^d}$ is in $\ell^{p,q,\omega}(\mathbb{Z}^{d+1})$ if $\{c(k_1, k_2)\omega(k_1, k_2)\}_{k_1 \in \mathbb{Z}, k_2 \in \mathbb{Z}^d}$ is in $\ell^{p,q}(\mathbb{Z}^{d+1})$. The corresponding weighted norms are defined as

$$\|f\|_{L^{p,q,\omega}(\mathbb{R}^{d+1})} := \|f\omega\|_{L^{p,q}(\mathbb{R}^{d+1})};$$

$$\|c\|_{\ell^{p,q,\omega}(\mathbb{Z}^{d+1})} := \|c\omega\|_{\ell^{p,q}(\mathbb{Z}^{d+1})}.$$

Especially, if ω is submultiplicative, then ω has a lower bound, hence, $L^{p,q,\omega}(\mathbb{R}^{d+1}) \subset L^{p,q}(\mathbb{R}^{d+1})$ and $\ell^{p,q,\omega}(\mathbb{Z}^{d+1}) \subset \ell^{p,q}(\mathbb{Z}^{d+1})$.

Note that, for a function f, $f^\vee$ denotes the reflection $f(-\cdot)$.

The scalar product is defined as

$$\langle f, g \rangle := \int_{\mathbb{R}^{d+1}} f(x)g(x)dx, \quad for \ f \in L^p(\mathbb{R}^{d+1}), g \in L^{p'}(\mathbb{R}^{d+1});$$

$$\langle a, b \rangle := \sum_{k \in \mathbb{Z}^{d+1}} a(k)b(k), \quad for \ a \in \ell^p(\mathbb{Z}^{d+1}), b \in \ell^{p'}(\mathbb{Z}^{d+1}).$$

The notation $\langle \cdot, \cdot \rangle$ is also used for the action of a distribution on a test function.

7.5.1 Samplings in Non-decaying Shift-Invariant Subspaces

For $\phi \in W(L^{1,1})(\mathbb{R}^{d+1})$, the non-decaying shift-invariant subspaces are defined as follows

$$V_{p,q,1/\omega}(\phi) := \left\{ \sum_{k_1 \in \mathbb{Z}} \sum_{k_2 \in \mathbb{Z}^d} c(k_1, k_2)\phi(\cdot - k_1, \cdot - k_2) : c \in \ell^{p,q,1/\omega}(\mathbb{Z}^{d+1}) \right\}.$$

Obviously, $V_{p,p}(\phi) = V_p(\phi)$. When $\omega = 1$, we omit the subscript $1/\omega$ and use $V_{p,q}(\phi)$ instead of $V_{p,q,1/\omega}(\phi)$, and we can prove that $V_{p,q,1/\omega}(\phi)$ is well-defined in the following part of this chapter.

The decaying weight $1/\omega$ controls the growth rate of the signals living in $V_{p,q,1/\omega}(\phi)$ where ϕ is a generator function(kernel). A biorthogonal kernel $\tilde{\phi}$ of ϕ is defined as

$$\langle \tilde{\phi}, \phi(\cdot - k_1, \cdot - k_2) \rangle = \delta(k_1, k_2), \ \forall k_1 \in \mathbb{Z}, k_2 \in \mathbb{Z}^d.$$

During the signal processing, this kernel is considered to be a reconstruction filter and its biorthogonal partner can be thought of as a sampling filter. The result of this way is a projection onto $V_{p,q,1/\omega}(\phi)$. When $\tilde{\phi}$ is also an element of $V_{p,q,1/\omega}(\phi)$, the projection becomes orthogonal. Under the circumstances, we call it a dual kernel of ϕ and use the notation ϕ_{dual} rather than $\tilde{\phi}$.

The following lemma shows that the convolution of a function h belong to $L^{p,q,1/\omega}(\mathbb{R}^{d+1})$ with a function f belong to $W(L^{1,1,\omega})(\mathbb{R}^{d+1})$ is continuous, that is the filtered signal is continuous. After that, we will introduce some Riesz-type bounds, which link the weighted discrete norm of the samples to the weighted continuous norms of the original signal and the interpolated signals. These Riesz-type bounds ensure that filtering and projection are meaningful.

Lemma 7.20 *Let* $1 \le p, q < +\infty$ *and let* ω *be a submultiplicative weighting function. If* $h \in L^{p,q,1/\omega}(\mathbb{R}^{d+1})$ *and* $f \in W(L^{1,1,\omega})(\mathbb{R}^{d+1})$, *then the convolution* $g = h * f$ *is continuous.*

Proof First, we prove that g is a well-defined function. Using submultiplicativity of ω and Hölder's inequality, one has

$$\frac{1}{\omega(x_1, x_2)} |(h * f)(x_1, x_2)|$$

$$= \frac{1}{\omega(x_1, x_2)} \left| \int_{\mathbb{R}} \int_{\mathbb{R}^d} h(y_1, y_2) f(x_1 - y_1, x_2 - y_2) dy_2 dy_1 \right|$$

$$= \frac{1}{\omega(x_1, x_2)} \left| \int_{\mathbb{R}} \int_{\mathbb{R}^d} \frac{h(y_1, y_2)}{\omega(y_1, y_2)} \omega(-y_1, -y_2) f(x_1 - y_1, x_2 - y_2) dy_2 dy_1 \right|$$

$$= \frac{1}{\omega(x_1, x_2)} \left| \int_{\mathbb{R}} \int_{\mathbb{R}^d} \frac{h(y_1, y_2)}{\omega(y_1, y_2)} \omega(-y_1 + x_1 - x_1, -y_2 + x_2 - x_2) f(x_1 - y_1, x_2 - y_2) dy_2 dy_1 \right|$$

$$\leq \frac{C_\omega}{\omega(x_1, x_2)} \left| \int_{\mathbb{R}} \int_{\mathbb{R}^d} \frac{h(y_1, y_2)}{\omega(y_1, y_2)} \omega(x_1 - y_1, x_2 - y_2) \omega(-x_1, -x_2) f(x_1 - y_1, x_2 - y_2) dy_2 dy_1 \right|$$

$$= C_\omega \left| \int_{\mathbb{R}} \int_{\mathbb{R}^d} \frac{h(y_1, y_2)}{\omega(y_1, y_2)} \omega(x_1 - y_1, x_2 - y_2) f(x_1 - y_1, x_2 - y_2) dy_2 dy_1 \right|$$

$$\leq C_\omega \left[\int_{\mathbb{R}} \left(\int_{\mathbb{R}^d} \left| \frac{h(y_1, y_2)}{\omega(y_1, y_2)} \right|^q dy_2 \right)^{\frac{p}{q}} dy_1 \right]^{\frac{1}{p}} \left[\int_{\mathbb{R}} \left(\int_{\mathbb{R}^d} |f(x_1 - y_1, x_2 - y_2) \right. \right.$$

$$\left. \left. \omega(x_1 - y_1, x_2 - y_2)|^{q'} dy_2 \right)^{\frac{p'}{q'}} dy_1 \right]^{\frac{1}{p'}}$$

$$= C_\omega \left\| \frac{h}{\omega} \right\|_{L^{p,q}(\mathbb{R}^{d+1})} \| f\omega \|_{L^{p',q'}(\mathbb{R}^{d+1})}$$

$$= C_\omega \|h\|_{L^{p,q,1/\omega}(\mathbb{R}^{d+1})} \|f\|_{L^{p',q',\omega}(\mathbb{R}^{d+1})}$$

$$= C_\omega \|h\|_{L^{p,q,1/\omega}(\mathbb{R}^{d+1})} \left[\int_{\mathbb{R}} \left(\int_{\mathbb{R}^d} |f(y_1, y_2) \omega(y_1, y_2)|^{q'} dy_2 \right)^{\frac{p'}{q'}} dy_1 \right]^{\frac{1}{p'}}$$

$$= C_\omega \|h\|_{L^{p,q,1/\omega}(\mathbb{R}^{d+1})} \tag{7.40}$$

$$\times \left[\sum_{l_1 \in \mathbb{Z}} \int_{[0,1]} \left(\sum_{l_2 \in \mathbb{Z}^d} \int_{[0,1]^d} |f(y_1 + l_1, y_2 + l_2) \omega(y_1 + l_1, y_2 + l_2)|^{q'} dy_2 \right)^{\frac{p'}{q'}} dy_1 \right]^{\frac{1}{p'}}$$

$$\leq C_\omega \|h\|_{L^{p,q,1/\omega}(\mathbb{R}^{d+1})} \left[\sum_{l_1 \in \mathbb{Z}} \text{esssup}_{y_1 \in [0,1]} \left(\sum_{l_2 \in \mathbb{Z}^d} \text{esssup}_{y_2 \in [0,1]^d} |f(y_1 + l_1, y_2 + l_2) \right. \right.$$

$$\left. \left. \omega(y_1 + l_1, y_2 + l_2) |^{q'} \right)^{\frac{p'}{q'}} \right]^{\frac{1}{p'}}$$

$$= C_\omega \|h\|_{L^{p,q,1/\omega}(\mathbb{R}^{d+1})} \|f\|_{W(L^{p',q',\omega})(\mathbb{R}^{d+1})}$$

$$\leq C_\omega \|h\|_{L^{p,q,1/\omega}(\mathbb{R}^{d+1})}$$

$$\times \sum_{l_1 \in \mathbb{Z}} \text{esssup}_{y_1 \in [0,1]} \sum_{l_2 \in \mathbb{Z}^d} \text{esssup}_{y_2 \in [0,1]^d} |f(y_1 + l_1, y_2 + l_2)\omega(y_1 + l_1, y_2 + l_2)|$$

$$= C_\omega \|h\|_{L^{p,q,1/\omega}(\mathbb{R}^{d+1})} \|f\|_{W(L^{1,1,\omega})(\mathbb{R}^{d+1})}. \tag{7.41}$$

Now we prove the continuity of g. Inequality (7.40) and the shift-invariance of convolution lead to

$$|g(x_1, x_2) - g(x_1 - x_0, x_2 - y_0)|$$

$$= |((h - h(\cdot - x_0, \cdot - y_0)) * f)(x_1, x_2)|$$

$$\leq C_\omega \, \omega(x_1, x_2) \|f\|_{W(L^{1,1,\omega})(\mathbb{R}^{d+1})} \|h - h(\cdot - x_0, \cdot - y_0)\|_{L^{p,q,1/\omega}(\mathbb{R}^{d+1})}$$

$$\leq C_\omega \, \omega(x_1, x_2) \|f\|_{W(L^{1,1,\omega})(\mathbb{R}^{d+1})} (2a)^{\frac{d}{q}+\frac{1}{p}} \sup_{y_1 \in [-a,a]} \sup_{y_2 \in [-a,a]^d} \frac{1}{\omega(y_1, y_2)} \sup_{y_1 \in [-a,a]} \sup_{y_2 \in [-a,a]^d}$$

$$|h(y_1, y_2) - h(y_1 - x_0, y_2 - y_0)|$$

$$\leq C_{\omega, x_1, x_2, f, a} \sup_{y_1 \in [-a,a]} \sup_{y_2 \in [-a,a]^d} |h(y_1, y_2) - h(y_1 - x_0, y_2 - y_0)|. \tag{7.42}$$

Therefore, when h is continuous, we can obtain that g is also continuous. It is easy to say that $\mathcal{C}_c(\mathbb{R}^{d+1}) \subset L^{p,q,\omega}(\mathbb{R}^{d+1}) \subset L^{p,q}(\mathbb{R}^{d+1}) \subset L^{p,q,1/\omega}(\mathbb{R}^{d+1})$ ($\mathcal{C}_c(\mathbb{R}^{d+1})$ is the set of continuous and compactly supported functions), $\mathcal{C}_c(\mathbb{R}^{d+1})$ is dense in $L^{p,q,1/\omega}(\mathbb{R}^{d+1})$. Hence, for all $h \in L^{p,q,1/\omega}(\mathbb{R}^{d+1})$, there exists $h_0 \in \mathcal{C}_c(\mathbb{R}^{d+1})$, such that h_0 converges to h, so we pick an interval $[-a, a]^{d+1}$ that contains the supports of both h and $h(\cdot - x_0, \cdot - x_0)$ for x_0 sufficiently small. In this case, the right side of (7.42) converges to zero. Thus g is continuous at the point (x_1, x_2). Note that $C_{\omega, x_1, x_2, f, a}$ is a constant that depends neither on h nor on x_0. The proof is completed. $\qquad\square$

Lemma 7.21 *Let* $1 \leq p, q < +\infty$ *and let* ω *be a submultiplicative weighting function. If* $\phi \in W(L^{1,1,\omega})(\mathbb{R}^{d+1})$ *and* $c \in \ell^{p,q,\omega}(\mathbb{Z}^{d+1})$, *then the function*

$$f = \sum_{k_1 \in \mathbb{Z}} \sum_{k_2 \in \mathbb{Z}^d} c(k_1, k_2)\phi(\cdot - k_1, \cdot - k_2)$$

belongs to $L^{p,q,\omega}(\mathbb{R}^{d+1})$ *and we have that*

$$\|f\|_{L^{p,q,\omega}(\mathbb{R}^{d+1})} \leq C_\omega \|c\|_{\ell^{p,q,\omega}(\mathbb{Z}^{d+1})} \|\phi\|_{W(L^{1,1,\omega})(\mathbb{R}^{d+1})}.$$

Furthermore, when $p = 1, q = 1$, *that is* $c \in \ell^{1,1,\omega}(\mathbb{Z}^{d+1})$, *the function* f *can belong to* $W_{1,1,\omega}(\mathbb{R}^{d+1})$ *and*

$$\|f\|_{W(L^{1,1,\omega})(\mathbb{R}^{d+1})} \leq C_\omega \|c\|_{\ell^{1,1,\omega}(\mathbb{Z}^{d+1})} \|\phi\|_{W(L^{1,1,\omega})(\mathbb{R}^{d+1})}.$$

Here C_ω *is the constant given in (7.39).*

Proof We fix $x_1 \in \mathbb{R}, k_1 \in \mathbb{Z}$, denote $c_{k_1}(k_2) = \{c(k_1, k_2) : k_2 \in \mathbb{Z}^d\}$, $\phi_{x_1-k_1}(x_2) = \phi(x_1 - k_1, x_2)$ and $\omega_{x_1-k_1}(x_2) = \omega(x_1 - k_1, x_2)$, then

$$\|f\|^p_{L^{p,q,\omega}(\mathbb{R}^{d+1})}$$

$$= \int_{\mathbb{R}} \left(\int_{\mathbb{R}^d} |\omega(x_1, x_2) f(x_1, x_2)|^q dx_2 \right)^{\frac{p}{q}} dx_1$$

$$= \int_{\mathbb{R}} \left(\int_{\mathbb{R}^d} \left| \omega(x_1, x_2) \sum_{k_1 \in \mathbb{Z}} \sum_{k_2 \in \mathbb{Z}^d} c(k_1, k_2) \phi(x_1 - k_1, x_2 - k_2) \right|^q dx_2 \right)^{\frac{p}{q}} dx_1$$

$$= \int_{\mathbb{R}} \left(\int_{\mathbb{R}^d} \left| \sum_{k_1 \in \mathbb{Z}} \sum_{k_2 \in \mathbb{Z}^d} \omega(x_1 - k_1 + k_1, x_2 - k_2 + k_2) \right. \right. \tag{7.43}$$

$$\times c(k_1, k_2) \phi(x_1 - k_1, x_2 - k_2) \Bigg|^q dx_2 \Bigg)^{\frac{p}{q}} dx_1$$

$$\leq C^p_\omega \int_{\mathbb{R}} \left(\int_{\mathbb{R}^d} \left| \sum_{k_1 \in \mathbb{Z}} \sum_{k_2 \in \mathbb{Z}^d} \omega(k_1, k_2) |c(k_1, k_2)| \right. \right. \tag{7.44}$$

$$\times \omega(x_1 - k_1, x_2 - k_2) |\phi(x_1 - k_1, x_2 - k_2)| \Bigg|^q dx_2 \Bigg)^{\frac{p}{q}} dx_1$$

$$= C^p_\omega \int_{\mathbb{R}} \left(\int_{\mathbb{R}^d} \left| \sum_{k_1 \in \mathbb{Z}} (c_{\omega,k_1} *_{sd} \phi_{\omega,x_1-k_1})(x_2) \right|^q dx_2 \right)^{\frac{p}{q}} dx_1 \tag{7.45}$$

$$= C^p_\omega \int_{\mathbb{R}} \left\| \sum_{k_1 \in \mathbb{Z}} c_{\omega,k_1} *_{sd} \phi_{\omega,x_1-k_1} \right\|^p_{L^q(\mathbb{R}^d)} dx_1$$

$$\leq C^p_\omega \int_{\mathbb{R}} \left(\sum_{k_1 \in \mathbb{Z}} \| c_{\omega,k_1} *_{sd} \phi_{\omega,x_1-k_1} \|_{L^q(\mathbb{R}^d)} \right)^p dx_1$$

$$\leq C^p_\omega \int_{\mathbb{R}} \left(\sum_{k_1 \in \mathbb{Z}} \| c_{\omega,k_1} \|_{\ell^q(\mathbb{R}^d)} \| \phi_{\omega,x_1-k_1} \|_{W(L^1)(\mathbb{R}^d)} \right)^p dx_1, \tag{7.46}$$

where we adopted in (7.45) again the notation

$$c_{\omega,k_1} = \{c_{\omega,k_1}(k_2) := \omega_{k_1}(k_2) |c_{k_1}(k_2)|, k_2 \in \mathbb{Z}^d\} \in \ell^q(\mathbb{R}^d),$$

$$\phi_{\omega,x_1-k_1}(\cdot) = \omega_{x_1-k_1}(\cdot - k_2) |\phi_{x_1-k_1}(\cdot - k_2)|,$$

and (7.46) is according to Proposition 7.2.

Let $d_\omega = \{d_\omega(k_1) = \|c_{\omega,k_1}\|_{\ell^q(\mathbb{R}^d)} : k_1 \in \mathbb{Z}\}$ and $h_\omega(x) = \|\phi_{\omega,x}\|_{W(L^1)(\mathbb{R}^d)}$, then using Proposition 7.2 again, we can get

$$\|f\|_{L^{p,q,\omega}(\mathbb{R}^{d+1})}$$

$$\leq C_\omega \left[\int_\mathbb{R} \left| \sum_{k_1 \in \mathbb{Z}} d_\omega(k_1) \cdot h_\omega(x_1 - k_1) \right|^p dx_1 \right]^{\frac{1}{p}}$$

$$= C_\omega \|d_\omega *_{sd} h_\omega\|_{L^p(\mathbb{R})}$$

$$\leq C_\omega \|d_\omega\|_{\ell^p(\mathbb{R})} \|h_\omega\|_{W(L^1)(\mathbb{R})}$$

$$= C_\omega \left(\sum_{k_1 \in \mathbb{Z}} \|c_{\omega,k_1}\|_{\ell^q(\mathbb{R}^d)}^p \right)^{\frac{1}{p}} \sum_{n \in \mathbb{Z}} \mathrm{esssup}_{x_1 \in [0,1]} \|\phi_{\omega,x_1+n}\|_{W(L^1)(\mathbb{R}^d)}$$

$$= C_\omega \left[\sum_{k_1 \in \mathbb{Z}} \left(\sum_{k_2 \in \mathbb{Z}^d} |\omega(k_1, k_2)c(k_1, k_2)|^q \right)^{\frac{p}{q}} \right]^{\frac{1}{p}} \cdot \sum_{n \in \mathbb{Z}} \mathrm{esssup}_{x_1 \in [0,1]}$$

$$\sum_{l \in \mathbb{Z}^d} \mathrm{esssup}_{x_2 \in [0,1]^d} |\omega(x_1 + n - k_1, x_2 + l - k_2)\phi(x_1 + n - k_1, x_2 + l - k_2)|$$

$$= C_\omega \|c\|_{\ell^{p,q,\omega}(\mathbb{R}^{d+1})} \|\phi\|_{W(L^{1,1,\omega})(\mathbb{R}^{d+1})}.$$

Thus, we obtain $\|f\|_{L^{p,q,\omega}(\mathbb{R}^{d+1})} \leq C_\omega \|c\|_{\ell^{p,q,\omega}(\mathbb{Z}^{d+1})} \|\phi\|_{W(L^{1,1,\omega})(\mathbb{R}^{d+1})}$.

When $p = 1, q = 1$, by using the submultiplicative property of ω and Definition 7.1, one has

$$\|f\|_{W(L^{1,1,\omega})(\mathbb{R}^{d+1})}$$

$$= \sum_{n_1 \in \mathbb{Z}} \mathrm{esssup}_{x_1 \in [0,1]} \sum_{n_2 \in \mathbb{Z}^d} \mathrm{esssup}_{x_2 \in [0,1]^d} |f(x_1 + n_1, x_2 + n_2)\omega(x_1 + n_1, x_2 + n_2)|$$

$$\leq C_\omega \sum_{n_1 \in \mathbb{Z}} \mathrm{esssup}_{x_1 \in [0,1]} \sum_{n_2 \in \mathbb{Z}^d} \mathrm{esssup}_{x_2 \in [0,1]^d} \left| \sum_{k_1 \in \mathbb{Z}} \sum_{k_2 \in \mathbb{Z}^d} \omega(k_1, k_2)c(k_1, k_2) \right.$$

$$\omega(x_1 + n_1 - k_1, x_2 + n_2 - k_2)\phi(x_1 + n_1 - k_1, x_2 + n_2 - k_2) \Big|$$

$$\leq C_\omega \sum_{n_1 \in \mathbb{Z}} \mathrm{esssup}_{x_1 \in [0,1]} \sum_{n_2 \in \mathbb{Z}^d} \mathrm{esssup}_{x_2 \in [0,1]^d} \sum_{k_1 \in \mathbb{Z}} \sum_{k_2 \in \mathbb{Z}^d} |\omega(k_1, k_2)||c(k_1, k_2)|$$

$$|\omega(x_1 + n_1 - k_1, x_2 + n_2 - k_2)||\varphi(x_1 + n_1 - k_1, x_2 + n_2 - k_2)|$$

$$\leq C_\omega \sum_{k_1 \in \mathbb{Z}} \sum_{k_2 \in \mathbb{Z}^d} |\omega(k_1, k_2)||c(k_1, k_2)| \left| \sum_{n_1 \in \mathbb{Z}} \mathrm{esssup}_{x_1 \in [0,1]} \sum_{n_2 \in \mathbb{Z}^d} \mathrm{esssup}_{x_2 \in [0,1]^d} \right.$$

$$\left. |\omega(x_1 + n_1 - k_1, x_2 + n_2 - k_2)||\phi(x_1 + n_1 - k_1, x_2 + n_2 - k_2)| \right|$$

$$= C_\omega \, \|c\|_{\ell^{1,1,\omega}(\mathbb{Z}^{d+1})} \|\phi\|_{W(L^{1,1,\omega})(\mathbb{R}^{d+1})}.$$

Therefore, one completes the proof.   □

Lemma 7.22 *Let* $1 \leq p, q < +\infty$ *and let* ω *be a submultiplicative weighting function. If* $f \in L^{p,q,\omega}(\mathbb{R}^{d+1})$ *and* $\varphi \in W(L^{1,1,\omega})(\mathbb{R}^{d+1})$*, then* $f * \varphi$ *is a well-defined continuous function. Its sampled sequence* $\{c(k_1, k_2) := (f * \varphi)(k_1, k_2)\}_{k_1 \in \mathbb{Z}, k_2 \in \mathbb{Z}^d}$ *belongs to* $\ell^{p,q,\omega}(\mathbb{Z}^{d+1})$ *and the bound*

$$\|c\|_{\ell^{p,q,\omega}(\mathbb{Z}^{d+1})} \leq C_\omega \, \|f\|_{L^{p,q,\omega}(\mathbb{R}^{d+1})} \|\varphi\|_{W(L^{1,1,\omega})(\mathbb{R}^{d+1})}$$

holds.

   Furthermore, if the function f *belongs to* $L^{p,q,1/\omega}(\mathbb{R}^{d+1})$*, then the sampled sequence* $c \in \ell^{p,q,1/\omega}(\mathbb{Z}^{d+1})$ *and*

$$\|c\|_{\ell^{p,q,1/\omega}(\mathbb{Z}^{d+1})} \leq C_\omega \, \|f\|_{L^{p,q,1/\omega}(\mathbb{R}^{d+1})} \|\varphi\|_{W(L^{1,1,\omega})(\mathbb{R}^{d+1})}.$$

Here C_ω *is the constant given in* (7.39)

Proof Since $L^{p,q,\omega}(\mathbb{R}^{d+1}) \subset L^{p,q}(\mathbb{R}^{d+1}) \subset L^{p,q,1/\omega}(\mathbb{R}^{d+1})$ and $\varphi \in W(L^{1,1,\omega})(\mathbb{R}^{d+1})$, it follows from Lemma 7.20 that $f * \varphi$ is a well-defined continuous function.

   When $f \in L^{p,q,\omega}(\mathbb{R}^{d+1})$, according to duality, we can express the ω-weighted norm of c as

$$\|c\|_{\ell^{p,q,\omega}(\mathbb{Z}^{d+1})} = \sup_{\|b\|_{\ell^{p',q',1/\omega}(\mathbb{Z}^{d+1})}=1} \langle c, b \rangle.$$

Then

$$\|c\|_{\ell^{p,q,\omega}(\mathbb{Z}^{d+1})}$$

$$= \sup_{\|b\|_{\ell^{p',q',1/\omega}(\mathbb{Z}^{d+1})}=1} \left| \sum_{k_1 \in \mathbb{Z}} \sum_{k_2 \in \mathbb{Z}^d} c(k_1, k_2) b(k_1, k_2) \right|$$

$$= \sup_{\|b\|_{\ell^{p',q',1/\omega}(\mathbb{Z}^{d+1})}=1} \left| \sum_{k_1 \in \mathbb{Z}} \sum_{k_2 \in \mathbb{Z}^d} b(k_1, k_2) \int_{\mathbb{R}} \int_{\mathbb{R}^d} f(x_1, x_2) \varphi(k_1 - x_1, k_2 - x_2) dx_2 dx_1 \right|$$

$$= \sup_{\|b\|_{\ell^{p',q',1/\omega}(\mathbb{Z}^{d+1})}=1} \left| \sum_{k_1 \in \mathbb{Z}} \sum_{k_2 \in \mathbb{Z}^d} \frac{b(k_1, k_2)}{\omega(k_1, k_2)} \int_{\mathbb{R}} \int_{\mathbb{R}^d} f(x_1, x_2) \omega(k_1 - x_1 + x_1, k_2 - x_2 + x_2) \right.$$

$$\left. \times \varphi(k_1 - x_1, k_2 - x_2) dx_2 dx_1 \right|$$

$$\le C_\omega \sup_{\|b\|_{\ell^{p',q',1/\omega}(\mathbb{Z}^{d+1})}=1} \left| \sum_{k_1 \in \mathbb{Z}} \sum_{k_2 \in \mathbb{Z}^d} \frac{b(k_1, k_2)}{\omega(k_1, k_2)} \int_{\mathbb{R}} \int_{\mathbb{R}^d} f(x_1, x_2) \omega(x_1, x_2) \varphi(k_1 - x_1, k_2 - x_2) \right.$$

$$\left. \times \omega(k_1 - x_1, k_2 - x_2) dx_2 dx_1 \right|$$

$$\le C_\omega \sup_{\|b\|_{\ell^{p',q',1/\omega}(\mathbb{Z}^{d+1})}=1} \int_{\mathbb{R}} \int_{\mathbb{R}^d} \left| f(x_1, x_2) \omega(x_1, x_2) \sum_{k_1 \in \mathbb{Z}} \sum_{k_2 \in \mathbb{Z}^d} \frac{b(k_1, k_2)}{\omega(k_1, k_2)} \varphi(k_1 - x_1, k_2 - x_2) \right.$$

$$\left. \times \omega(k_1 - x_1, k_2 - x_2) \right| dx_2 dx_1$$

$$= C_\omega \sup_{\|b\|_{\ell^{p',q',1/\omega}(\mathbb{Z}^{d+1})}=1} \left\| f\omega \sum_{k_1 \in \mathbb{Z}} \sum_{k_2 \in \mathbb{Z}^d} \frac{b(k_1, k_2)}{\omega(k_1, k_2)} \varphi(k_1 - \cdot, k_2 - \cdot) \omega(k_1 - \cdot, k_2 - \cdot) \right\|_{L^{1,1}(\mathbb{R}^{d+1})}$$

$$\le C_\omega \sup_{\|b\|_{\ell^{p',q',1/\omega}(\mathbb{Z}^{d+1})}=1} \| f\omega \|_{L^{p,q}(\mathbb{R}^{d+1})}$$

$$\times \left\| \sum_{k_1 \in \mathbb{Z}} \sum_{k_2 \in \mathbb{Z}^d} \frac{b(k_1, k_2)}{\omega(k_1, k_2)} \varphi(k_1 - \cdot, k_2 - \cdot) \omega(k_1 - \cdot, k_2 - \cdot) \right\|_{L^{p',q'}(\mathbb{R}^{d+1})} \tag{7.47}$$

$$\le C_\omega \sup_{\|b\|_{\ell^{p',q',1/\omega}(\mathbb{Z}^{d+1})}=1} \| f\omega \|_{L^{p,q}(\mathbb{R}^{d+1})} \left\| \frac{b}{\omega} \right\|_{\ell^{p',q'}(\mathbb{Z}^{d+1})} \| \varphi\omega \|_{W(L^{1,1})(\mathbb{R}^{d+1})} \tag{7.48}$$

$$= C_\omega \sup_{\|b\|_{\ell^{p',q',1/\omega}(\mathbb{Z}^{d+1})}=1} \| f \|_{L^{p,q,\omega}(\mathbb{R}^{d+1})} \| b \|_{\ell^{p',q',1/\omega}(\mathbb{Z}^{d+1})} \| \varphi \|_{W(L^{1,1,\omega})(\mathbb{R}^{d+1})}$$

$$= C_\omega \| f \|_{L^{p,q,\omega}(\mathbb{R}^{d+1})} \| \varphi \|_{W(L^{1,1,\omega})(\mathbb{R}^{d+1})},$$

where (7.47) follows by Proposition 7.3 and (7.48) follows by Proposition 7.1. Thus, we obtain

$$\| c \|_{\ell^{p,q,\omega}(\mathbb{Z}^{d+1})} \le C_\omega \| f \|_{L^{p,q,\omega}(\mathbb{R}^{d+1})} \| \varphi \|_{W(L^{1,1,\omega})(\mathbb{R}^{d+1})}.$$

When $f \in L^{p,q,1/\omega}(\mathbb{R}^{d+1})$, by duality, Hölder's inequality, and Lemma 7.21 one has

$$\| c \|_{\ell^{p,q,1/\omega}(\mathbb{Z}^{d+1})}$$

$$= \sup_{\|b\|_{\ell^{p',q',\omega}(\mathbb{Z}^{d+1})}=1} \langle c, b \rangle$$

$$
= \sup_{\|b\|_{\ell^{p',q'},\omega(\mathbb{Z}^{d+1})}=1} \sum_{k_1\in\mathbb{Z}} \sum_{k_2\in\mathbb{Z}^d} b(k_1,k_2) \int_{\mathbb{R}} \int_{\mathbb{R}^d} f(x_1,x_2)\varphi(k_1-x_1,k_2-x_2)dx_2 dx_1
$$

$$
\leq \sup_{\|b\|_{\ell^{p',q'},\omega(\mathbb{Z}^{d+1})}=1} \left[\int_{\mathbb{R}} \left(\int_{\mathbb{R}^d} \left| \frac{f(x_1,x_2)}{\omega(x_1,x_2)} \right|^q dx_2 \right)^{\frac{p}{q}} dx_1 \right]^{\frac{1}{p}}
$$

$$
\times \left[\int_{\mathbb{R}} \left(\int_{\mathbb{R}^d} \left| \sum_{k_1\in\mathbb{Z}} \sum_{k_2\in\mathbb{Z}^d} \omega(x_2,x_2) b(k_1,k_2)\varphi(k_1-x_1,k_2-x_2) \right|^{q'} dx_2 \right)^{\frac{p'}{q'}} dx_1 \right]^{\frac{1}{p'}}
$$

$$
= \|f\|_{L^{p,q,1/\omega}(\mathbb{R}^{d+1})} \sup_{\|b\|_{\ell^{p',q'},\omega(\mathbb{Z}^{d+1})}=1} \left\| \sum_{k_1\in\mathbb{Z}} \sum_{k_2\in\mathbb{Z}^d} b(k_1,k_2)\varphi^{\vee}(x_1-k_1,x_2-k_2) \right\|_{L^{p',q'},\omega(\mathbb{R}^{d+1})}
$$

$$
\leq \|f\|_{L^{p,q,1/\omega}(\mathbb{R}^{d+1})} \sup_{\|b\|_{\ell^{p',q'},\omega(\mathbb{Z}^{d+1})}=1} C_\omega \|b\|_{\ell^{p',q'},\omega(\mathbb{Z}^{d+1})} \|\varphi^{\vee}\|_{W(L^{1,1,\omega})(\mathbb{R}^{d+1})}
$$

$$
= C_\omega \|f\|_{L^{p,q,1/\omega}(\mathbb{R}^{d+1})} \|\varphi\|_{W(L^{1,1,\omega})(\mathbb{R}^{d+1})}.
$$

Thus, we obtain $\|c\|_{\ell^{p,q,1/\omega}(\mathbb{Z}^{d+1})} \leq C_\omega \|f\|_{L^{p,q,1/\omega}(\mathbb{R}^{d+1})} \|\varphi\|_{W(L^{1,1,\omega})(\mathbb{R}^{d+1})}.$ □

Lemma 7.21 and 7.22 are extensions of [168]. Now we show the similar bounds for the decaying weighting function $1/\omega$.

Lemma 7.23 *Let* $1 \leq p, q < +\infty$ *and let* ω *be a submultiplicative weighting function. If* $\varphi \in W(L^{1,1,\omega})(\mathbb{R}^{d+1})$ *and* $c \in \ell^{p,q,1/\omega}(\mathbb{Z}^{d+1})$, *then the function*

$$
f = \sum_{k_1\in\mathbb{Z}} \sum_{k_2\in\mathbb{Z}^d} c(k_1,k_2)\varphi(\cdot - k_1, \cdot - k_2)
$$

belongs to $L^{p,q,1/\omega}(\mathbb{R}^{d+1})$ *and we have that*

$$
\|f\|_{L^{p,q,1/\omega}(\mathbb{R}^{d+1})} \leq C_\omega \|c\|_{\ell^{p,q,1/\omega}(\mathbb{Z}^{d+1})} \|\varphi\|_{W(L^{1,1,\omega})(\mathbb{R}^{d+1})},
$$

where C_ω *is the constant given in* (7.39).

Proof By duality, we can express the $1/\omega$-weighted norm of f as

$$
\|f\|_{L^{p,q,1/\omega}(\mathbb{R}^{d+1})} = \sup_{\|g\|_{L^{p',q'},\omega(\mathbb{R}^{d+1})}=1} \langle f, g \rangle.
$$

Using Hölder's inequality and Lemma 7.22, one has

$$\|f\|_{L^{p,q,1/\omega}(\mathbb{R}^{d+1})}$$

$$= \sup_{\|g\|_{L^{p',q',\omega}(\mathbb{R}^{d+1})}=1} \left\langle \sum_{k_1\in\mathbb{Z}}\sum_{k_2\in\mathbb{Z}^d} c(k_1,k_2)\varphi(\cdot-k_1,\cdot-k_2), g\right\rangle$$

$$= \sup_{\|g\|_{L^{p',q',\omega}(\mathbb{R}^{d+1})}=1} \sum_{k_1\in\mathbb{Z}}\sum_{k_2\in\mathbb{Z}^d} \frac{c(k_1,k_2)}{\omega(k_1,k_2)}\omega(k_1,k_2)\langle\varphi(\cdot-k_1,\cdot-k_2), g\rangle$$

$$\leq \sup_{\|g\|_{L^{p',q',\omega}(\mathbb{R}^{d+1})}=1} \left[\sum_{k_1\in\mathbb{Z}}\left(\sum_{k_2\in\mathbb{Z}^d}\left|\frac{c(k_1,k_2)}{\omega(k_1,k_2)}\right|^q\right)^{\frac{p}{q}}\right]^{\frac{1}{p}} \left[\sum_{k_1\in\mathbb{Z}}\left(\sum_{k_2\in\mathbb{Z}^d}|\omega(k_1,k_2)\right.\right.$$

$$\left.\left.\langle\varphi(\cdot-k_1,\cdot-k_2), g\rangle|^{q'}\right)^{\frac{p'}{q'}}\right]^{\frac{1}{p'}}$$

$$= \sup_{\|g\|_{L^{p',q',\omega}(\mathbb{R}^{d+1})}=1} \|c\|_{\ell^{p,q,1/\omega}(\mathbb{Z}^{d+1})}\|g * \varphi^\vee\|_{\ell^{p',q',\omega}(\mathbb{R}^{d+1})}$$

$$\leq \sup_{\|g\|_{L^{p',q',\omega}(\mathbb{R}^{d+1})}=1} \|c\|_{\ell^{p,q,1/\omega}(\mathbb{Z}^{d+1})}C_\omega\|g\|_{L^{p',q',\omega}(\mathbb{R}^{d+1})}\|\varphi^\vee\|_{W(L^{1,1,\omega})(\mathbb{R}^{d+1})}$$

$$= C_\omega\|c\|_{\ell^{p,q,1/\omega}(\mathbb{Z}^{d+1})}\|\varphi\|_{W(L^{1,1,\omega})(\mathbb{R}^{d+1})}.$$

Thus, we obtain $\|f\|_{L^{p,q,1/\omega}(\mathbb{R}^{d+1})} \leq C_\omega\|c\|_{\ell^{p,q,1/\omega}(\mathbb{Z}^{d+1})}\|\varphi\|_{W(L^{1,1,\omega})(\mathbb{R}^{d+1})}.$ □

Now we give one of the main results in this section, the following theorem captures both the sampling and the reconstruction of signals in $L^{p,q,1/\omega}(\mathbb{R}^{d+1})$.

Theorem 7.10 *Let $\varphi \in W(L^{1,1,\omega})(\mathbb{R}^{d+1})$ and $\tilde{\varphi} \in W(L^{1,1,\omega})(\mathbb{R}^{d+1})$ be two biorthogonal functions such that $\langle\varphi, \tilde{\varphi}(\cdot-k_1,\cdot-k_2)\rangle = \delta(k_1,k_2), \forall k_1 \in \mathbb{Z}, k_2 \in \mathbb{Z}^d$. We also assume that the weighting function ω is submultiplicative. Then, the linear operator*

$$P_\varphi f := \sum_{k_1\in\mathbb{Z}}\sum_{k_2\in\mathbb{Z}^d} \langle f, \tilde{\varphi}(\cdot-k_1,\cdot-k_2)\rangle\varphi(\cdot-k_1,\cdot-k_2)$$

is a projector that continuously maps $L^{p,q,1/\omega}(\mathbb{R}^{d+1})$ into the subspace $V_{p,q,1/\omega}(\varphi) \subset L^{p,q,1/\omega}(\mathbb{R}^{d+1})$.

Proof If P_φ is a projector, then $P_\varphi^2 = P_\varphi$. Especially, for all $f \in L^{p,q,1/\omega}(\mathbb{R}^{d+1})$, since $\langle\varphi, \tilde{\varphi}(\cdot-k_1,\cdot-k_2)\rangle = \delta(k_1,k_2)$, we can get

$$P_\varphi^2 f = \sum_{k_1\in\mathbb{Z}}\sum_{k_2\in\mathbb{Z}^d} \langle P_\varphi f, \tilde{\varphi}(\cdot-k_1,\cdot-k_2)\rangle\varphi(\cdot-k_1,\cdot-k_2)$$

$$= \sum_{k_1 \in \mathbb{Z}} \sum_{k_2 \in \mathbb{Z}^d} \left\langle \sum_{l_1 \in \mathbb{Z}} \sum_{l_2 \in \mathbb{Z}^d} \langle f, \tilde{\varphi}(\cdot - l_1, \cdot - l_2) \rangle \varphi(\cdot - l_1, \cdot - l_2), \tilde{\varphi}(\cdot - k_1, \cdot - k_2) \right\rangle$$

$$\times \varphi(\cdot - k_1, \cdot - k_2)$$

$$= \sum_{k_1 \in \mathbb{Z}} \sum_{k_2 \in \mathbb{Z}^d} \sum_{l_1 \in \mathbb{Z}} \sum_{l_2 \in \mathbb{Z}^d} \langle f, \tilde{\varphi}(\cdot - l_1, \cdot - l_2) \rangle \delta(k_1 - l_1, k_2 - l_2) \varphi(\cdot - k_1, \cdot - k_2)$$

$$= \sum_{k_1 \in \mathbb{Z}} \sum_{k_2 \in \mathbb{Z}^d} \langle f, \tilde{\varphi}(\cdot - k_1, \cdot - k_2) \rangle \varphi(\cdot - k_1, \cdot - k_2)$$

$$= P_\varphi f.$$

It is well-known that P_φ is bounded in the $L^{p,q,1/\omega}$-norm then it is also continuous. Let $c(k_1, k_2) = \langle f, \tilde{\varphi}(\cdot - k_1, \cdot - k_2) \rangle = (f * \tilde{\varphi}^\vee)(k_1, k_2)$ be the expansion coefficients of $P_\varphi f$, thus we can invoke Lemma 7.22 to deduce that

$$\|c\|_{\ell^{p,q,1/\omega}(\mathbb{Z}^{d+1})} \leq C_\omega \|\tilde{\varphi}\|_{W(L^{1,1,\omega})(\mathbb{R}^{d+1})} \|f\|_{L^{p,q,1/\omega}(\mathbb{R}^{d+1})}.$$

Using this bound and Lemma 7.23, we obtain

$$\|P_\varphi f\|_{L^{p,q,1/\omega}(\mathbb{R}^{d+1})} \leq C_\omega \|c\|_{\ell^{p,q,1/\omega}(\mathbb{Z}^{d+1})} \|\varphi\|_{W(L^{1,1,\omega})(\mathbb{R}^{d+1})}$$

$$\leq \underbrace{C_\omega^2 \|\tilde{\varphi}\|_{W(L^{1,1,\omega})(\mathbb{R}^{d+1})} \|\varphi\|_{W(L^{1,1,\omega})(\mathbb{R}^{d+1})}}_{constant} \|f\|_{L^{p,q,1/\omega}(\mathbb{R}^{d+1})},$$

which shows the boundedness of the operator P_φ. Therefore, one completes the proof. □

In order to obtain another main theorem, we need the following results.

Lemma 7.24 *Let $1 \leq p, q < +\infty$ and let ω be a submultiplicative weighting function that additionally satisfies the Gelfand-Raikov-Shilov(GRS) condition*

$$\lim_{n \to \infty} \omega(nk)^{\frac{1}{n}} = 1, \quad \forall k \in \mathbb{Z}^d.$$

Suppose that the generator φ is in $W(L^{1,1,\omega})(\mathbb{R}^{d+1})$ and that $\{\varphi(\cdot - k_1, \cdot - k_2)\}_{k_1 \in \mathbb{Z}, k_2 \in \mathbb{Z}^d}$ is a Riesz basis for $V_2(\varphi)$. Then, the dual generator φ_{dual} is in $W(L^{1,1,\omega})(\mathbb{R}^{d+1})$ as well.

Proof Let us put $\psi(x_1, x_2) := \overline{\varphi(-x_1, -x_2)}$ and define the autocorrelation sequence of φ as

$$a(k_1, k_2) := \int_\mathbb{R} \int_{\mathbb{R}^d} \varphi(x_1, x_2) \overline{\varphi(x_1 - k_1, x_2 - k_2)} dx_2 dx_1, \quad for \ k_1 \in \mathbb{Z}, k_2 \in \mathbb{Z}^d,$$

then $a(k_1, k_2) = (\varphi * \psi)(k_1, k_2)$. We first show that $a \in \ell^{1,1,\omega}(\mathbb{Z}^{d+1})$, denote

$$\varphi_{\omega,x_1,x_2}(l_1, l_2) = \omega(x_1 + l_1, x_2 + l_2)|\varphi(x_1 + l_1, x_2 + l_2)|,$$

$$\psi_{\omega,-x_1,-x_2}(k_1 - l_1, k_2 - l_2) = \omega(k_1 - l_1 - x_1, k_2 - l_2 - x_2)|\psi(k_1 - l_1 - x_1, k_2 - l_2 - x_2)|.$$

Then using Proposition 7.2, one has

$$\|a\|_{\ell^{1,1,\omega}(\mathbb{Z}^{d+1})}$$

$$= \|\varphi * \psi\|_{\ell^{1,1,\omega}(\mathbb{Z}^{d+1})}$$

$$\leq C_\omega \sum_{k_1 \in \mathbb{Z}} \sum_{k_2 \in \mathbb{Z}^d} \left| \int_{\mathbb{R}} \int_{\mathbb{R}^d} \omega(x_1, x_2)|\varphi(x_1, x_2)|\omega(k_1 - x_1, k_2 - x_2) \right.$$

$$\left. |\psi(k_1 - x_1, k_2 - x_2)|dx_2 dx_1 \right|$$

$$= C_\omega \sum_{k_1 \in \mathbb{Z}} \sum_{k_2 \in \mathbb{Z}^d} \left| \sum_{l_1 \in \mathbb{Z}} \int_{[0,1]} \sum_{l_2 \in \mathbb{Z}^d} \int_{[0,1]^d} \omega(x_1 + l_1, x_2 + l_2)|\varphi(x_1 + l_1, x_2 + l_2)| \right.$$

$$\left. \omega(k_1 - (x_1 + l_1), k_2 - (x_2 + l_2))|\psi(k_1 - (x_1 + l_1), k_2 - (x_2 + l_2))|dx_2 dx_1 \right|$$

$$\leq C_\omega \int_{[0,1]} \int_{[0,1]^d} \sum_{k_1 \in \mathbb{Z}} \sum_{k_2 \in \mathbb{Z}^d} \left| \sum_{l_1 \in \mathbb{Z}} \sum_{l_2 \in \mathbb{Z}^d} \varphi_{\omega,x_1,x_2}(l_1, l_2) \right.$$

$$\left. \psi_{\omega,-x_1,-x_2}(k_1 - l_1, k_2 - l_2) \right| dx_2 dx_1$$

$$= C_\omega \int_{[0,1]} \int_{[0,1]^d} \left\| \varphi_{\omega,x_1,x_2} * \psi_{\omega,-x_1,-x_2} \right\|_{\ell^{1,1}(\mathbb{Z}^{d+1})} dx_2 dx_1$$

$$\leq C_\omega \int_{[0,1]} \int_{[0,1]^d} \left\| \varphi_{\omega,x_1,x_2} \right\|_{\ell^{1,1}(\mathbb{Z}^{d+1})} \left\| \psi_{\omega,-x_1,-x_2} \right\|_{\ell^{1,1}(\mathbb{Z}^{d+1})} dx_2 dx_1$$

$$= C_\omega \int_{[0,1]} \int_{[0,1]^d} \left\| \varphi_{\omega,x_1,x_2} \right\|^2_{\ell^{1,1}(\mathbb{Z}^{d+1})} dx_2 dx_1 \qquad (7.49)$$

$$\leq C_\omega \, \text{esssup}_{x_1 \in [0,1]} \text{esssup}_{x_2 \in [0,1]^d} \left\| \varphi_{\omega,x_1,x_2} \right\|^2_{\ell^{1,1}(\mathbb{Z}^{d+1})}$$

$$\leq C_\omega \left(\sum_{k_1 \in \mathbb{Z}} \text{esssup}_{x_1 \in [0,1]} \sum_{k_2 \in \mathbb{Z}^d} \text{esssup}_{x_2 \in [0,1]^d} |\varphi_{\omega,x_1,x_2}(k_1, k_2)| \right)^2$$

$$= C_\omega \|\varphi\|^2_{W(L^{1,1,\omega})(\mathbb{R}^{d+1})} < \infty.$$

Since the symmetry of ω, $\|\varphi_{\omega,x_1,x_2}\|_{\ell^{1,1}(\mathbb{Z}^{d+1})} = \|\psi_{\omega,-x_1,-x_2}\|_{\ell^{1,1}(\mathbb{Z}^{d+1})}$, this verifies (7.49).

According to [206], if $\{\varphi(\cdot - k_1, \cdot - k_2)\}_{k_1 \in \mathbb{Z}, k_2 \in \mathbb{Z}^d}$ is a Riesz basis for $V_2(\varphi)$, then the dual generator φ_{dual} exists and is unique. In this case, φ_{dual} determined by the expension

$$\varphi_{dual} = \sum_{k_1 \in \mathbb{Z}} \sum_{k_2 \in \mathbb{Z}^d} b(k_1, k_2) \varphi(\cdot - k_1, \cdot - k_2),$$

where the coefficient sequence b is given in the Fourier domain by

$$\hat{b}(\xi_1, \xi_2) = \frac{1}{\sum_{k_1 \in \mathbb{Z}} \sum_{k_2 \in \mathbb{Z}^d} |\hat{\varphi}(\xi_1 + k_1, \xi_2 + k_2)|^2} = \frac{1}{\hat{a}(\xi_1, \xi_2)}.$$

Here, $\hat{a}$ and $\hat{b}$ are the Fourier series associated with the sequences of coefficients a and b, respectively. Because $a \in \ell^{1,1,\omega}(\mathbb{Z}^{d+1})$ and ω satisfies the GRS condition, then using Proposition 7.3, we can deduce that $b \in \ell^{1,1,\omega}(\mathbb{Z}^{d+1})$. Finally, from Lemma 7.21, we have that

$$\|\varphi_{dual}\|_{W(L^{1,1,\omega})(\mathbb{R}^{d+1})} \leq C_\omega \|b\|_{\ell^{1,1,\omega}(\mathbb{Z}^{d+1})} \|\varphi\|_{W(L^{1,1,\omega})(\mathbb{R}^{d+1})} < \infty.$$

In conclusion, the dual kernel φ_{dual} also belongs to $W(L^{1,1,\omega})(\mathbb{R}^{d+1})$.   □

Now we give another main result in this section.

Theorem 7.11 *Let $1 \leq p, q < +\infty$ and let ω be a submultiplicative weighting function satisfying the Gelfand-Raikov-Shilov(GRS) condition. Assume that $\varphi \in W(L^{1,1,\omega})(\mathbb{R}^{d+1})$ and that $\{\varphi(\cdot - k_1, \cdot - k_2)\}_{k_1 \in \mathbb{Z}, k_2 \in \mathbb{Z}^d}$ is a Riesz basis for $V_2(\varphi)$. Then, there exist constants $C_{\omega,\varphi} < \infty$ and $\tilde{C}_{\omega,\varphi} > 0$ such that*

$$\tilde{C}_{\omega,\varphi} \|c\|_{\ell^{p,q,1/\omega}(\mathbb{Z}^{d+1})} \leq \left\| \sum_{k_1 \in \mathbb{Z}} \sum_{k_2 \in \mathbb{Z}^d} c(k_1, k_2) \varphi(\cdot - k_1, \cdot - k_2) \right\|_{1,p,q,1/\omega(\mathbb{R}^{d+1})}$$

$$\leq C_{\omega,\varphi} \|c\|_{\ell^{p,q,1/\omega}(\mathbb{Z}^{d+1})},$$

where $c \in \ell^{p,q,1/\omega}(\mathbb{Z}^{d+1})$.

Proof Let $f := \sum_{k_1 \in \mathbb{Z}} \sum_{k_2 \in \mathbb{Z}^d} c(k_1, k_2) \varphi(\cdot - k_1, \cdot - k_2)$. Using Lemma 7.23, we can get the right-hand-side inequality

$$\|f\|_{L^{p,q,1/\omega}(\mathbb{R}^{d+1})} \leq C_\omega \|c\|_{\ell^{p,q,1/\omega}(\mathbb{Z}^{d+1})} \|\varphi\|_{W(L^{1,1,\omega})(\mathbb{R}^{d+1})}$$

$$= C_{\omega,\varphi} \|c\|_{\ell^{p,q,1/\omega}(\mathbb{Z}^{d+1})},$$

where the constant $C_{\omega,\varphi}$ is equal to $C_\omega \|\varphi\|_{W(L^{1,1,\omega})(\mathbb{R}^{d+1})}$.

The left-hand-side inequality can be obtained by noting that

$$c(k_1, k_2) - \langle f, \varphi_{dual}(\cdot - k_1, \cdot - k_2) \rangle = (f * \varphi_{dual}^\vee)(k_1, k_2), \quad \forall k_1 \subset \mathbb{Z}, k_2 \subset \mathbb{Z}^d.$$

Since Lemma 7.24, we can get $\varphi_{dual} \in W(L^{1,1,\omega})(\mathbb{R}^{d+1})$, then $\varphi_{dual}^\vee \in W(L^{1,1,\omega})(\mathbb{R}^{d+1})$. This allows us to invoke Lemma 7.22 to get

$$\begin{aligned}
\|c\|_{\ell^{p,q,1/\omega}(\mathbb{Z}^{d+1})} &\leq C_\omega \|f\|_{L^{p,q,1/\omega}(\mathbb{R}^{d+1})} \|\varphi_{dual}^\vee\|_{W(L^{1,1,\omega})(\mathbb{R}^{d+1})} \\
&= C_\omega \|f\|_{L^{p,q,1/\omega}(\mathbb{R}^{d+1})} \|\varphi_{dual}\|_{W(L^{1,1,\omega})(\mathbb{R}^{d+1})} \\
&= \tilde{C}_{\omega,\varphi} \|f\|_{L^{p,q,1/\omega}(\mathbb{R}^{d+1})},
\end{aligned}$$

where the constant $\tilde{C}_{\omega,\varphi}$ is equal to $C_\omega \|\varphi_{dual}\|_{W(L^{1,1,\omega})(\mathbb{R}^{d+1})}$. Therefore, one completes the proof. □

7.5.2 Ideal Samplings

7.5.2.1 Samplings in Weighted Mixed Sobolev Spaces

First, we introduce the definition of weighted mixed Sobolev spaces.

Definition 7.14 (Weighted Mixed Sobolev Spaces) For $s \in \mathbb{R}$, $1 \leq p, q < +\infty$, and for a weighting function ω, the weighted mixed Sobolev spaces $L^s_{p,q,\omega}(\mathbb{R}^{d+1})$ is defined as

$$L^s_{p,q,\omega}(\mathbb{R}^{d+1}) := \left\{ f \in \mathcal{S}'(\mathbb{R}^{d+1}) : \mathcal{F}^{-1}\left\{ (1 + \|\cdot\|^2)^{\frac{s}{2}} \hat{f} \right\} \in L^{p,q,\omega}(\mathbb{R}^{d+1}) \right\}.$$

The corresponding weighted norms are defined as

$$\|f\|_{L^s_{p,q,\omega}(\mathbb{R}^{d+1})} := \left\| \mathcal{F}^{-1}\left\{ (1 + \|\cdot\|^2)^{\frac{s}{2}} \hat{f} \right\} \right\|_{L^{p,q,\omega}(\mathbb{R}^{d+1})}.$$

Let $f \in L^s_{p,q,\omega}(\mathbb{R}^{d+1})$, then it can be written as the convolution $f = f_s * \varphi_s$, here $f_s := \mathcal{F}^{-1}\left\{ (1 + \|\cdot\|^2)^{\frac{s}{2}} \hat{f} \right\} \in L^{p,q,\omega}(\mathbb{R}^{d+1})$, and $\varphi_s := \mathcal{F}^{-1}\left\{ (1 + \|\cdot\|^2)^{-\frac{s}{2}} \right\}$. When $s > 0$, we call φ_s a Bessel potential kernel. By [104], there exists a constant C_s such that

$$\varphi_s(x) \leq C_s e^{-\pi\|x\|}, \quad \forall \|x\| > \frac{1}{\pi}.$$

and when $s > d + 1$, $\varphi_s \in L^\infty(\mathbb{R}^{d+1})$. The following lemma gives a stronger property of φ_s.

Lemma 7.25 *Let* $1 \leq p, q < +\infty$, $s > d + 1$, *and* $\omega(x) = (1 + \|x\|^2)^{\frac{\alpha}{2}}$ *for some* $\alpha \geq 0$. *Then, the Bessel potential kernel* $\varphi_s \in W(L^{1,1,\omega})(\mathbb{R}^{d+1})$.

Proof By the properties of φ_s, we have

$$\|\varphi_s\|_{W(L^{1,1,\omega})(\mathbb{R}^{d+1})}$$

$$= \|\varphi_s\|_{W(L^{1,\omega})(\mathbb{R}^{d+1})}$$

$$= \sum_{k \in \mathbb{Z}^{d+1}} \mathrm{esssup}_{x \in [0,1]^{d+1}} |\varphi_s(x+k)\omega(x+k)|$$

$$= \sum_{k \in \mathbb{Z}^{d+1}} \mathrm{esssup}_{x \in [0,1]^{d+1}} |\varphi_s(x+k)| \left(1 + \|x+k\|^2\right)^{\frac{\alpha}{2}}$$

$$\leq \sum_{\|k\| \leq \sqrt{d+1}+\frac{1}{\pi}} \mathrm{esssup}_{x \in [0,1]^{d+1}} \|\varphi_s\|_{L^\infty(\mathbb{R}^{d+1})} \left[1 + (\|x\| + \|k\|)^2\right]^{\frac{\alpha}{2}}$$

$$+ C_s \sum_{\|k\| > \sqrt{d+1}+\frac{1}{\pi}} \mathrm{esssup}_{x \in [0,1]^{d+1}} e^{-\pi(\|k\|-\|x\|)} \left[1 + (\|x\| + \|k\|)^2\right]^{\frac{\alpha}{2}}. \quad (7.50)$$

Let

$$A = \sum_{\|k\| \leq \sqrt{d+1}+\frac{1}{\pi}} \mathrm{esssup}_{x \in [0,1]^{d+1}} \|\varphi_s\|_{L^\infty(\mathbb{R}^{d+1})} \left[1 + (\|x\| + \|k\|)^2\right]^{\frac{\alpha}{2}}$$

and

$$B = C_s \sum_{\|k\| > \sqrt{d+1}+\frac{1}{\pi}} \mathrm{esssup}_{x \in [0,1]^{d+1}} e^{-\pi(\|k\|-\|x\|)} \left[1 + (\|x\| + \|k\|)^2\right]^{\frac{\alpha}{2}}.$$

Then, we can get

$$A \leq \|\varphi_s\|_{L^\infty(\mathbb{R}^{d+1})} \sum_{\|k\| \leq \sqrt{d+1}+\frac{1}{\pi}} \left[1 + (\sqrt{d+1} + \|k\|)^2\right]^{\frac{\alpha}{2}}$$

$$= \sharp \left\{ k \in \mathbb{Z}^{d+1} : \|k\| \leq \sqrt{d+1} + \frac{1}{\pi} \right\} \cdot C_s \|\varphi_s\|_{L^\infty(\mathbb{R}^{d+1})} \left[1 + (2\sqrt{d+1} + \frac{1}{\pi})^2\right]^{\frac{\alpha}{2}}$$

$$\leq +\infty,$$

and

$$B \leq C_s \sum_{k \in \mathbb{Z}^{d+1}} \mathrm{esssup}_{x \in [0,1]^{d+1}} e^{\pi \|x\|} \left[1 + (\|x\| + \|k\|)^2\right]^{\frac{\alpha}{2}} e^{-\pi \|k\|}$$

$$= C_s e^{\pi \sqrt{d+1}} \sum_{k \in \mathbb{Z}^{d+1}} \left[1 + (\sqrt{d+1} + \|k\|)^2\right]^{\frac{\alpha}{2}} e^{-\pi \|k\|}$$

$$= C_s e^{\pi \sqrt{d+1}} \sum_{k \in \mathbb{Z}^{d+1}} \left[1 + d + 1 + 2\sqrt{d+1}\|k\| + \|k\|^2\right]^{\frac{\alpha}{2}} e^{-\pi \|k\|}$$

$$\leq C_s e^{\pi \sqrt{d+1}} \sum_{k \in \mathbb{Z}^{d+1}} \left[1 + d + 1 + d + 1 + \|k\|^2 + \|k\|^2\right]^{\frac{\alpha}{2}} e^{-\pi \|k\|}$$

$$= C_s e^{\pi \sqrt{d+1}} \sum_{k \in \mathbb{Z}^{d+1}} \left[3 + 2d + 2\|k\|^2\right]^{\frac{\alpha}{2}} e^{-\pi \|k\|}$$

$$= C_s e^{\pi \sqrt{d+1}} \sum_{k \in \mathbb{Z}^{d+1}} 2^{\frac{\alpha}{2}} \left(\frac{3}{2} + d + \|k\|^2\right)^{\frac{\alpha}{2}} e^{-\pi \|k\|}$$

$$= C_{s,\alpha} e^{\pi \sqrt{d+1}} \sum_{k \in \mathbb{Z}^{d+1}} \left(\frac{3}{2} + d + \|k\|^2\right)^{\frac{\alpha}{2}} e^{-\pi \|k\|} \tag{7.51}$$

$$< +\infty,$$

where (7.50) follows from $\|x + k\| \geq \|k\| - \|x\|$. The sum in (7.51) is finite because the polynomial growth $\left(\frac{3}{2} + d + \|k\|^2\right)^{\frac{\alpha}{2}}$ will be dominated by the exponential decay $e^{-\pi \|k\|}$ when $\|k\|$ is large enough. Therefore, one completes the proof. $\square$

The following theorem 7.12 tells us that every element of $L^s_{p,q,1/\omega}(\mathbb{R}^{d+1})$ is an $L^{p,q,\omega}(\mathbb{R}^{d+1})$ function when $s > 0$ and the weighting function $\omega(x) = (1 + \|x\|^2)^{\frac{\alpha}{2}}$ for some $\alpha \geq 0$.

Theorem 7.12 *Suppose that $1 \leq p, q < +\infty$, $s > 0$, and $\omega(x) = (1 + \|x\|^2)^{\frac{\alpha}{2}}$ for some $\alpha \geq 0$. Then, $L^s_{p,q,1/\omega}(\mathbb{R}^{d+1}) \subset L^{p,q,1/\omega}(\mathbb{R}^{d+1})$.*

Proof We first show that $\varphi_s \in L^{1,1,\omega}(\mathbb{R}^{d+1}) = L^{1,\omega}(\mathbb{R}^{d+1})$. Indeed, from the listed properties of φ_s, we have that

$$\|\varphi_s\|_{L^{1,\omega}(\mathbb{R}^{d+1})} = \int_{\mathbb{R}^{d+1}} |\varphi_s(x)\omega(x)| dx$$

$$= \int_{\|x\| \leq \frac{1}{\pi}} |\omega(x)||\varphi_s(x)| dx + \int_{\|x\| > \frac{1}{\pi}} |\omega(x)||\varphi_s(x)| dx$$

$$\leq \|\varphi_s\|_{L^1(\mathbb{R}^{d+1})} \sup_{\|x\|\leq\frac{1}{\pi}} |\omega(x)| + C_s \int_{\|x\|>\frac{1}{\pi}} (1+\|x\|^2)^{\frac{\alpha}{2}} e^{-\pi\|x\|} dx$$

$$< \infty.$$

Thus, $\varphi_s \in L^{1,\omega}(\mathbb{R}^{d+1})$. Let $f \in L^s_{p,q,1/\omega}(\mathbb{R}^{d+1})$. Using duality, the submultiplicativity of ω, Hölder's, and Young's inequalities, we have that

$$\|f\|_{L^{p,q,1/\omega}(\mathbb{R}^{d+1})}$$

$$= \sup_{\|g\|_{L^{p',q',\omega}(\mathbb{R}^{d+1})}=1} \langle f, g \rangle$$

$$= \sup_{\|g\|_{L^{p',q',\omega}(\mathbb{R}^{d+1})}=1} \langle f_s * \varphi_s, g \rangle$$

$$= \sup_{\|g\|_{L^{p',q',\omega}(\mathbb{R}^{d+1})}=1} \left\langle \frac{f_s}{\omega}, \omega \cdot (g * \varphi_s) \right\rangle$$

$$= \sup_{\|g\|_{L^{p',q',\omega}(\mathbb{R}^{d+1})}=1} \int_{\mathbb{R}} \int_{\mathbb{R}^d} \frac{f_s(x_1, x_2)}{\omega(x_1, x_2)} [\omega(g * \varphi_s)](x_1, x_2) dx_1 dx_2$$

$$\leq \sup_{\|g\|_{L^{p',q',\omega}(\mathbb{R}^{d+1})}=1} \left[\int_{\mathbb{R}} \left(\int_{\mathbb{R}^d} \left| \frac{f_s(x_1, x_2)}{\omega(x_1, x_2)} \right|^q dx_2 \right)^{\frac{p}{q}} dx_1 \right]^{\frac{1}{p}}$$

$$\times \left[\int_{\mathbb{R}} \left(\int_{\mathbb{R}^d} |[\omega(g * \varphi_s)](x_1, x_2)|^{q'} dx_2 \right)^{\frac{p'}{q'}} dx_1 \right]^{\frac{1}{p'}}$$

$$= \sup_{\|g\|_{L^{p',q',\omega}(\mathbb{R}^{d+1})}=1} \left\| \frac{f_s}{\omega} \right\|_{L^{p,q}(\mathbb{R}^{d+1})} \|\omega(g * \varphi_s)\|_{L^{p',q'}(\mathbb{R}^{d+1})}$$

$$= \sup_{\|g\|_{L^{p',q',\omega}(\mathbb{R}^{d+1})}=1} \|f_s\|_{L^{p,q,1/\omega}(\mathbb{R}^{d+1})} \|g * \varphi_s\|_{L^{p',q',\omega}(\mathbb{R}^{d+1})}$$

$$= \sup_{\|g\|_{L^{p',q',\omega}(\mathbb{R}^{d+1})}=1} \|f_s\|_{L^{p,q,1/\omega}(\mathbb{R}^{d+1})} C_\alpha \|g\|_{L^{p',q',\omega}(\mathbb{R}^{d+1})} \|\varphi_s\|_{L^{1,1,\omega}(\mathbb{R}^{d+1})}$$

$$= C_\alpha \|f_s\|_{L^{p,q,1/\omega}(\mathbb{R}^{d+1})} \|\varphi_s\|_{L^{1,1,\omega}(\mathbb{R}^{d+1})}$$

$$< \infty,$$

which implies that $f \in L^{p,q,1/\omega}(\mathbb{R}^{d+1})$. Therefore, one completes the proof. □

Now, we give the main result of this section.

Theorem 7.13 *Let* $1 \leq p, q < +\infty$, $s > d$, *and* $\omega(x) = (1 + \|x\|^2)^{\frac{\alpha}{2}}$ *for some* $\alpha \geq 0$. *Then, the sampling operator* $f \mapsto f|_{\mathbb{Z}^{d+1}}$ *is bounded from* $L^s_{p,q,1/\omega}(\mathbb{R}^{d+1}) \cap C(\mathbb{R}^{d+1})$ *to* $\ell^{p,q,1/\omega}(\mathbb{Z}^{d+1})$, *i.e., there exists a constant* $C_{\alpha,s}$ *depending only on* α *and* s *such that*

$$\|f\|_{\ell^{p,q,1/\omega}(\mathbb{Z}^{d+1})} \leq C_{\alpha,s} \|f\|_{L^s_{p,q,1/\omega}(\mathbb{R}^{d+1})}, \quad \forall f \in L^s_{p,q,1/\omega}(\mathbb{R}^{d+1}) \cap C(\mathbb{R}^{d+1}).$$

Proof According to Lemma 7.25, we can get $f \overset{a.e.}{=} f_s * \varphi_s$, where $f_s \in L^{p,q,1/\omega}(\mathbb{R}^{d+1})$ and $\varphi_s \in W(L^{1,1,\omega})(\mathbb{R}^{d+1})$. By Lemma 7.20, we know that the convolution $f_s * \varphi_s$ is continuous everywhere. Since f is continuous, we deduce that $f = f_s * \varphi_s$ everywhere. Now we can write $f(k_1, k_2) = (f_s * \varphi_s)(k_1, k_2)$, for all $k_1 \in \mathbb{Z}, k_2 \in \mathbb{Z}^{d+1}$. Using Lemma 7.22, we have that

$$\|f\|_{\ell^{p,q,1/\omega}(\mathbb{Z}^{d+1})} \leq C_\alpha \|\varphi_s\|_{W(L^{1,1,\omega})(\mathbb{R}^{d+1})} \|f_s\|_{L^{p,q,1/\omega}(\mathbb{R}^{d+1})}$$
$$= \underbrace{C_\alpha \|\varphi_s\|_{W(L^{1,1,\omega})(\mathbb{R}^{d+1})}}_{C_{\alpha,s}} \|f\|_{L^s_{p,q,1/\omega}(\mathbb{R}^{d+1})}.$$

Therefore, one completes the proof. $\square$

7.5.2.2 Spline Interpolation

We consider the spline interpolation about a function $f(x)$ from its samples $\{f(k)\}_{k \in \mathbb{Z}^{d+1}}$. Let

$$f_{int}(x) = \sum_{k \in \mathbb{Z}^{d+1}} c(k)\varphi(x - k),$$

where φ is the chosen kernel (such as a B-spline) with desirable properties (such as localization, smoothness, etc.) and the coefficients

$$c(k) = \sum_{n \in \mathbb{Z}^{d+1}} f(n)h(k - n).$$

Here $\{h(k)\}_{k \in \mathbb{Z}^{d+1}}$ is a digital interpolation filter whose discrete-domain Fourier transform is given by

$$\hat{h}(\xi) := \frac{1}{\sum_{k \in \mathbb{Z}^{d+1}} \varphi(k)e^{-2\pi j\langle \xi, k \rangle}}, \tag{7.52}$$

where $\sum_{k \in \mathbb{Z}^{d+1}} \varphi(k)e^{-2\pi j\langle \xi, k \rangle} \neq 0$, for almost all $\xi \in \mathbb{R}^{d+1}$. Let φ_{int} be the interpolation which given by

$$\varphi_{int} = \sum_{k \in \mathbb{Z}^{d+1}} h(k)\varphi(\cdot - k).$$

Then φ_{int} satisfies the interpolation condition

$$\varphi_{int}(k) = \delta(k), \quad \forall k \in \mathbb{Z}^{d+1},$$

and

$$f_{int}(x) = \sum_{k \in \mathbb{Z}^{d+1}} f(k)\varphi_{int}(x - k),$$

hence we can get $f_{int}(k) = f(k)$.

When $f \in L^s_{p,q,1/\omega}(\mathbb{R}^{d+1}) \cap C(\mathbb{R}^{d+1})$ and φ has desirable properties, we get the following result.

Theorem 7.14 *Let* $1 \leq p, q < +\infty$, $s > d$, *and* $\omega(x) = (1 + \|x\|^2)^{\frac{\alpha}{2}}$ *for some* $\alpha \geq 0$. *Assume that* $\varphi \in W(L^{1,1,\omega})(\mathbb{R}^{d+1})$, $\varphi(k) \in \ell^{1,1,\omega}(\mathbb{Z}^{d+1})$, *and that* $\sum_{k \in \mathbb{Z}^{d+1}} \varphi(k)e^{-2\pi j \langle \xi, k \rangle}$ *is nonzero for almost all* ξ. *If* $f \in L^s_{p,q,1/\omega}(\mathbb{R}^{d+1}) \cap C(\mathbb{R}^{d+1})$, *then the interpolated function* $f_{int}(x_1, x_2) = \sum_{k_1 \in \mathbb{Z}} \sum_{k_2 \in \mathbb{Z}^d} c(k_1, k_2)\varphi(x_1 - k_1, x_2 - k_2)$ *is included in* $L^{p,q,1/\omega}(\mathbb{R}^{d+1})$, *and we have that*

$$C_{\alpha,\varphi} \|f_{int}\|_{L^{p,q,1/\omega}(\mathbb{R}^{d+1})} \leq \|f[\cdot]\|_{\ell^{p,q,1/\omega}(\mathbb{Z}^{d+1})} \leq C_{\alpha,s} \|f\|_{L^s_{p,q,1/\omega}(\mathbb{R}^{d+1})}.$$

Proof On the one hand, since the weight $\omega(x) = (1 + \|x\|^2)^{\frac{\alpha}{2}}$ is submultiplicative and satisfies the GRS condition, $\varphi(k) \in \ell^{1,1,\omega}(\mathbb{Z}^{d+1})$ and $\hat{h}(\xi) = 1/\sum_{k \in \mathbb{Z}^{d+1}} \varphi(k)e^{-2\pi j \langle \xi, k \rangle}$, it follows from the weighted version of Wiener's lemma [103] that the interpolation filter h is a sequence in $\ell^{1,1,\omega}(\mathbb{Z}^{d+1})$. Combining this with Lemma 7.21 and $\varphi_{int} = \sum_{k \in \mathbb{Z}^{d+1}} h(k)\varphi(\cdot - k)$, we can get the interpolant φ_{int} also belongs to $W(L^{1,1,\omega})(\mathbb{R}^{d+1})$. According to Lemma 7.23 and $f_{int}(x) = \sum_{k \in \mathbb{Z}^{d+1}} f(k)\varphi_{int}(x - k)$, one has

$$\|f_{int}\|_{L^{p,q,1/\omega}(\mathbb{R}^{d+1})} \leq \underbrace{C_\alpha \|\varphi_{int}\|_{W(L^{1,1,\omega})(\mathbb{R}^{d+1})}}_{1/C_{\alpha,\varphi}} \|f[\cdot]\|_{\ell^{p,q,1/\omega}(\mathbb{Z}^{d+1})},$$

which proves the left-hand-side inequality. On the other hand, the right-hand-side inequality follows directly from Theorem 7.13.

Therefore, one completes the proof.   □

References

1. S. T. Ali, J. P. Antoine, J. P. Gazeau, *Coherent States, Wavelets, and Their Generalizations.* Theoretical and mathematical physics (Springer, New York, 2000)
2. A. Aldroubi, H. Feichtinger, Exact iterative reconstruction algorithm for multivariate irregularly sampled functions in spline-like spaces: The L^p theory. Proc. Amer. Math. Soc. **126**(9), 2677–2686 (1998)
3. A. Aldroubi, K. Gröchenig, Beurling-Landau-type theorems for non-uniform sampling in shift invariant spline spaces. J. Fourier Anal. Appl. **6**(1), 93–103(2000)
4. A. Aldroubi, K. Gröchenig, Nonuniform sampling and reconstruction in shift-inviant spaces. SIAM Rev. **43**(4), 585–620(2001)
5. F. Albiac, N. Kalton, *Topics in Banach Space Theory.* Graduate Texts in Mathematics 233(Springer Cham, 2016)
6. A. Aldroubi, Non-uniform weighted average sampling and reconstruction in shift-invariant and wavelet spaces. Appl. Comput. Harmon. Anal. **13**(2), 151–161(2002)
7. Z. Amiri, R. A. Kamyabi-Gol, Distance between continuous frames in Hilbert space. J. Korean. Math. Soc. **54**, 215-225(2017)
8. R. Alaifari, M. Wellershoff, Uniqueness of STFT phase retrieval for bandlimited functions. Appl. Comput. Harmon. Anal. **50**, 34-48(2021)
9. R. Alaifari, F. Bartolucci, M. Wellershoff, Phase retrieval of bandlimited functions for the wavelet transform. Appl. Comput. Harmon. Anal. **64**, 102–117(2023)
10. A. Aldroubi, M. Unser, Sampling procedure in function spaces and asymptotic equivalence with Shannon's sampling theory. Numer. Funct. Anal. Optim. **15**(1), 1–21(1994)
11. S. Banach, *Théorie des opérations linéaires* (Instytut Matematyczny Polskiej Akademi Nauk, 1932)
12. R. Balan, Equivalence relations and distances between Hilbert frames. Proc. Amer. Math. Soc. **127**, 2353–2366 (1999)
13. R. Balian, Un principe d'incertitude fort en théorie du signal ou en mécanique quantique. C. R. Acad. Sci., Paris, Sér. **292**(20), 1357–1361 (1981)
14. B. V. Rajarama Bhat, T. Bhattacharyya, *Dilations, Completely Positive Maps and Geometry.* Texts and Readings in Mathematics 84 (Springer, Singapore, 2024)
15. R. Balan, B. G. Bodmann, P. G. Casazza, D. Edidin, Fast algorithms for signal reconstruction without phase. In: *Wavelets XII* (Vol. 6701, pp. 712–722) (Proceedings of SPIE, 2007)
16. R. Balan, B. G. Bodmann, P. G. Casazza, D. Edidin, Painless reconstruction from magnitudes of frame vectors. J. Fourier Anal. Appl. **15**, 488–501 (2009)
17. T. Banerjee, B. R. Bhat, M. Kumar, C^*-extreme points of positive operator valued measures and unital completely positive maps. Commun. Math. Phys. **388**, 1235–1280 (2021)

18. S. Botelho-Andrade, P. G. Casazza, D. Cheng, J. Haas, T. T. Tran, The quantum detection Problem: A survey. Springer Proc. Math. Stat. **255**, 337–352 (2017)

19. T. Bemrose, P. G. Casazza, D. Cheng, J. Haas, H. Van Nguyen, Computing the distance between frames and between subspaces of a Hilbert space. In: *Frames and Other Bases in Abstract and Function Spaces: Novel Methods in Harmonic Analysis* (Vol. 1, pp. 81–99) (Birkhäuser, Cham, 2017)

20. S. Botelho-Andrade, P. G. Casazza, D. Cheng, T. T. Tran, The solution to the frame quantum detection problem. J. Fourier. Anal. Appl. **25**, 2268–2323 (2019)

21. R. Balan, P. G. Casazza, D. Edidin, On signal reconstruction from the absolute value of the frame coefficients. In: *Wavelets XI* (Vol. 5914, pp. 362–369) (Proceedings of SPIE, 2005)

22. R. Balan, P. Casazza, D. Edidin, On signal reconstruction without phase. Appl. Comput. Harmon. Anal. **20**(3), 345–356 (2006)

23. A. Benedek, A. P. Calderón, R. Panzone, Convolution operators on Banach space valued functions. Proc. Nat. Acad. Sci. USA. **48**, 356–365 (1962)

24. R. Balan, P. G. Casazza, D. Edidin, Equivalence of reconstruction from the absolute value of the frame coefficients to a sparse representation problem. IEEE Signal Process. Lett. **14**(5), 341–345 (2007)

25. C. de Boor, R. DeVore, A. Ron, The structure of finitely generated shift-invariant spaces in $L^2(\mathbb{R}^d)$. J. Funct. Anal. **119**(1), 37–78 (1994)

26. B. Bekka, Square integrable representations, von Neumann algebras and an application to Gabor analysis. J. Fourier Anal. Appl. **10**, 325–349 (2004)

27. J. J. Benedetto, P. J. S. G. Ferreira, *Modern Sampling Theory, Mathematics and Applications.* Applied and Numerical Harmonic Analysis (ANHA) (Birkhäuser, Boston, 2001)

28. J. J. Benedetto, M. Fickus, Finite normalized tight frames. Adv. Comput. Math. **18**, 357–385 (2003)

29. I. Bojarovska, A. Flinth, Phase retrieval from Gabor measurements. J. Fourier Anal. Appl. **22**, 542–567 (2016)

30. H. Bercovici, C. Foias, L. Kerchy, B. Sz.-Nagy, *Harmonic Analysis of Operators on Hilbert Space.* Universitext (Springer, Berlin, 2010)

31. K. Beanland, D. Freeman, R. Liu. Upper and lower estimates for Schauder frames and atomic decompositions. Fund. Math. **231**, 161–188 (2015)

32. M. Bownik, N. Kaiblinger, Minimal generator sets for finitely generated shift-invariant spaces of $L^2(\mathbb{R})$. J. Math. Anal. Appl. **313**(1), 342–352 (2006)

33. J. J. Benedetto, A. Kebo, The role of frame force in quantum detection. J. Fourier Anal. Appl. **14**, 443–474 (2008)

34. P. Busch, P. Lahti, J. P. Pellonpää, K. Ylinen, *Quantum Measurement.* Theoretical and Mathematical Physics (Springer, Berlin, 2016)

35. M. Bownik, The structure of shift-invariant spaces of $L^2(\mathbb{R}^n)$. J. Funct. Anal. **177**(2), 282–309 (2000)

36. A. Benedek, R. Panzone, The space L^p with mixed norm. Duke Math. J. **28**, 301–324 (1961)

37. M. Bownik, K. Ross, The structure of translation-invariant spaces on locally compact Abelian groups. J. Fourier Anal. Appl. **21**(4), 849–884 (2015)

38. J. W. Bunce, The similarity problem for representations of C^*-algebras. Proc. Am. Math. Soc. **81**(1981), 409–414 (1981)

39. A. Buchholz, Norm of convolution by operator-valued functions on free groups. Proc. Amer. Math. Soc. **127**, 1671–1682 (1999)

40. L. J. Bunce, J. D. M. Wright, The Mackey-Gleason problem for vector measures on projections in von Neumann algebras. J. London Math. Soc. **49**, 133–149 (1994)

41. A. Bhandari, A. Zayed, Shift-invariant and sampling spaces associated with the fractional Fourier transform domain, IEEE Trans. Signal Process. **60**(3), 1627–1637 (2012)

42. P. G. Casazza, Finite dimensional decompositions in Banach spaces, Cont. Math. **52**, 1–31 (1986)

43. P. G. Casazza, Approximation properties. In: *Handbook of the Geometry of Banach Spaces* (vol 1, pp. 271–316) (Elsevier Science B.V., 2021)

44. J. Cahill, P. G. Casazza, I. Daubechies, Phase retrieval in infinite-dimensional Hilbert spaces. Trans. Am. Math. Soc. Ser. B. **3**(3), 63–76 (2016)

45. P. G. Casazza, S. J. Dilworth, E. Odell, Th. Schlumprecht, A. Zsák, Coeffcient quantization for frames in Banach spaces. J. Math. Anal. Appl. **348**, 66–86 (2008)

46. C. Cheng, J. Fu, On the rings of projective characters of Abelian groups and Dihedral groups, preprint (2016)

47. P. Casazza, M. Fickus, J. Tremain, E. Weber, The Kadison-Singer problem in mathematics and engineering: a detailed account. In: *Operator Theory, Operator Algebras, and Applications (Contemp. Math.)* (Vol. 414, pp. 299–355) (American Mathematical Society, 2006)

48. M. D. Choi, Completely positive linear maps on complex matrices. Linear. Algebra. Appl. **10**, 285–290 (1975)

49. E. Christensen, On non-selfadjoint representations of operator algebras. Amer. J. Math., **103**, 817–834 (1981)

50. C. Cheng, A character theory for projective representations of finite groups. Linear Algebra Appl. **469**, 230–242 (2015)

51. O. Christensen, *An Introduction to Frames and Riesz Bases, Second Edition*. Applied and Numerical Harmonic Analysis (ANHA) (Birkhäuser Cham, 2016)

52. P. Casazza, D. Han, D. Larson, Frames for Banach spaces In: *The Functional and Harmonic Analysis of Wavelets and Frames (Contemp. Math.)* (Vol. 247, pp. 149–182) (American Mathematical Society, 1999, 247)

53. P. G. Casazza, N. J. Kalton, Unconditional bases and unconditional finite-dimensional decompositions in Banach spaces. Israel J. Math. **95**, 349–373 (1996)

54. P. G. Casazza, G. Kutyniok, (Eds.), *Finite Frames, Theory and Applications*. Applied and Numerical Harmonic Analysis (ANHA) (Springer Science + Business Media, New York, 2013)

55. D. Carando, S. Lassalle, Duality, reflexivity and atomic decompositions in Banach spaces. Studia Math. **191**, 67–80 (2009)

56. D. Carando, S. Lassalle, P. Schmidberg, The reconstruction formula for Banach frames and duality. J. Approx. Theory. **163**, 640–651 (2011)

57. A. Connes, Classification of injective factors. Cases $II_1, II_\infty, III_\lambda, \lambda \neq 1$. Ann. Math. **104**, 73–115 (1976)

58. P. G. Casazza, J. C. Tremain,The Kadison-Singer problem in mathematics and engineering. Proc. Natl. Acad. Sci. U. S. A. **103**, 2032–2039 (2006)

59. A. M. Davie, The approximation problem for Banach spaces, Bull. London Math. Soc. **5**, 261–266 (1973)

60. I. Daubechies, *Ten Lectures on Wavelets* (Society for industrial and applied mathematics (SIAM), Philadelphia, 1992)

61. K. R. Davidson, A. Dor-On, O. M. Shalit, B. Solel, Dilations, Inclusions of Matrix Convex Sets, and Completely Positive Maps. Int. Math. Res. Not. **13**, 4069–4130 (2017)

62. D. L. Donoho, M. Elad, Optimally sparse representations in general non orthogonal dictionaries via ℓ_1 minimization. Proc. Natl. Acad. Sci. USA, **100**, 2197–2202 (2003)

63. I. Daubechies, A. Grossmann, Frames in the Bargmann space of entire functions. Comm. Pure Appl. Math. **41**, 151–164 (1988)

64. I. Daubechies, A. Grossmann, Y. Meyer, Painless nonorthogonal expansions. J. Math. Phys. **27**, 1271–1283 (1986)

65. D. Dutkay, D. Han, D. Larson, A duality principle for groups, J. Funct. Anal. **257**, 1133–1143 (2009)

66. J. Diestel, H. Jarchow, A. Tonge, *Absolutely Summing Operators*. Cambridge Studies in Advanced Mathematics 43 (Cambridge University Press, Cambridge, 1995)

67. I. Daubechies, H. Landau, Z. Landau, Gabor time-frequency lattices and the Wexler-Raz identity. J. Fourier Anal. Appl. **1**, 437–478 (1995)

68. R. G. Douglas, On majorization, factorization, and range inclusion of operators on Hilbert space. Proc. Am. Math. Soc. **17**, 413–415 (1966)

69. R. J. Duffin, A. C. Schaeffer, A class of nonharmonic Fourier series. Trans. Am. Math. Soc. **72**(2), 341–66 (1952)

70. X. Dai, Q. Sun, *The abc-Problem for Gabor Systems*. Mem. Am. Math. Soc. **244**, 1152 (2016)

71. J. Diestel, J. J. Uhl, Jr, *Vector Measures*. Mathematical Surveys and Monographs 15 (American Mathematical Society, 1977)

72. J. Duoandikoetxea, *Fourier Analysis* (American Mathematical Society, Rhode Island, 2000)

73. K. Dykema, Interpolated free group factors. Pacific J. Math. **163**, 123–135 (1994)

74. Y. C. Eldar, G. D. Forney, Optimal tight frames and quantum measurement. IEEE Trans. Inform. Theory. **48**, 599–610 (2002)

75. J. Eisner, D. Freeman, Continuous Schauder frames for Banach spaces. J Fourier Anal. Appl. **26**, 66 (2020)

76. Y. C. Eldar, N. Hammen, D. Mixon, Recent advances in phase retrieval. IEEE Signal Process. Mag. **33**(5), 158–162 (2016)

77. P. Enflo, A counterexample to the approximation problem in Banach spaces. Acta Math. **130**, 309–317 (1973)

78. E. Effros, N. Ozawa, Z. Ruan, On injectivity and nuclearity for operator spaces, Duke Math. J. **110**, 489–521 (2001)

79. E. Effros, Z. Ruan, *Operator spaces*. London Mathematical Society Monographs New Series 23 (Oxford University Press, 2000)

80. D. L. Fernandez, Vector-valued singular integral operators on L^p-spaces with mixed norms and applications. Pac. J. Math. **129**(2), 257–275 (1987)

81. H. G. Feichtinger, K. Gröchenig, A unified approach to atomic decompositions via integrable group representations. In: *Function Spaces and Applications*. Lecture Notes in Mathematics (Vol. 1302, pp. 52–73) (Springer, Berlin, 1988)

82. H.G. Feichtinger, K. Grochenig, Banach spaces related to integrable group representations and their atomic decompositions, I. J. Funct. Anal. **86**, 305–340 (1989)

83. H. G. Feichtinger, K. Grochenig, Banach spaces related to integrable group representations and their atomic decompositions, II. Monatsh. fur Math. **108**, 129–148 (1989)

84. M. Fabian, P. Habala, P. Hájek, V. M. Santalucía, J. Pelant, V. Zizler, *Functional Analysis and Infinite-Dimensional Geometry*. CMS Books in Mathematics 8 (Springer-Verlag, New York, 2001)

85. T. Figiel, W. B. Johnson, The approximation property does not imply the bounded approximation property. Proc. Amer. Math. Soc. **41**, 197–200 (1973)

86. G. B. Folland, *Real Analysis: Modern Techniques and Their Applications*. Pure and Applied Mathematics: A Wiley Series of Texts, Monographs and Tracts (John Wiley & Sons, New York, 1984)

87. J. L. Francia, F. J. Ruiz , J. L. Torrea, Calderón–Zygmund theory for operator-valued kernels. Adv. Math. **62**(1), 7–48 (1986)

88. D. Freeman, D. Speegle, The discretization problem for continuous frames. Adv. Math. **345**, 784–813 (2019)

89. D. Gabor, Theory of communications. J. Inst. Elec. Eng. **93**, 429–457 (1946)
90. K. Gröchenig, C. Heil, Modulation spaces and pseudodifferential operators. Integral Equ. Oper. Theory. **34**, 439–457 (1999)
91. J-P. Gabardo, D. Han, Subspace Weyl-Heisenberg frames. J. Fourier Anal. Appl. **7**, 419–433 (2001)
92. J-P. Gabardo, D. Han, Frame representations for group-like unitary operator systems, J. Oper. Theory. **49**, 223–244 (2003)
93. J-P. Gabardo, D. Han, Frames associated with measurable spaces, Adv. Comput. Math. **18**, 127–14 (2003)
94. G. Godefroy, N. J. Kalton, Lipschitz-free Banach spaces. Studia Math. **159**, 121–141 (2003)
95. H. G. Feichtinger, W. Kozek, F. Luef, Gabor analysis over finite Abelian groups. Appl. Comput. Harmon. Anal. **26**, 230–248 (2009)
96. A. Goldberger, S. Kang, K. A. Okoudjou, Towards a classification of incomplete Gabor POVMs in $\mathbb{C}^d$. Linear Multilinear Algebra. **70**(22), 7536–7557 (2022)
97. A. M. Gleason, Measures on the closed subspaces of a Hilbert space. In: *The Logico-Algebraic Approach to Quantum Mechanics*(Vol. 1, pp. 123–133) (Springer Netherlands, Dordrecht, 1975)
98. V. P. Gupta, P. Mandayam and V. S. Sunder, *The Functional Analysis of Quantum Information Theory* Lecture Notes in Physics 902 (Springer International Publishing, Switzerland, 2015)
99. G. Godefroy, N. Ozawa, Free Banach spaces and the approximation properties. Proc. Amer. Math. Soc. **142**(5), 1681–1687 (2014)
100. A. Grothendieck, *Produits tensoriels topologiques et espaces nucléaires*. Mem. Amer. Math. Soc. **16**, (1955)
101. K. Gröchenig, Describing functions: atomic decompositions versus frames, Monatsh. Math. **112**, 1–42 (1991)
102. K. Gröchenig, *Foundations of Time-Frequency Analysis*. Applied and Numerical Harmonic Analysis (ANHA) (Springer Science + Business Media, New York, 2001)
103. K. Gröchenig, Weight functions in time-frequency analysis. Pseudodifferential Operators: Partial Differential Equations and Time-Frequency Analysis. **52**, 343–366 (2007)
104. L. Grafakos, *Modern Fourier Analysis*. Graduate Texts in Mathematics 250 (Springer, New York, 2008)
105. P. Grohs, M. Rathmair, Stable Gabor phase retrieval and spectral clustering. Commun. Pure Appl. Math. **72**(5), 981–1043 (2019)
106. S. Goh, A. Ron, Z. Shen, *Gabor and Wavelet Frames*. Lecture notes series 10 (World Scientific, Singapore, 2007)
107. Q. Gu, D. Han, When a characteristic function generates a Gabor frame. Appl. Comp. Harm. Anal. **24**(3), 290–309 (2008)
108. K. Gröchenig, J. Stöckler, Gabor frames and totally positive functions. Duke Math. J. **162**(6), 1003–1031 (2013)
109. U. Haagerup, An example of a nonnuclear C^*-algebra, which has the metric approximation property, Invent. Math. **50**, 279–293 (1978)
110. D. W. Hadwin, Dilations and Hahn decompositions for linear maps. Canad. J. Math. **33**, 826–839 (1981)
111. U. Haagerup, Solutions of the similarity problem for cyclic representations of C^*-algebras, Ann. Math. **118**, 215–240 (1983)
112. D. Han, A note on the density theorem for projective unitary representations. Proc. Amer. Math. Soc. **145**, 1739–1745 (2017)
113. C. Heil, An introduction to weighted Wiener amalgams. In: *Wavelets and Their Applications* (pp. 183–216) (Allied Publishers, New Delhi, 2003)

114. C. Heil, *Harmonic Analysis and Applications*. Applied and Numerical Harmonic Analysis (Birkhäuser, Boston, 2006)

115. C. Heil, History and evolution of the Density Theorem for Gabor frames. J. Fourier Anal. Appl. **13**, 113-166(2007)

116. C. Heil, The density theorem and the homogeneous approximation property for Gabor frames. In: *Representations, Wavelets, and Frames: A Celebration of the Mathematical Work of Lawrence Baggett*. (Applied and Numerical Harmonic Analysis, pp.71–102) (Birkhäuser, Boston, 2008)

117. C. Heil, *A Basis Theory Primer, Expanded Edition*. Applied and Numerical Harmonic Analysis (Springer Science + Business Media, New York, 2011)

118. D. Han, Q. Hu, R. Liu. Injective continuous frames and quantum detections. Banach J. Math. Anal. **15**, 12 (2021)

119. D. Han, Q. Hu, R. Liu, Generalized Choi-Kraus dilations of linear maps between matrix algebras. Oper. Matrices. **16**, 1219–1237 (2022)

120. D. Han, Q. Hu, D. R. Larson, R. Liu, Dilations for operator-valued quantum measures. Adv. Math. **438**, 109476 (2024)

121. D. Han, Q. Hu, R. Liu, H. Wang, Quantum injectivity of multi-window Gabor frames in finite dimensions. Ann. Funct. Anal. **13**, 4 (2022)

122. D. Han, D. R. Larson, *Frames, Bases and Group Representations*. Mem. Am. Math. Soc. **147**, 697 (2000)

123. D. Han, T. Juste, Y. Li, W. Sun, Frame phase-retrievability and exact phase-retrievable frames. J. Fourier Anal. Appl. **25**(6), 3154–3173 (2019)

124. D. Han, D. R. Larson, R. Liu, Dilations of operator-valued measures with bounded *p*-variations and framings on Banach spaces. J. Funct. Anal. **274**, 1466–1490 (2018)

125. D. Han, D. Larson, B. Liu, R. Liu, *Operator-Valued Measures, Dilations, and the Theory of Frames*. Mem. Am. Math. Soc. **229**, 1075 (2014)

126. D. Han, D. R. Larson, R. Liu, B. Liu, Dilations for systems of imprimitivity acting on Banach spaces. J. Funct. Anal. **266**, 6914–6937 (2014)

127. D. Han, D. R. Larson, B. Liu, R. Liu, Dilations of frames, operator-valued measures and bounded linear maps. In: *Operator Methods in Wavelets, Tilings, and Frames (Contemp. Math.)* (Vol. 626, pp. 33–53) (American Mathematical Society, 2014)

128. D. Han, D. R. Larson, R. Liu, B. Liu, Structural properties of homomorphism dilation systems. Chin. Ann. Math. Ser. B. **41**, 585–600 (2020)

129. J. W. Helton, I. Klep, S. McCullough, M. Schweighofer, *Dilations, Linear Matrix Inequalities, the Matrix Cube Problem and Beta Distributions*. Mem. Amer. Math. Soc. **257**, 1232 (2019)

130. J. Holub, Pre-frame operators, Besselian frame and near-Riesz bases in Hilbert spaces. Proc. Amer. Math. Soc. **122**, 779–785 (1994)

131. A. S. Holevo, *Statistical Structure of Quantum Theory*. Lecture Notes in Physics Monographs 67 (Springer Berlin, Heidelberg, 2003)

132. C. Heil, D. Walnut, Continuous and discrete wavelet transforms. SIAM Rev. **31**, 628–666 (1989)

133. D. Han, Y. Wang, Lattice tiling and the Weyl-Heisenberg frames, Geom. Funct. Anal. **11**, 742–758 (2001)

134. D. Han, Y. Wang, The existence of Gabor bases. In: *Wavelets, Frames and Operator Theory(Contemp. Math.)* (Vol. 345, pp. 183–192) (American Mathematical Society, 2004)

135. R. C. James, Bases and reflexivity of Banach spaces. Ann. Math. **52**, 518–527(1950)

136. A. Janssen, Duality and biorthogonality for Weyl-Heisenberg frames. J. Fourier Anal. Appl. **1**, 403–436 (1995)

137. A. Janssen, Representations of Gabor frame operators. In: *Twentieth Century Harmonic Analysis - A Celebration*. NATO Science Series II: Mathematics, Physics and Chemistry (Vol. 33, pp. 73–101) (Kluwer Academic, Dordrecht, 2001)

138. A. Janssen, Zak transforms with few zeros and the tie. In: *Advances in Gabor Analysis*. (Applied and Numerical Harmonic Analysis, pp. 31–70) (Birkhäuser, Boston, 2003)

139. A.J. Jerri, The Shannon sampling theorem-its various extensions and applications: a tutorial review. Proc. IEEE. **65**(11), 1565–1596 (1977)

140. R. Q. Jia, Shift-invariant spaces and linear operator equations. Israel Math. J. **103**(1), 259–288 (1998)

141. R. Q. Jia, Stability of the shifts of a finite number of functions, J. Approx. Theory. **95**(2), 194–202 (1998)

142. R. Q. Jia, C. A. Micchelli, Using the refinement equations for the construction of pre-wavelets II: Powers of two. In: Curves and Surfaces (pp. 209–246) (Academic Press, New York, 1991)

143. R. Q. Jia, C. A. Micchelli, On linear independence for integer translates of a finite number of functions. Proc. Edinburgh Math. Soc. **36**, 69–85 (1992)

144. M. Junge, N. J. Nielsen, Z. J. Ruan, Q. Xu, $\mathscr{COL}_p$ spaces – the local structure of non-commutative L_p spaces. Adv. Math. **187**, 257–319 (2004)

145. M. Junge, Z. Ruan, Approximation properties for noncommutative L_p-spaces associated with discrete groups. Duke Math. J. **117**, 313–341 (2003)

146. W. B. Johnson, H. P. Rosenthal, M. Zippin, On bases, finite dimensional decompositions and weaker structures in Banach spaces. Israel J. Math. **9**, 488–506 (1971)

147. R. V. Kadison, On the orthogonalization of operator representations, Amer. J. of Math. **77**, 600–620 (1955)

148. G. Karpilovsky, *The Schur Multiplier*. London Mathematical Society Monographs (Oxford University Press, 1987)

149. Y. Katznelson, *An Introduction to Harmonic Analysis* (Cambridge University Press, Cambridge, 1968)

150. H. Kim, R. Kim, J. Lim, Characterization of the closedness of the sum of two shift-invariant spaces. Appl. Comput. Harmon. Anal. **320**(1), 381–395 (2006)

151. V. Kaftal, P. W. Ng, S. Zhang, Commutators and linear spans of projections in certain finite C^*-algebras. J. Funct. Anal. **266**, 1883–1912 (2014)

152. J. Lindenstrauss, *Extension of Compact Operators*. Mem. Amer. Math. Soc. **48**, (1964)

153. R. Liu, On shrinking and boundedly complete Schauder frames of Banach spaces. J. Math. Anal. Appl. **365**, 385–398 (2010)

154. Li L, Juste T, Brennan J, Cheng C, Han D. Phase retrievable projective representation frames for finite abelian groups. J. Fourier Anal. Appl., 2017, 25.

155. F. Low, Complete sets of wave packets. In: *A Passion for Physics Essays in Honor of Geoffrey Chew* (pp. 17–22) (World Scientific, Singapore, 1985)

156. R. Liu, Z. Ruan, Cb-frames for operator spaces. J. Funct. Anal. **270**(11), 4280–4296 (2016)

157. R. Liu, J. Shen, B. Zheng, Operators with the Lipschitz bounded approximation property. Sci. China Math. **66**, 1545–1554 (2023)

158. Yu. Lyubarskii, Frames in the Bargmann space of entire functions. In: *Entire and Subharmonic Functions* (Vol. 11, pp. 167–180) (American Mathematical Society, 1992)

159. R. Liu, B. Zheng, A characterization of Schauder frames which are near-Schauder bases. J. Fourier Anal. Appl. 16, 791–803 (2010)

160. G. W. Mackey, Imprimitivity for representations of locally compact groups. Proc. Natl. Acad. Sci. U. S. A. **35**, 537–545 (1949)

161. G. W. Mackey, *Unitary Group Representations in Physics, Probability, and Number Theory*. Mathematics Lecture Notes Series 55 (Benjamin-Cummings Publishing Company, 1989)

162. F. Marvasti, *Nonuniform Sampling, Theory and Practice*. Information Technology: Transmission, Processing and Storage (PSTE) (Kluwer Academic/Plenum Publishers, New York, 2001)

163. Y. Meyer, *Wavelets and Operators*. (Translated by D. H. Salinger) Cambridge Studies in Advanced Mathematics 37 (Cambridge University Press, Cambridge, 1992)

164. F. J. Murray, J. von Neumann, On rings of operators, IV. Ann. Math. **44**, 716–808 (1943)

165. M. A. Nielsen, I. L. Chuang,, *Quantum Computation and Quantum Information: 10th Anniversary Edition* (Cambridge University Press, Cambridge, 2010)

166. J. von Neumann, *Mathematical Foundations of Quantum Mechanics: New Edition* (Princeton University Press, Princeton, 2018)

167. M.Z. Nashed, Q. Sun, Sampling and reconstruction of signals in a reproducing kernel subspace of $L^p(\mathbb{R}^d)$. J. Funct. Anal. **258**(7), 2422–2452 (2010)

168. H. Q. Nguyen, M. Unser, A sampling theory for non-decaying signals. Appl. Comput. Harmon. Anal. **43**(1), 76–93 (2017)

169. E. Odell, H. P. Rosenthal, A double-dual characterization of separable Banach spaces containing ℓ_1. Israel J. Math. **20**(3–4), 375–384 (1975)

170. T. Oikhberg, E. Ricard, Operator spaces with few completely bounded maps. Math. Ann. **328**, 229–259 (2004)

171. V. Paulsen, *Completely Bounded Maps and Operator Algebras*. Cambridge Studies in Advanced Mathematics 78 (Cambridge University Press, Cambridge, 2002)

172. K. R. Parthasarathy, *Mathematical Foundation of Quantum Mechanics*. Texts and Readings in Mathematics 35 (Hindustan Book Agency, Gurgaon 2005)

173. V. Paulsen, A survey of completely bounded maps. Notes at the Banff International Research Station (2007)

174. A. Pełczyński, Any separable Banach space with the bounded approximation property is a complemented subspace of a Banach space with a basis. Studia Math. **40**, 239–243 (1971)

175. D. Petz, *Quantum Information Theory and Quantum Statistics*. Theoretical and Mathematical Physics (TMP) (Springer Berlin, Heidelberg, 2008)

176. G. Pisier, *Similarity Problems and Completely Bounded Maps: Second, Expanded Edition*. Lecture Notes in Mathematics 1618 (Springer Berlin, Heidelberg, 2021)

177. G. Pisier, *Introduction to Operator Space Theory*. London Mathematical Society Lecture Note Series 294 (Cambridge University Press, Cambridge, 2003)

178. G. Pisier, *Tensor Products of C*-Algebras and Operator Spaces: The Connes–Kirchberg Problem*. London Mathematical Society Student Texts 96 (Cambridge University Press, Cambridge, 2020)

179. G. Pisier, Q. Xu, Non-commutative L_p-spaces. In: *Handbook of the Geometry of Banach Spaces* (Vol. 2, pp. 1459–1517). (Elsevier Science B.V., 2003)

180. F. Radulescu, The fundamental group of the von Neumann algebra of a free group with infinitely many generators is $\mathbb{R}_+ \setminus \{0\}$. J. Amer. Math. Soc. **5**, 517–532 (1992)

181. F. Radulescu, Random matrices, amalgamated free products and subfactors of the von Neumann algebra of a free group, of noninteger index. Invent. Math. **115**, 347–389 (1994)

182. D. Robert, M. Combescure, *Coherent States and Applications in Mathematical Physics: Second Edition*. Theoretical and Mathematical Physics(TMP) (Springer Nature Switzerland AG, 2021)

183. A. Ron, Z. Shen, Frames and stable bases for shift-invariant subspaces of $L^2(\mathbb{R})$. Canad. J. Math. **47**(5), 1051–1094 (1995)

184. A. Ron, Z. Shen, Weyl-Heisenberg frames and Riesz bases in $L_2(\mathbb{R}^d)$. Duke Math. J. **89**, 237–282 (1997)

185. A. Ron , Z. Shen, Generalized shift-invariant systems. Constr. Approx. **22**(1), 1–45 (2005)

186. Z. Ruan, Subspaces of C*-algebras. J. Funct. Anal. **76**, 217–230 (1988)

187. W. Rudin, *Functional Analysis* (McGraw Hill Book Company, New York, 1973)

188. R. Ryan, *Introduction to Tensor Products of Banach Spaces*. Springer Monographs in Mathematics (Springer-Verlag, London, 2003)

189. M. Schechter, *Principles of Functional Analysis* (Academic Press, New York, 1971)

190. K. Seip, Density theorems for sampling and interpolation in the Bargmann-Fock space I. J. Reine Angew. Math. **429**, 91–106 (1992)

191. C. Shannon, Communication in the presence of noise. Proc IRE. **37**(1), 10–21 (1949)

192. R. Smith, Completely bounded maps between C^*-algebras. J. London Math. Soc. **27**, 157–166 (1983)

193. Z. Shang, W. Sun, X. Zhou, Vector sampling expansions in shift invariant subspaces. J. Math. Anal. Appl. **325**(2), 898–919 (2007)

194. Q. Sun, Nonuniform average sampling and reconstruction of signals with finite rate of innovation. SIAM J. Math. Anal. **38**(5), 1389–1422 (2006)

195. K. Seip, R. Wallstén, Sampling and interpolation in the Bargmann-Fock space II. J. Reine Angew. Math. **429**, 107–113 (1992)

196. A. Szankowski, Subspaces without the approximation property. Israel J. Math. **30**, 123-129 (1978)

197. A. Szankowski, $B(\mathcal{H})$ does not have the approximation property. Acta Math. **146**, 89–108 (1981)

198. S. J. Szarek, A Banach space without a basis which has the bounded approximation property. Acta Math. **159**, 81–98 (1987)

199. W. Sun, X. Zhou, Sampling theorem for multiwavelet subspaces. Chinese Sci. Bull. **44**(14), 1283–1286 (1999)

200. W. Sun, X. Zhou, Sampling theorem for wavelet subspaces: error estimate and irregular sampling. IEEE Trans. Signal Process. **48**(1), 223–226 (2000)

201. M. Talagrand, *Pettis Integral and Measure Theory*. Mem. Am. Math. Soc. **51**, 307 (1984)

202. M. Takesaki, *Theory of Operator Algebras: I, II* (Springer, New York, 2003)

203. R. Torres, E. Ward, Leibniz's Rule, Sampling and wavelets on mixed Lebesgue spaces. J. Fourier Anal. Appl. **21**(5), 1053–1076 (2015)

204. M. Unser, A. Aldroubi, M. Eden, B-spline signal processing: part I-theory. IEEE Trans. Signal Process. **41**(2), 821–833 (1993)

205. M. Unser, Splines: a perfect fit for signal and image processing. IEEE Signal Process. Mag. **16**(6), 22–38 (1999)

206. M. Unser, Sampling-50 years after Shannon. Proc. IEEE. **88**(4), 569–587 (2000)

207. R. Venkataramani, Y. Bresler, Sampling theorems for uniform and periodic nonuniform MIMO sampling of multiband signals. IEEE Trans. Signal Process. **51**(12), 3152–3163 (2003)

208. P. P. Vaidyanathan, B. Vrcelj, Biorthogonal partners and applications. IEEE Trans. Signal Process. **49**, 1013–1027 (2001)

209. E. Ward, *New Estimates in Harmonic Analysis for Mixed Lebesgue Spaces*. Ph.D. Thesis (University of Kansas, 2010)

210. S. F. Waldron, *An Introduction to Finite Tight Frames*. Applied and Numerical Harmonic Analysis (ANHA) (Springer Science + Business Media, New York, 2018)

211. N. Weaver, *Lipschitz Algebras: Second Edition* (World Scientific Publishing Company, 2018)

212. Y. Wang, Z. Xu, Phase retrieval for sparse signals. Appl. Comput. Harmon. Anal. **37**, 531–544 (2014)

213. F. Zhang, *Matrix Theory: Basic Results and Techniques*. Universitext (Springer Science + Business Media, New York, 2011)

214. X. Zhou, W. Sun, On the sampling theorem for wavelet subspaces. J. Fourier Anal. Appl. **5**(4), 347–354 (1999)

215. Q. Zhang, W. Sun, Invariance of shift invariant spaces. Sci. China Seri. A. **55**(7), 1395–1401 (2012)
216. Q. Zhang, W. Sun, The structure of finitely generated shift-invariant subspaces in super Hilbert spaces. Numer. Func. Anal. Opt. **34**(3), 349–364 (2013)
217. Q. Zhang, W. Sun, Vector-stability of refinable vectors. Appl. Anal. **92**(10), 2215–2228(2013)
218. Q. Zhang, W. Sun, Reconstruction of Splines from Nonuniform Samples. Acta. Math. Sin.-English Ser. **35**, 245–256 (2019)

The manufacturer's authorised representative in the EU is Springer
Nature Customer Service Centre GmbH, Europaplatz 3, 69115 Heidelberg,
Germany. If you have any concerns regarding our products, please
contact ProductSafety@springernature.com

Printed and bound by CPI Group (UK) Ltd, Croydon, CR0 4YY
23/04/2026
02095593-0012